Sustainable Performance in the Digital Age

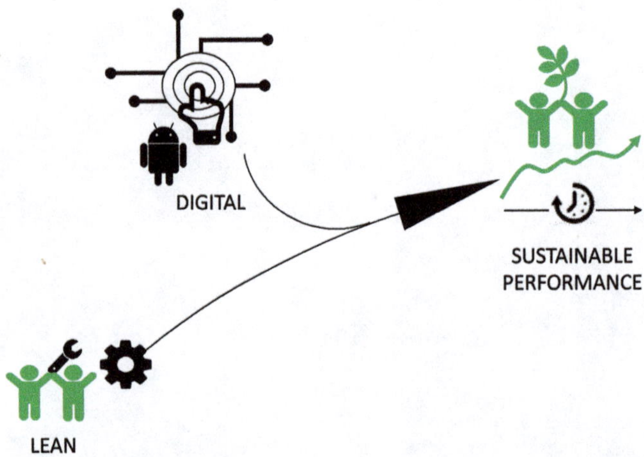
DIGITAL
SUSTAINABLE
PERFORMANCE
LEAN

Carine Vinardi

Sustainable Performance in the Digital Age

When Lean Meets Digital

Springer

Carine Vinardi
Aix-en-Provence, France

ISBN 978-3-032-01251-7 ISBN 978-3-032-01252-4 (eBook)
https://doi.org/10.1007/978-3-032-01252-4

This Springer imprint is published by the registered company Springer Nature Switzerland AG
The registered company address is: Gewerbestrasse 11, 6330 Cham, Switzerland

Acknowledgments The subject that inspired me throughout my career, my past books and my thesis too has been the sustainable efficiency and performance of organizations, whatever their size, primarily when it addresses what makes sense for people.

After writing on performing drivers such as Lean System, Multicultural Teams and *Hoshin kanri*,[1] unsurprisingly, this, my fourth book, falls in that category too.

In this extra professional project, I would like to thank, my daughter Mariane and my husband Richard particularly. Once again, they have given me their unfailing support in this venture because of its burden on already scarce personal resources.

Special and heartfelt thanks also go to my kind but demanding reviewers: my husband Richard, Bertrand Rose and Benoit Eynard (both Professors), for their precious contributions once more.

My gratitude also goes to all who agreed to give me their time to answer my questions and who participated in the book through their interviews: Luc Julia, Philippe Roger, Daniel Hager, Nicolas Siegler, Fabrice Barthelemy, Bruno Echardour, Christian Cornille, Bertile Burel, Delphine Asseraf, Stéphanie Perchet, David Briggs, Richard Vinardi, Eric Ciampi, Olivier Drean, Marie-Ange Giuseppin, José Carlos Pires, Eric Le Bourdon, Vincent Di Lorenzo and François Moallic.

Thank you also to a long list of people who have contributed through reflection, contacts, and references.

Thank you to Francesco Cipollone, who undertook the translation for the second time, and the delicate and demanding challenge of negotiating around the French original while hoping to satisfy you, the English language reader.

And lastly, my thanks also go to Springer Nature, my editor, who trusted me once again, with this publication.

Competing Interests The author has no competing interests to declare that are relevant to the content of this manuscript.

[1] A more precise definition is found in the Glossary (See No. 3).

Contents

Chapter 1
Introduction

The role of Digital technology, once considered episodic, has become wholly part of our daily lives since the general adoption of the Internet between 1995 and 2000. In 2024, **65% of the world's population** used the Internet, and workers rely on it for an average of **40% of their work time** (Istrate et al., 2024).

Digital technology is also considered essential for the economy as it even has its own industry. Some compelling figures for 2024:

- The value of the worldwide digital economy is estimated at **15 trillion Euros** and could **double by 2030** (OECD, 2024).
- For the industry sector alone, the estimated value of digital transformation is expected to be **$767 bn in 2026** (Arumugam, 2023).
- The worldwide market for Artificial intelligence or AI (defined in Chap. 2) amounts to between 140 and 240 billion dollars (Sternchuss, 2024). This could be **multiplied by 10** or even reach 1.5 trillion Euros by 2030! (OECD, 2024).

In this context, there is often a peculiar mix of high enthusiasm for the answers digital technology promises for the future, together with considerable apprehension about uncontrolled or careless use. Humans still have to learn "to tame" digital technology. The man-digital interface has yet to be optimized to achieve its full potential. Already at the beginning of the last century, Albert Einstein said this about the first computers:

> "Computers are incredibly fast, accurate, and stupid. Human beings are incredibly slow, inaccurate, and brilliant. **Together, they are powerful beyond imagination**."

In the professional arena, that means taking tested methods such as Lean or an Operational Excellence alternative (which I will detail later), and learning how to

C. Vinardi, *Sustainable Performance in the Digital Age*,
https://doi.org/10.1007/978-3-032-01252-4_1

update and enhance them with digital solutions and their analytical power. This is what this book is about.

If many organizations have implemented Lean, only a few have truly achieved it globally and sustainably. As these organizations integrate digital technology, the same syndromes that were seen with Lean reappear: commitment is sometimes motivated by the "Fear Of Being Obsolete" (FOBO) or by the "Fear Of Missing Out" (FOMO). Before understanding how Lean and Digital technology are complementary, it is best to begin with the conditions for success together with the pitfalls to be avoided.

The best way to start is actually by being committed for the right reasons and having a work plan that is consistent for the organization. As we will see, the implementation of an adapted digital strategy is decisive, as stated by G. Westerman, Leading Expert on Digital transformation & Innovation culture for MIT Sloane Initiative on Digital Economy:

> "When digital transformation is done right, it is like a caterpillar turning into a butterfly. **When done wrong, all you have is really a fast caterpillar.**"

Companies such as "GAFAM" (for Google, Apple, Facebook, Amazon and Microsoft) have built their business model on digital development by capitalizing on data. Other companies, the great majority by the way, produce real objects or provide services that are not, in essence, digital but try to benefit from the worldwide trend. All these organizations **all have in common the goal of a sustainable improvement of their performance**, and even more so for those in the private sector with its competitive business landscape.

Since performance is the subject matter throughout this book, it is appropriate to clarify its definition. Simply put, performance is a measure of the capacity to produce results by a person, a team, or some equipment (or more largely, by a company or an organization when looking at overall contribution).

Measuring results and, therefore, the level of performance means putting results into perspective, which is given by the defined objective, the resources used, or in comparison to the result of a similar organization. Performance can therefore be analyzed on three levels:

- Result/Objective: **effectiveness**
- Result/Committed resources: **efficiency**
- Result/Result of another organization: ***benchmarking***[1]

Performance inside an organization or a company therefore requires **having and developing all the human resources** (people, skills, organization), with the **material resources** (given, matters, equipment, software...) and combining **both with the right *process***[2] to get the **best results**.

[1] A more precise definition is found in the Glossary (See No. 1).

[2] A more precise definition is found in the Glossary (See No. 2).

Where can the results be seen? This question brings us to the aim of every enterprise and organization: to satisfy current and future **customer** expectations, whether they are called “patients” in the medical sphere or “citizens” in the public sector. That means satisfying the current and future expectations of **employees, owners, or investors** (shareholders for the private sector or public organizations) without forgetting those of its **suppliers.**

The overall performance of an organization is, therefore, a measure of all the performance results, including **customer, financial, social, and partnership performance results across all business lines.** In the **transaction** sector relating to services (mainly providing non-physical property), performance involves the processing of information from various inputs up to the customer.

In the **industry** sector, mainly relating to goods being tangible, physical products, performance involves the processing of materials into products from various inputs up to the customer.

Note that the transaction sector can also provide physical outputs (files and documents) just as the industry sector can provide non-tangible services. This is why I say “mainly”.

The search for performance is already well established, so it helps to look at performance levers that are proven, such as Lean. Afterward, we will answer how they still make sense today.

We will also see that these approaches, just like digital technologies, must be adapted in their implementation to produce the expected performance. Just as for cooking, in addition to the recipe, the quality of each product, the equipment, the utensils, the method, and the “dexterity” of the cook (i.e., of people and their skills) all count and make a difference. With common principles (the recipes), the implementation must be adapted **to company cultures** (national, activities, businesses…) and **their maturity**. The result is not the same when these parameters are changed.

Lastly, it is important to say that this book addresses **performance sustainability on two levels.**

The first level of sustainability looks at **performance in time.** An approach or technology is considered **durable progress** only if it is **for the many** and if **it is used appropriately**. Whereas Lean is now established and accessible, it is not so for all digital solutions, namely because of cost. In addition, as already noted by some, digital deployment is primarily focused on cost-cutting (Deloitte, 2023). Yet at the time, it had already been seen that Lean is not sustainable when used exclusively for cost-cutting, and sometimes it is even detrimental to employees (Vinardi, 2019). So it is important to stress how to reconcile company objectives with lasting performance.

The second level of sustainability is about being **respectful of the environment and the planet**. Regarding digital solutions, a positive impact on the planet is far from being achieved. Some estimate a reduction in Greenhouse Gas (GHG) emissions of − 20% by 2050 thanks to digital technology including − 4% from AI, by 2030 (George et al., 2022). We will see how some digital solutions contribute to reducing environmental impacts.

However, this reduction seems very far from compensating worldwide contributions by the whole of the digital industry (i.e., all those involved in the Value Chain of digital solutions). Several studies have shown that the **digital industry** already accounts for **4% of worldwide GHG emissions** and that it could reach **10% by 2030**, including Data Centers which already amount to 2–2.2% of emissions, i.e. as much as the aircraft industry worldwide, whose contribution is estimated at 2.1% (Jackson & Hodgkinson, 2022; Jungblut, 2024).

By analyzing performance on its two levels of sustainability, i.e., **in time** and **the environment,** additional filters can be considered to measure levers of performance (in addition to those relating to **customers, partnerships, financial and social performance**) to appreciate what Lean and digital technologies can bring, individually or together.

To move progressively, Chap. 2 will clarify several definitions and terms to make the reading of the book easier. In the next chapter, we will look back at the past, with some key dates to understand where digital and Lean implementations have come from and if it is more appropriate to talk about digital evolution or digital revolution. Chapter 4 will clarify the importance of the context and the constraint fields past and present in our approach to performance and in future trends. We will end by defining the filters to understand performance sustainability on its two levels of Lean or digital approaches (independently considered) in Chap. 5. We will then look at the combined performance in Chap. 6. After the performance conditions are defined, in Chap. 7 we will examine the implementation and prioritization stages. We will look in detail at several different companies in Chap. 8. Lastly, before concluding in the last chapter, Chap. 9 will take a prospective look at some technologies that should impact the performance sustainability of our organizations.

Writing this book is motivated by wanting to share experience and recommendations in the light of a scientific perspective. The primary scientific references used to support the arguments in this book are in English, French, and Spanish, as well as some prominent publications on the subject. As far as feedback of experience is concerned, I have also endeavored to share the experience I have acquired during my career, both as a Leader and Advisor, on designing and implementing both approaches. It will often confirm and sometimes infirm academic references. Lastly, I will also share **nineteen invaluable supporting contributions** from my interviews with leaders, consultants or experts who agreed to share their experiences with me in full transparency.

Among them, we have **thirteen top and senior executives** with diverse experience in the subject and whose **companies vary in culture and size**.

Daniel HAGER	**Chairman of the supervisory board of HAGER GROUP,** former CEO from 2008 to 2023 HAGER GROUP: International Industrial company (with €3.2 bn revenues and 13,000 employees in 2023). Producing electrical distribution solutions
Nicolas SIEGLER	**Deputy CEO for MAIF and CEO for MAIF INTERNATIONAL** MAIF: One of the main French Mutual Insurance Company (with €4.7 bn revenues in 2023). Operating in transactional business

(continued)

(continued)

Fabrice BARTHELEMY	**CEO of TARKETT** since 2019 TARKETT: International Industrial company (with €3.3 bn revenues and 12,000 employees in 2023), World Leader producing flooring and sports surface
Christian CORNILLE	**CEO of MECACHROME** since 2020 MECACHROME: Medium Size Industrial International company (with €605 m revenues and 4800 employees in 2023). World Leader producing powertrains
Bruno ECHARDOUR	**CIO of LA POSTE GROUP** since 2016 LA POSTE GROUP: Now International and Transactional Retail company in postal/bank/insurance services, digital and service tools (with €34.1 bn revenues, 232,000 employees, 17,000 retail outlets in 2023). Coming from the French National Postal Company, it is a public owned company
Bertile BUREL	**CO-FOUNDER of WONDERBOX** since 2004 WONDERBOX: Medium Size Retail company (with €350 m revenues and 700 people in 2023), French Leader of Gift-boxes
Delphine ASSERAF	**DEPUTY CEO of Harvest Group/CEO of Harvest France SAS** HARVEST: Small Size Tech company in Digital software solutions for wealth managers, (with €80 m revenues, 500 people in 2023)
Philippe ROGER	**ORANGE Consulting CEO from 2014 to 2023** International Consulting firm (In House and External) belonging to Orange, one of the Leader in Europe for supporting Digital transformation
Stéphanie PERCHET	**CO-FOUNDER of ADOPTIN,** born in May 2024 ADOPTIN: Start-up, Software—Application in Change Management
David BRIGGS	**COO of NAAREA** NAAREA: French Deep Tech Start Up (300 FTE), developing Nuclear Small Modular Reactor (SMR), founded in 2020, first production delivery in 2030
Richard VINARDI	**Strategic Industrial Director for L'Occitane-En-Provence** since 2011 L'OCCITANE-EN-PROVENCE: International and Industrial/Transactional company with retail activities. Cosmetic business (with €2.5 bn revenues, 10,000 people in 2023)
Vincent DI LORENZO	**Operations VP for BAUMERT** IBAUMET: ndustrial Small Size Enterprise (with €28 m, 180 people in 2023), expert in customized technical doors for high-risk sites protection
Dr François MOALLIC	**Head of R&D for TMI BU of CERATIZIT/PLANSEE GROUP** PLANSEE: International and Industrial Company (with €1.5 bn revenues, 7000 people). Producing cutting tools & hard material solutions

Five Senior Partners or Experts (In House or External consulting):

Eric CIAMPI	**SENIOR PARTNER—Head of Operations Practice in France & Paris Office Leader for OLIVER WYMAN** OLIVIER WYMAN: Worldwide American Strategy and Management consulting firm (with €3.3 bn revenues for 7000 professionals in 2023)

(continued)

(continued)

José Carlos PIRES	**SENIOR PARTNER North America for KAIZEN INSTITUTE** KAIZEN INSTITUTE: Worldwide recognized consulting Leader (with €100 m revenues and 800 consultants worldwide in 2023) primarly focused on Opex/Lean. They completed their offer with concurrent Digital support
Olivier DREAN	**SENIOR DIRECTOR, in charge of North America Manufacturing Consulting for DASSAULT SYSTEM** DASSAULT SYSEM: Worldwide digital consulting firm (with €5.2 bn revenues, 22,000 people in 2023)
Marie-Ange GUISEPPIN	**HEAD OF INDUSTRY & DEFENSE for INETUM** INETUM: Worldwide IT Consulting Leader (with €2.5 bn revenues, 28,000 consultants, worldwide in 2023) primarly focused on digital services and solutions. They completed with an offer in Opex/Lean support
Eric LE BOURDON	**OPEX Director for LABEYRIE FINE FOODS** LABEYRIE: International and Industrial Leader Fine Foods company (with €1 bn revenues and 3400 people in 2023)

Lastly, I will share the experience of **Luc Julia,** who is one of the world's most well-known and recognized experts on AI. Co-creator of Siri in 2010, former CTO and Senior Vice President of Innovation for Samsung, he has also launched several start-ups. I had the chance to interview him at the end of 2024.

Luc JULIA	**CHIEF SCIENTIFIC OFFICER for RENAULT Group since 2021** RENAULT: French multinational automotive manufacturer with 4 brands (with €52.3 bn and 105,500 people in 2023)

References

Arumugam, A. (2023). Digitalization and its impact in manufacturing sector—Leveraging digital technology to improve efficiency and productivity. Available at SSRN: https://ssrn.com/abstract=4858067

Deloitte, Global Shared Services and Outsourcing Survey. (2023). Executive summary. https://www2.deloitte.com/us/en/pages/operations/articles/shared-services-survey.html

George, M., O'Regan, K., Holst, A. (2022). *Digital solutions can reduce global emissions by up to 20%. Here's how*, World Economic Forum. https://www.weforum.org/agenda/2022/05/how-digital-solutions-can-reduce-global-emissions/

Istrate, R., Tulus, V., Grass, R. N., et al. (2024). The environmental sustainability of digital content consumption. *Nature Communications, 15*, 3724. https://doi.org/10.1038/s41467-024-47621-w

Jackson, T., Hodgkinson, I. R. (2022) *What is 'dark data' and how is it adding to all of our carbon footprints?* World Economic Forum. https://www.weforum.org/agenda/2022/10/dark-data-is-killing-the-planet-we-need-digital-decarbonisation/

Jungblut, S.-I. (2024) *Our digital carbon footprint: What's the environmental impact of the online world?* Reset Digital for Good. https://en.reset.org/our-digital-carbon-footprint-environmental-impact-living-life-online-12272019/

OECD Digital Economy Outlook. (2024). *Embracing the technology frontier* (Vol. 1). OECD Publishing, Paris. https://doi.org/10.1787/a1689dc5-en

Sternchuss, A. (2024). Les chiffres 2023-2024 du marché de l'IA dans le monde. https://lehub.bpifrance.fr/les-chiffres-2023-2024-du-marche-de-lia-dans-le-monde/

Vinardi, C. (2019). Thesis on industrial engineering. *Les défis du Lean à l'ère de la mondialisation et de l'industrie 4.0*. Université de Technologie de Compiègne, Français. NNT:2019COMP2500, tel-03201462

Chapter 2
Definitions and Terminology

Abstract To write as clearly as possible, it is important to have a common language for definitions and terminology. The Lean system (which includes Hoshin kanri (A more precise definition is found in the Glossary (See No. 3))) and parts of digital systems (with a specific focus on Artificial Intelligence, Industries 4.0 and 5.0) are described here, holistically and then in detail for their various key components. In addition, a glossary is included at the end of chapters, giving a quick summary. This chapter is about clarifying the "what" question and analyzing criteria concerning performance. The meaning, the raison d'être of its approaches, and their implementation for their part are addressed from Chap. 5 onwards.

2.1 The Lean System Including *Hoshin kanri*[1]

In the introduction, I referred to Operational Excellence (OPEX). It is sometimes used to describe performance systems including *Lean* (and other associated approaches such as 6sigma, WCM, Agile…). However, I have chosen not to use this term in the book because:

- All approaches and performance systems originate from the Toyota Production System (TPS). Toyota was indeed the first, not to invent, but to formalize a system that produces results based on the separate and combined experience of components that already existed (this chapter).
- Lean terminology appeared several years before OPEX when it was intended to talk about what made Toyota successful without necessarily referring to the company or its TPS acronym (Vinardi, 2019).
- OPEX has its roots in Lean. The term OPEX appeared because Lean was often misunderstood or reduced to its constituent parts or a few principles focusing on costs, namely in the United States. (Found et al., 2018).

[1] A more precise definition is found in the Glossary (See No. 3).

C. Vinardi, *Sustainable Performance in the Digital Age*,
https://doi.org/10.1007/978-3-032-01252-4_2

- The notion of "Excellence" is often more attractive and makes sense for organizations when they associate their name with it, creating a "Business Excellence or Performance System". However, the definition of an "Excellence System" is not universal (Found et al., 2018), and no company in particular stood out when researching for this book.

  Consequently, **I suggest using the term "Lean"** but note that it includes:

"Doing better, quicker, cheaper, and sustainable for customers and employees" by maximizing value to the customer, which is also called "Added Value" (AV), and by minimizing the waste, which is also called "Non-Value added" (NVA).

For the customer, AV is produced **on what is called the Front Line, respectful to the employees**: in factory workshops, R&D laboratories, in customer service centers, in a service organization, the welcome desk of an administration, etc.

The notion of a "Lean system" is certainly the most appropriate because Lean is based on three areas (A, B, C) which are developed coherently:

- A. an **organization and individual behavior**, making sure they are suited,
- B. specific **management practices,**
- C. selective and systematic use of **methods and tools** (Vinardi, 2019).

Lean starts with and focuses on the customer (with their needs and expectations, whether implicit or explicit). This point is often forgotten when implementing Lean, especially when it is used to reduce costs. Without detailing the various components already described in a separate work (Vinardi, 2013, 2019, 2023), nonetheless, it is important to state the fundamentals and the key aspects of the subject in this book. I will therefore share the key *Lean* Leadership aspects (A and B) and also the key Lean Tactical aspects (C).

2.1.1 Key Leadership Elements

There are at least three fundamental aspects that make sense anytime, anyplace.

- <u>**Standard Visual Management**</u> (Vinardi, 2019)
  - The **standard is a statement of the best common rules to carry out a task (in terms of life, work, or management), a product, or a service, aiming for the best options for performance, safety, quality, time, and cost**. In other words, they are the common rules to follow for best results (service or product, for example) at a given time. When optimized, it is using only the resources that

are necessary and adapted (including the notion of efficiency). Satisfying and being respectful of employees means taking into account their safety and work conditions. The standard must allow a task to be carried out painlessly without wounding (i.e., as ergonomic as possible) and without creating unnecessary physical and psychological stress.

- **Everyone, including craftsmen who work alone, has a standard, but it is not always documented.** Documentation is truly significant when several people carry out the same task, when newcomers must be trained quickly, or when knowledge must be shared.

  For millennia, word of mouth (tacit knowledge) was the rule, but printing and more recently, electronics have enabled the written documentation of knowledge wherever possible (explicit and tacit knowledge will be considered later in the context of deployment in Chap. 7). Documentation must be done with and for users: it must be as visual as possible, as simple as possible, and easily and frequently updated where required. After sharing, the standard needs to be implemented and then improved when any implementation difficulties are solved. It is the **Golden Triangle,** as illustrated in Fig. 2.1.

- **The standard gives a reference for Daily Performance Management (DPM), based on indicators to obtain results for an objective and therefore the comparison can be made** (Fig. 2.2). **This is a Key Performance Indicator (KPI). This indicator may be very simple (e.g., yes/no for a rule) or indicate an expected value or an interval range.** In that case, the result is compliant with the expected values when it lies between a min and max, as shown in Fig. 2.3.

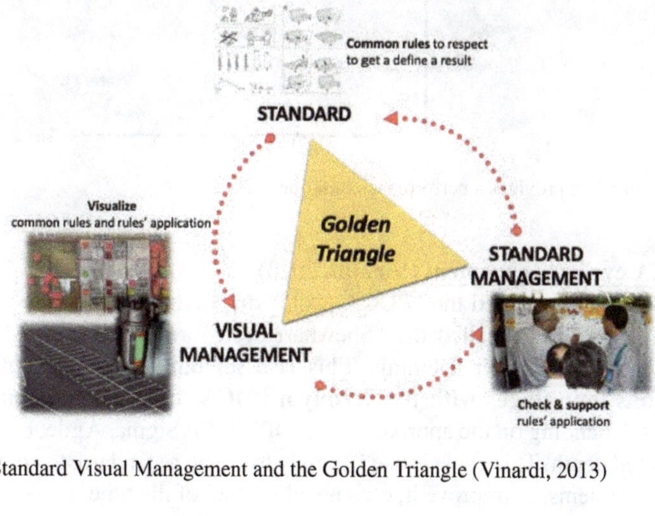

Fig. 2.1 Standard Visual Management and the Golden Triangle (Vinardi, 2013)

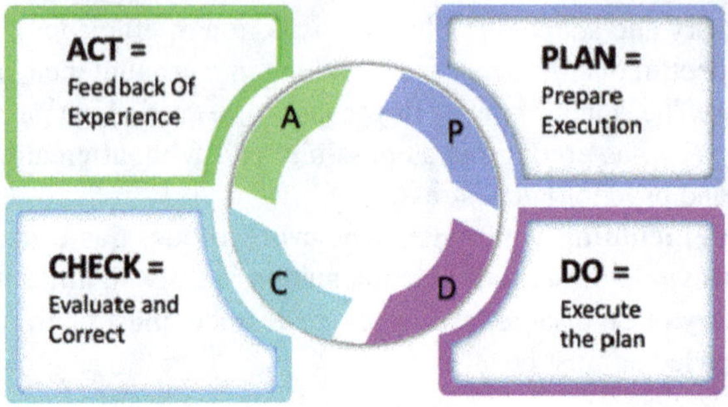

Fig. 2.2 PDCA cycle is applicable everywhere and at all times

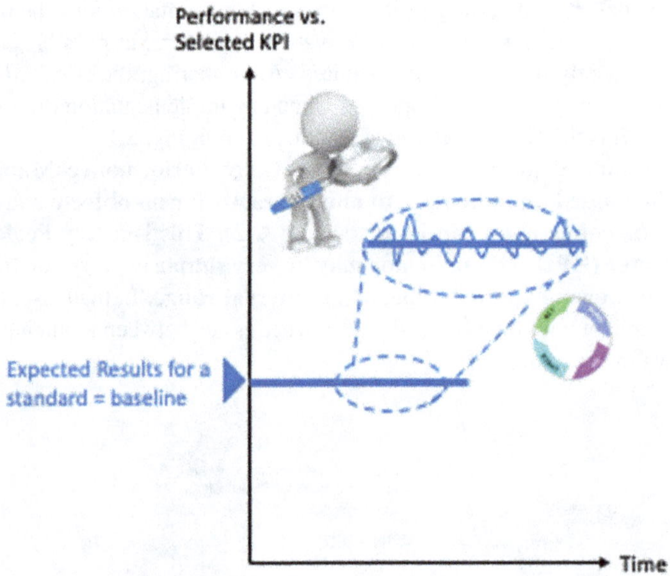

Fig. 2.3 The standard provides a performance baseline

– **PDCA cycle management** (Vinardi, 2023)

This cycle is called the "PDCA cycle" or "Deming wheel." In the beginning, it was rather called the "Shewhart cycle," referring to Mr. Shewhart, who worked with Mr. Deming. **This is a formal training/learning cycle and has four stages with the acronym PDCA**. It has different names and stages depending on the approach (e.g., OPEX, SixSigma, Agile, etc.) and the expected depth. This cycle is used namely to create a standard, to correct it and solve problems, to improve it, etc., i.e. almost all of the time. It doesn't matter which stage to start at, even if the four stages are explained in the "PDCA" order. Note that this logical cycle may be applied to very simple actions or complex projects alike. The parts described below can be adjusted accordingly:

- **P for PLAN**

  Define the subject, the problem to solve, and the improvement to make.
  Determine the initial state with the data collection and objectives with KPIs
  Analyze the gap to be filled and the main success conditions.
  Define assumptions and work methods through a plan (protocol) and a schedule.
  Inform and train the people involved if needed.

- **D for DO**

  Execute the plan.
  Identify deviance (what can distort results).

- **C for CHECK**

  Validate the results
  Correct the plan if necessary

- **A for ACT**

  Establish the conditions for success and reasons for failure
  Formalize lessons learned
  Standardize and deploy new standards for success

- **Problem Solving and Escalation management**
  - A **problem can be seen as the gap with an expected reference**. In performance coordination, a problem is considered an unfavorable gap in the results that the standard is supposed to yield. If there is no standard, the first problem to solve before anything else is to write the common rules for the best reachable result. The **objective of problem solving is indeed to reach the reference value** as shown in Fig. 2.4.
  - There are **several simple methods in problem solving to break down a problem** (when it is complex) and to analyze it (Vinardi, 2013), (Vinardi, 2019). Teams that work on the same problem must apply factual and rigorous analysis together in a common mindset. **To perform, people must work together.** When several causes and therefore solutions appear, prioritizing and testing them is recommended. Each is rated for its impact, ease of implementation, etc. A **priority matrix** is obtained by combining both, as represented in Fig. 2.5, where **priorities and initial quick wins are identified, thus motivating the teams.**
  - **Problem solving requires skills** to implement the methods of definitions and analysis. These skills are in addition to the subject at hand (mechanics, biology, sales, …). **The capacity of the teams to be autonomous is a factor for effectiveness and also for commitment, as we will see later. Problems can also be escalated, according to importance,** whenever an individual or a team cannot solve it at their level, either for lack of means (money, time) or for lack of skills.

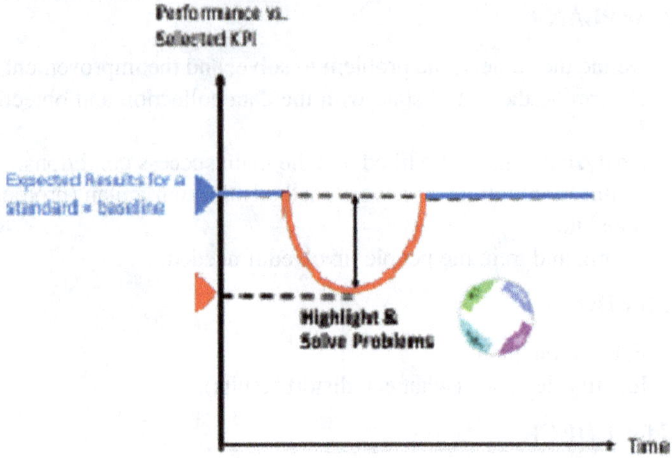

Fig. 2.4 Identifications of problems showing the gap to fill against the baseline (*Source* Vinardi, 2023)

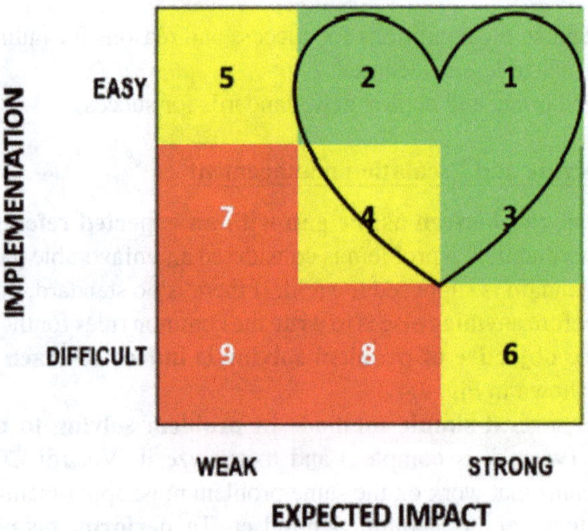

Fig. 2.5 Priority matrix (*Source* Vinardi, 2023)

Concerning performance, the team Leader should be at the heart of the escalation *process*[2] to understand why it is a priority problem to be escalated, how to answer it (in a cascading way, because answers cascade down too), and how to give the team more autonomy for the same problems in the future (development). Lastly, two requirements are needed for the process[3] to be fast and powerful: **transparency and**

[2] A more precise definition is found in the Glossary (See No. 2).

[3] A more precise definition is found in the Glossary (See No. 2).

the smallest number of hierarchical levels to **avoid interpretation filters, and it also helps for speed.**

Solving problems is not exclusively about being **effective but also about being creative,** particularly if new solutions to new situations are needed. Experience relating to issues past and present must be a starting point, as well as finding new solutions for new problems. Lastly, solving problems means learning, in particular, learning how not to repeat them. **A performing organization is a learning one** (Vinardi, 2023).

Apart from **DPM which was mentioned above, two other management rhythms** play their part in the Lean System. The visual coordination of standards, to solve and escalate problems, and the PDCA cycle are both useful in two particular situations:

- **Continuous Improvement (CI)**

  Here, the **standard works and delivers the expected result**. The organization needs to improve it (both the standard and the expected result). There are ideas to develop the standard (conditions, parameters, accessories, for example) without compromising it. The **improvement here is incremental.**
- **Breakthrough Improvement (BI)**

  In this second case, the standard works and delivers too. Let's imagine the organization needs considerably better results than those currently obtained. **The risk is great, but it aims at a higher gain too.**

  - In some cases, a wider generalization of the standard in the organization helps to reach it. A **fast large-scale action can have the effect of over-driving the impact.** Even for simple cases, any large-scale deployment is always complex whatever the solution, as we will see in Chaps. 5 and 7.
  - **Even for simple cases, any large-scale deployment is always complex whatever the solution, as we will see in** Chaps. 5 and 7.

A general overall representation of Fig. 2.4 is shown in Fig. 2.6. The context of BI is typically in strategic situations. This kind of improvement is preferably put forward by the teams concerned, but the decision to carry them out must be aligned with the *vision*[4] and the *Strategic Objectives*[5] of the company or organization.

The definition and implementation of **BIs broadly fit in what is called *Hoshin Kanri*[6]** (Vinardi, 2023). It is used to define and deploy an organization's strategic plan in a real and powerful way, across all levels and for all its teams. This approach was formalized in the 1960s in Japan before it was adopted by Toyota, as it was for TPS, which it is part of. Despite its results when correctly implemented, it did not receive the same attention as the other components of TPS, in particular, because it is **very demanding on Top Management** (Vinardi, 2023).

---

[4] A more precise definition is found in the Glossary (See No. 4).

[5] A more precise definition is found in the Glossary (See No. 5).

[6] A more precise definition is found in the Glossary (See No. 3).

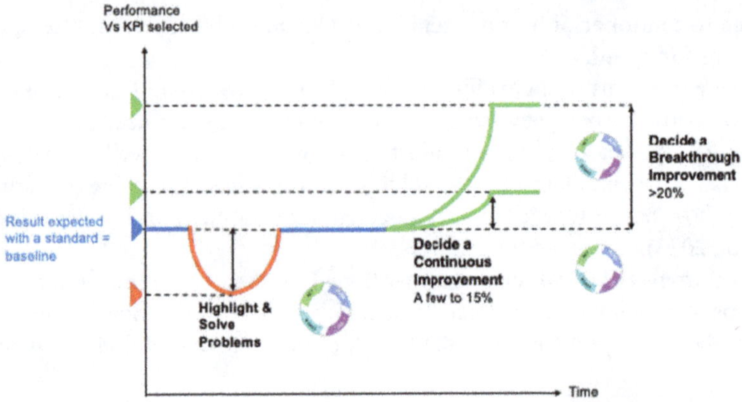

Fig. 2.6 Baseline and management rhythms against time (*Source* Vinardi, 2023)

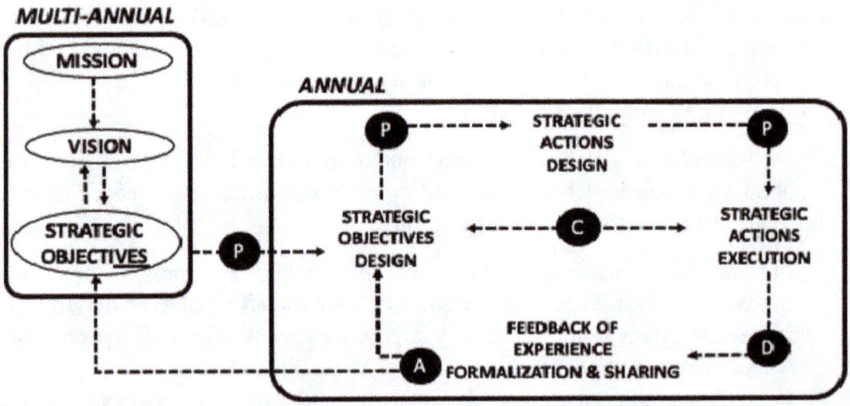

Fig. 2.7 *PDCA* in the multi-year and single-year stages of ***Hoshin kanri***.[9] (*Source* Vinardi, 2019)

Figure 2.7 shows how *vision*[7] translates into concrete actions. One of the secrets of the approach is in particular to carry out the **same exercise with all the teams coherently, that is by ensuring actions are suited at the right level** (Vinardi, 2023). When referring to the *Lean* System, it implies *Hoshin Kanri*[8] de facto.

The **three management rhythms (DPM, CI, and BI) all have the same basic aspects** (Standards, PDCA, and problem solving). In addition, speed is decisive in problem solving because instabilities are often provoked and **implementation problems arise when a change is made.** Moreover, leaders often address these **three rhythms together with all issues to be managed at the same time,** as in Fig. 2.8.

[7] A more precise definition is found in the Glossary (See No. 4).

[8] A more precise definition is found in the Glossary (See No. 3).

[9] A more precise definition is found in the Glossary (See No. 3).

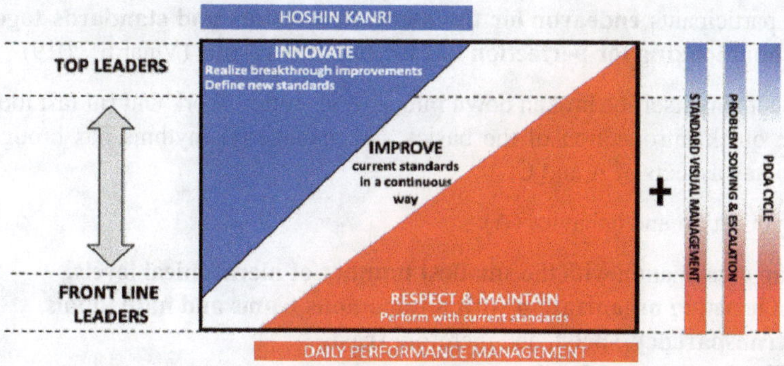

Fig. 2.8 An example of Nemoto diagram (*Source* Vinardi, 2023)

- The **teams closest to the front line mainly implement the standards and the associated problem solving (for 70 to 90% of the time),** a little on CI (10 to 20%), and only very little on BI.
- On the other hand, **Top Leaders are in charge of preparing the future with BI, for which they should devote 50 to 70% of their time**. 20 to 30% of the time should be dedicated to developing the potential of CI, and do just a little DPM at their level, on systemic problems coming up from the front line.

Figure 2.8 shows the ideal time management for leaders of the best-performing companies (apply the right Lean System in the right way…). It is named after its author Masao Nemoto. As a former top executive of Toyota, he had various operational functions as a company Frontline Leader to the Toyota Board, where he was also in charge of the deployment of *Hoshin kanri*[10] (Vinardi, 2023).

Note that the successful interworking of intermediate teams depends on matching *Bottom Up*[11] information and suggestions from the Front Line with *Top Down*[12] decisions from Top Leaders. **They must regularly go and check with the Front Line** to see how the strategic plan is deployed, to get a "true" status report, and to enquire and understand how they can help (Vinardi, 2023).

Let us also note that coordination and problem solving of CIs and BIs can be done in **project mode or Action Work Out, both following the PDCA cycle.**

- **Project mode** is appropriate for a **complex action** (in general, the action is broken down into several simple actions). The implementation is measured **in weeks, months, or even years with validation milestones (or stages)** and relies on a multidisciplinary team that meets at defined intervals, in particular at milestones.
- **Action Work Out** is appropriate for a relatively **simple action**, deployed with a group of people that work on the **same shared objective, with the same short period (a few days maximum), same location, space, and purpose.** All

[10] A more precise definition is found in the Glossary (See No. 3).

[11] A more precise definition is found in the Glossary (See No. 6).

[12] A more precise definition is found in the Glossary (See No. 7).

the participants **endeavor for the associated results and standards together, without looking for perfection but for tangible results** (Vinardi, 2019).

A project may itself be broken down into several Action Work Out (in fast loops).

The quick introduction of the basics and managerial rhythms has brought up several key aspects of A and C:

- Organization and behavior (A):
  - an organization with the **smallest number of hierarchical levels,**
  - a **learning organization with autonomous teams and individuals,**
  - **transparency, speed,** and therefore **trust,**
  - **Top Leaders on the shop.**
- Methods and tools (C): to define and improve the standards as well as coordinating them effectively; to define, characterize, and solve problems visually. The approach using PDCA, even if it is a learning cycle, can also be regarded as a method. Are there other methods and key tools?

2.1.2 Key Tactical Elements

Let us take another look at the essence of Lean, seen at the beginning of this chapter: "doing better, quicker, cheaper, and sustainable for customers and employees", maximizing the Added Value for the customer while minimizing waste (Non-Value added).

Lean methods and tools maximize the Added Value and eliminate the Non-Value added (waste). First, let us point out three categories of waste: the **"3 *MUs*"** standing for the first syllable of these Japanese words:

- ***MURA***: waste **related to uncontrolled variations** in the *process*[13] due to customer demand. It is normal that customer demand varies, however, the organization's human and material resources are not extensible ad infinitum, and they have a price. This client-related cause has a wave-like effect internally if there are no methods or tools to create filters, so it seizes the teams' energy without necessarily satisfying customers.
- ***MURI***: waste **related to resources**. Examples for people would be a task or *process*[14] that does not make sense, one that is illogical or lacks of ergonomy (resulting in a waste of time, stress, or physical pain), or keying in data twice, correcting flaws, or stopping and starting. Equipment, for example, can result in frequent stops and starts (which machines in general, just like humans, do not like) or usage in unsuitable conditions (under or overuse).

[13] A more precise definition is found in the Glossary (See No. 2).

[14] A more precise definition is found in the Glossary (See No. 2).

- ***MUDA***: waste **directly related to the product or the service**. For example, there are defects and scraps, excessive quality, unnecessary transfers or transport, inventories, moves, or waiting time.

To tackle *MURA* and reduce variations, Leveling is used to distribute and balance the workload so that deadlines are more reliable and aligned with customer demand and greater ease of work for the teams. A Safety Buffer, for example, helps to deal with customer emergencies as well as for internal ones. These aspects are represented in a simple way, in Fig. 2.9 (Vinardi, 2013).

To address the *MURA, MURI,* and *MUDA* categories correctly, it is necessary to understand the ***process***[15] of how individuals work together, how they share data, documents, materials, or parts, and how the equipment produces the deliverables from the inputs.

- **Value Stream Mapping** (VSM) is a key method for sharing everything starting from simple and visual maps, identifying the gaps, analyzing, and prioritizing the elimination of waste (Vinardi, 2013). An example illustrates a manufacturing activity in Fig. 2.10.

Lastly, I would like to mention **two key methods and also two pillars,** which include various tools and methods that can be set up according to maturity, specifically **to tackle *MUDA* in an organization.** However, they are **to be done only after the impacts of the first two categories of waste are understood.**

- **5S** is an approach with 5 stages (Sort, Set, Shine with inspection, Standardize, Sustain). The implementation of 5S is used to create and maintain a safe working environment, with the best operating conditions, suited to the expected use, and without waste of time. This makes sense everywhere: in a workshop, in an office, with equipment, a computer, or a shared folder (Vinardi, 2023).
- **Total Productive Maintenance (TPM),** especially for activities with a lot of equipment, because it must reduce or eliminate breakdowns (waste) and must

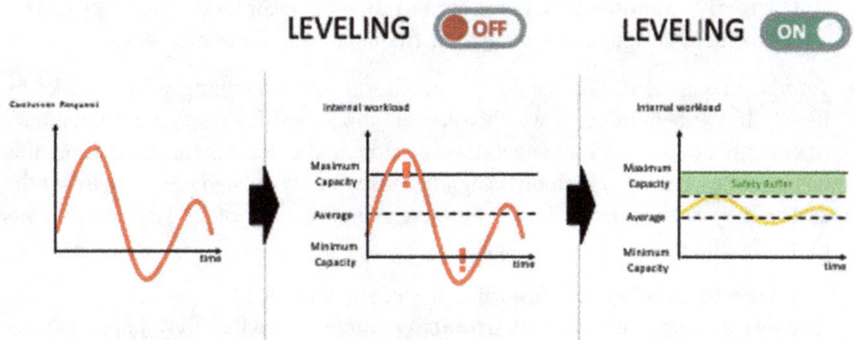

Fig. 2.9 Leveling impact (*Source* Vinardi, 2013)

[15] A more precise definition is found in the Glossary (See No. 2).

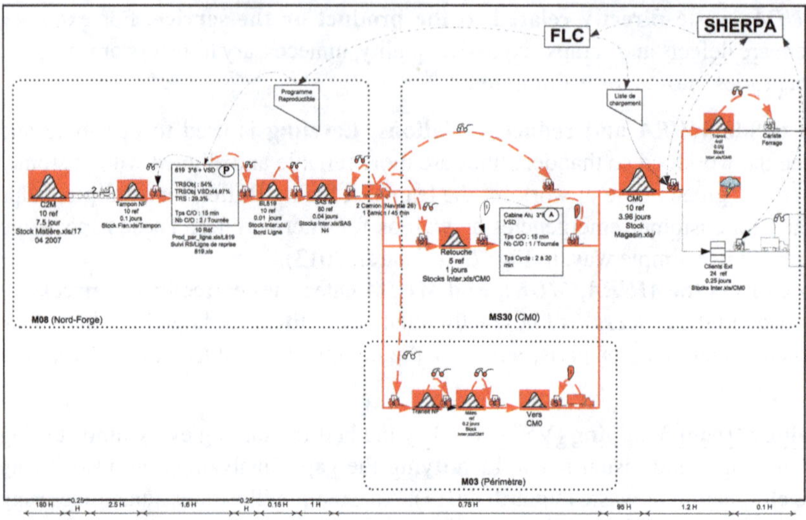

Fig. 2.10 Example of VSM current state after analysis (Vinardi, 2013)

help prevention. Depending on the organization, it has several parts. Here are the essential ones:

- **5S** is mentioned above and is the basis for any optimized equipment.
- **Autonomous maintenance** is the organization of activities carried out by maintenance operators and employees themselves, who are trained by the maintenance teams. Because they are physically nearest to the equipment, operators can identify weak signals (abnormal temperatures, odors, or unusual vibrations, for example) before failures happen.
- **Preventive maintenance** is the set of tasks to avoid stops and breakdowns. Whenever possible, these tasks are carried out without stoppages.
- **Corrective maintenance** is the organization of repair tasks when they cannot be avoided, designed to reduce their frequency and duration.

- ***Jidoka***: In Japanese, this means "to transform into something which works by itself". In the beginning, it was just the organization of tasks that a human and/or equipment had to do to detect defects, stop, and correct them. Over time, this has been broadened to include design because quality-related costs are lower the earlier they are tackled (Fig. 2.11). Here is a summary of its key components (Vinardi, 2023):
  - **integrate quality at design** for a product or a service,
  - **control, stop, and correct problems** where and when they happen (filter defects that are not eliminated during design),
  - **set up a final inspection** (final filter if the design and intermediary filters are lacking).

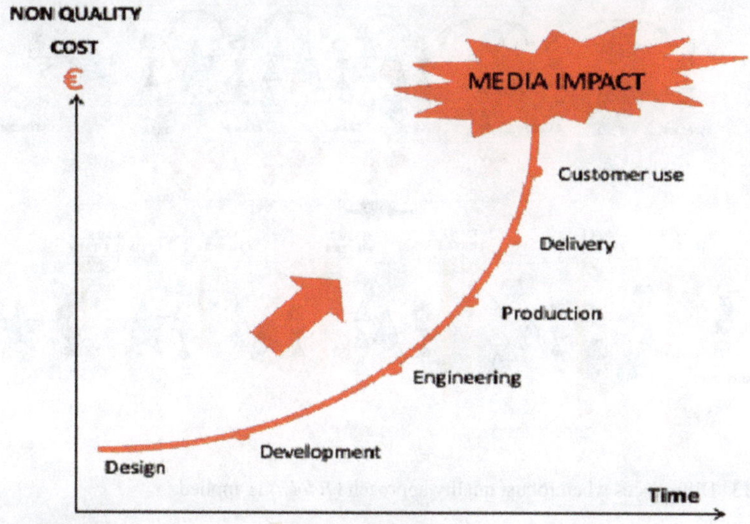

Fig. 2.11 Non quality costs (*Source* Vinardi, 2013)

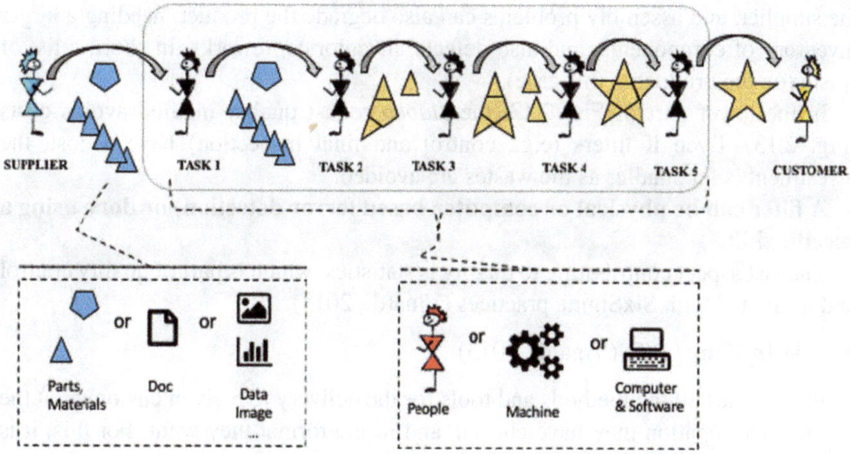

Fig. 2.12 Example of *process*[17] with 5 tasks

Figure 2.12 is an example with five tasks to be done in a manufacturing *process*[16]. This example can be extrapolated where the materials or components are documents, data, or images for a service activity. Also, the example shows people carrying out tasks, but they can also be machines, computers, or a combination of the three.

[16] A more precise definition is found in the Glossary (See No. 2).

[17] A more precise definition is found in the Glossary (See No. 2).

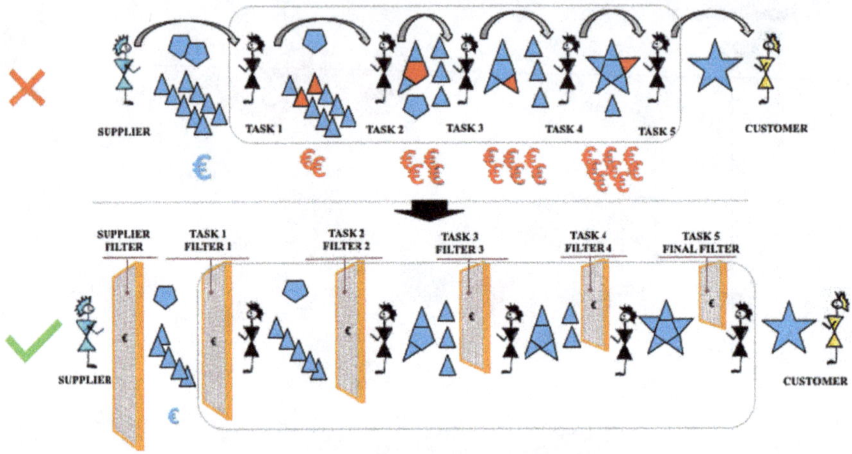

Fig. 2.13 Differences when robust quality approach (*Jidoka*) is applied

The upper part of Fig. 2.13 shows that without a quality method incorporated in the design across all value chains, some defective parts (in red) may arrive from the supplier, and assembly problems can also degrade the product, needing a larger inventory of components, and also defects, inventories, reworks, in short, a list of waste (on the product in this case).

In the lower part of Fig. 2.13, the *Jidoka* robust quality method avoids costs (Fig. 2.13). Even if filters (e.g., control and final inspection) have a cost, the investment is far smaller as the wastes are avoided.

A **filter can be physical or computer-based (error detection) or done using a specific skill.**

The last aspect contributing to quality is statistics, which is part of quality control and is aligned with SixSigma practices (Vinardi, 2013).

- **Just In Time (JIT)** (Vinardi, 2013)

  It combines all the methods and tools for the delivery to a given customer, at the time and location they have chosen, and in the format they want. For this, it is necessary to:

  - **Synchronize** the launch of the product or service by the customer at the right time so that it arrives at the desired moment. This is called Pull Flows.
  - **Harmonize** with the customer by producing the product or the service at least as quickly as it takes for the customer "to consume" the previous one. This is called **Takt Time.**
  - **Minimize flow stoppages,** i.e. minimizing the handover and handling of products or data so reducing problems and therefore waste (errors, deterioration, time, etc.). This is called **Continuous Flow.**

– **Ensure flows are independent from the start-up to the delivery** of the service or product to a customer to prevent disturbances if one flow impacts another. This is called **Autonomous Flow.**

In order to illustrate, I have chosen the aspects of Just In Time with the workings of a restaurant serving fresh produce for its customer orders.

Figure 2.14 illustrates the Pull Flow and Continuous Flow, Fig. 2.15 illustrates Takt Time, and Fig. 2.16, the Autonomous Flow.

This is of course just an illustration and it is always good to check suitability. For example, must we keep a batch flow at the bar (Fig. 2.14)? Must we keep the autonomous flow with the waiter and cook, for each table? (Fig. 2.15).

By implementing these principles, the right questions are asked to move fast and speed up the flow (Figs. 2.15 and 2.16).

Here ends the quick tour of the key methods and tools in Lean. It included smoothing, VSM, 5S, TPM, *Jidoka,* and JIT. They all rest on three Leadership fundamentals (Standard Visual Management, PDCA, and Problem Solving and Escalation) and must be integrated into the three management rhythms (Daily Performance Management, Continuous and Breakthrough Improvement Management).

Coordination to solve problems and to set improvements can have long loops (projects) and/or fast loops (Action Work Out).

The Action Work Out, also called fast loop, is also found in Agile, but with another name. This approach was gradually adopted by IT/Software organizations, which often were in charge of Digital deployment. I will present what it is about, with their similarities and differences.

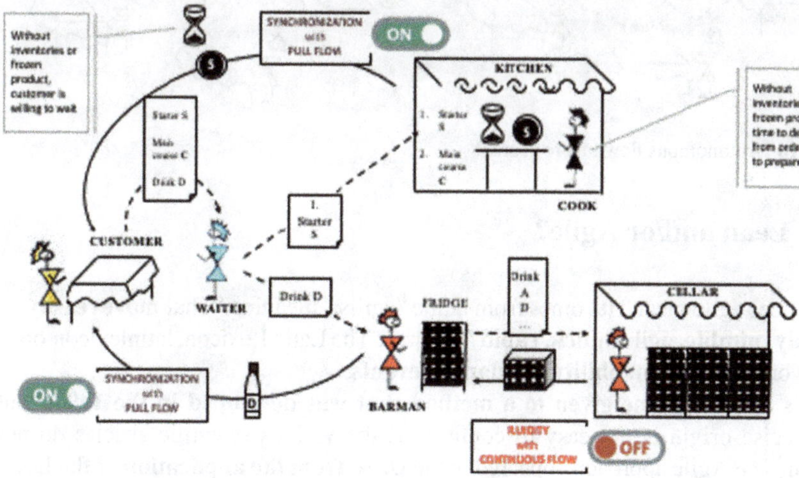

Fig. 2.14 Pull and continuous flow in a restaurant

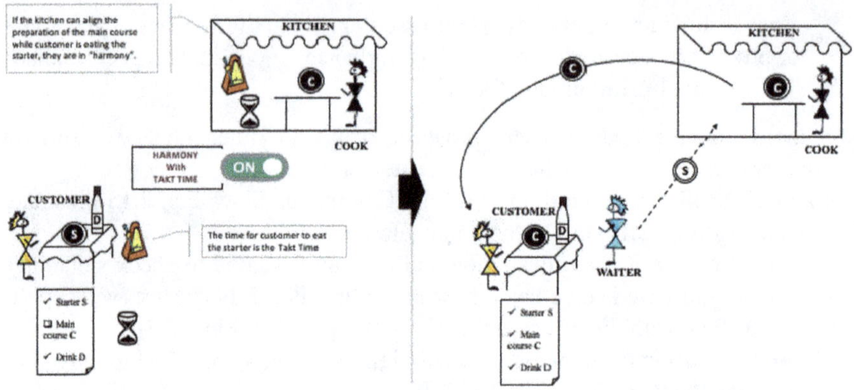

Fig. 2.15 Takt time in a restaurant

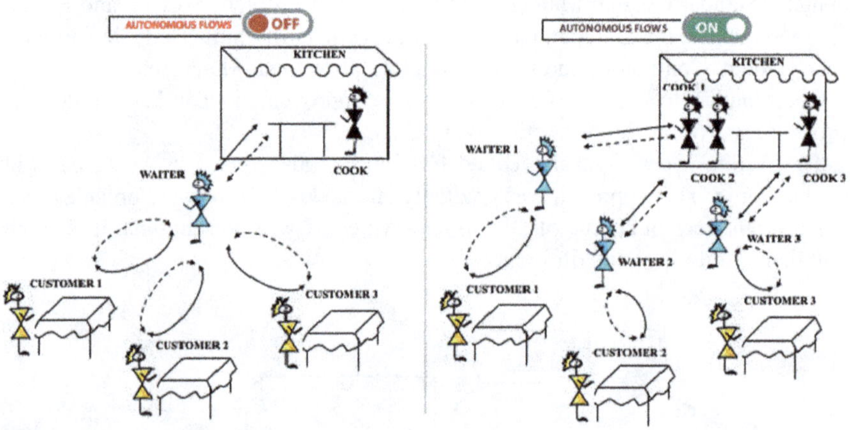

Fig. 2.16 Autonomous flow in a restaurant

2.2 Lean and/or Agile?

What does agile mean? It comes from latine "agilis," meaning "That **moves easily or quickly; nimble, agile, quick, rapid**" (Source: The Latin Lexicon, latinlexicon.org). The word suggests **an ability to adapt to events.**

It is also the **name given to a method that was developed in the 1990s**, but the precise origin is not easy to confirm, as the various scientific articles do not concur. The Agile approach appeared in the USA, **from the application of the Lean approach to the software industry**. Renowned developers met together and noted how their projects were systematically delivered late and did not meet the real needs of the customer. So they thought up a solution, named it "Agile project management", which had a short iterative *process*[18] and involved the customer at regular intervals (Barrand & Moïchine, 2024).

The initial method was designed in opposition to the traditional project mode (linear sequential approach, either waterfall type or V cycle, linear but with more iterations).

"Project culture" made way for "Product culture". The main objectives were to reduce the Time to Market and the associated costs, to value the clients by involving them in the *process*,[19] to act meaningfully, and to optimize. These objectives were very commendable, even if they are not unique to Agile.

The framework of the IT project deployment and the way of thinking of software application teams strongly influenced the Agile Approach. As the "customers" do not necessarily know how to express clear technical specifications for software people and the needs often change, it was a question of focusing a multidisciplinary team on a single subject with the customer, for a short period and delivering a usable solution even if imperfect. The end of each sequence was marked by a choice to either stop and be content with the result or start a new sequence again. Iterative cycles were short and repeated: **this is exactly the (Lean) Action Work Out mentioned previously…However, as "workshop mode" was not always included in all Lean deployment, IT people were often not exposed to it**.

In 2001, the "Manifesto for Agile Software Development" stated four core principles and twelve connected declinations (Cervone, 2011):

- Individuals and interactions over processes and tools.
- Working software over comprehensive documentation.
- Customer collaboration over contract negotiation.
- Responding to change over following a plan.

Twelve variations are associated with these basic principles, and the manifesto is often cited as a reference element. However, it is far removed from the first publications describing well-defined methods in computer engineering. And increasingly, we are also witnessing a form of dilution of the sense of agility in a broader application and theoretical framework (Berthou, 2022).

Anyway, Agile started to be "famous" in 2010s with the boom of Start-ups. As the Americans had done by creating the terminology Lean, initially launched for "marketing" reasons to avoid talking about the Toyota System, the term Agile appeared in the same logic to differentiate itself from Lean. Another explanation is that Lean itself is often misunderstood in its composition and articulation, with a good number of companies, especially American ones, reducing it to tools and/or the notion of Just-in-Time. The creation of a new term is not surprising. Agile also includes a specific English vocabulary with "Scrums" and "Sprints":

- **"Sprint" is an iteration** lasting 1 to 4 weeks during which the team performs a given job and delivers an increment (usable deliverable) to meet a given objective.

[18] A more precise definition is found in the Glossary (See No. 2).

[19] A more precise definition is found in the Glossary (See No. 2).

- **"Scrum" is a framework** which guides teams in the iterative and incremental delivery of a product or service. During the sprint, the team checks in daily with each other in the form of a 15-min meeting known as a scrum. This meeting is facilitated by a ScrumMaster (Sliger, 2011).

There are various other frameworks (Spotify, SAFe, LeSS, SoS…) with alternatives in terms of vocabulary. The form and terminology of Agile are specific, **but the content already exists widely or can be integrated into a complete and well-understood Lean system**. The duration of coordination loops is already a topic that depends on the activity in the Lean world.

Therefore, Agile and Lean have many aspects in common, but one of the differences is their focus:

- Lean is often **concerned with *process*[20] improvement** (where "the *process*.[21] leads to results"
- Agile only **focuses on service or product improvement**.

When there is a physical flow (manufacturing a product, for example), focusing on the *process*[22] leading to the result makes sense because the (visible) physical flow guides information flows and the organization (Vinardi, 2013). On the other hand, when there is no physical flow (when providing a service, for example), **the organization strongly guides the *process*, and therefore merely focusing on the *process*[23] is not enough to obtain the results. This is one of the factors used to adapt Lean in transactional activities** (Vinardi, 2013).

Moreover, just as for Lean, Agile is based on "Test and Learn", which is part of the US and Japanese culture, where learning is based on "Empirical first" or "Learning by doing first" (Vinardi, 2016, 2019).

Agile has some specific prerequisites:

- Prior evaluation of workload (even if it is adjustable),
- A broad budget framework because detailed estimates cannot be given.

It is recommended to have autonomous teams of 10–12 people that are multi-disciplinary and that include the client or their representative. However, there are autonomous teams in the Lean approach (Vinardi, 2013), so this is not specific.

? Is Agile in its principles the ultimate evolution of Lean as some authors put forward?

We must be careful. Firstly, let's take the traditional project mode that Agile called into question. Even if it is more structured and rigid in places, according to the type

[20] A more precise definition is found in the Glossary (See No. 2).

[21] A more precise definition is found in the Glossary (See No. 2).

[22] A more precise definition is found in the Glossary (See No. 2).

[23] A more precise definition is found in the Glossary (See No. 2).

of project, Agile can be used pragmatically, simply and flexibly, as is the case in problem solving. It is not a question of using a sledgehammer to crack a nut.

Also, **remember that Agile was developed in a software environment (with an "interaction-based" IT team, and where the Client is in the *process*[24]), as opposed to a hardware environment with its often costly infrastructure (with its "industrial-based" IT, where *processes*[25] structure the organization).** Therefore, when designing an airplane, or any project with **a large budget, architecture requirements, and where errors are costly,** it is advisable to **keep a structuring project approach, with a development coordination based on rapid loops.**

That was the case in the car and construction industry. To illustrate this, let's take a French success story: the building of the Millau Viaduct in 2004. It was a technical and architectural feat (the world's highest bridge at the time).

Contrary to the norm of the profession, the project strictly complied with the timeline (three years after laying the first stone) and budget, all of this, despite strong headwinds. (Godfrain, 2006). It was not deployed in Agile mode but in traditional project mode with iteration loops. Without going through all the reasons for the success, we will look at the technical part (Godfrain, 2006):

- **Concurrent engineering and manufacturing** (not a purely linear approach): preparation is done as early as possible (with prefabricated parts) and use of metal for the deck instead of concrete, for example.
- **Usage of available techniques**: to ensure the vertical and horizontal positioning (Permanent monitored using a geostationary satellite), for example.
- **Use of tried and tested technologies** even if not at the same scale (limited challenge) with a very large design stage.

For the social and environmental part (Godfrain, 2006):

- **Autonomous entity-based** organization.
- **Simple and efficient quality control**, using iteration loops.
- **The working conditions and the respect for the environment** at all stages of the project were recognized by the local authorities and the workforce.

Naturally, the point is not about saying that the Agile mode is not meaningful, but rather to state when it is best to use it. It is useful for projects that are not yet scoped (and that may not need to be) without depending on complex systems in terms of integration. When problems are known, it is better to incrementally improve, while in complex situations where problems are not known, a probe-sense-respond approach based on experimentation and the exploration of new solutions is preferable (Furlan, 2023).

Lastly, it would be more fitting to avoid creating new terminology but apply the Agile mode in the existing performance system (Table 2.1).

[24] A more precise definition is found in the Glossary (See No. 2).

[25] A more precise definition is found in the Glossary (See No. 2).

Table 2.1 The main advantages and disadvantages of Agile compared to the traditional project approach (sources including Cervone, 2011; Kischelewski et al. 2024; Năftănăilă, 2008)

Agile ADVANTAGES	Agile DISADVANTAGES
• **The client is at the center** of the problem to be solved. As for Lean, but it is imperative for Agile projects. • **No need for detailed specifications** to begin. • **Risks are minimized** by focusing on short iterations of clearly defined deliverables.	• Client involvement, namely for external clients, during the whole cycle (several weeks) and **across several projects can be difficult or impossible.** • **If the client does not have a clear sense of the product's direction,** the members of the development team will tend to behave in the same way, and the final product can be significantly different to what is expected.
• **Deliverables are given at regular** short intervals. • **The test plan is reviewed after** each Sprint instead of waiting for the project test phase. • **Changes in the definition** of the deliverable are always possible. • Project management is simpler.	• **The integration of deliverables in complex or large scale systems may be a challenge.** In addition, research is lacking comprehensive overviews about the implementation of large-scale agile. • **It might be difficult for novice agile project managers** to enforce some of the most important facets of agile methods, such as "no changes during a sprint".
• **Direct communication** with partners in the development *process*[2] is emphasized helps to avoid creating copious project documentation.	• **Frequent communication does not mean that no documentation** is required and without adequate oversight, a project could degenerate into "undisciplined hacking" because documentation is not maintained.

The coexistence of agile and non-agile units, inter-team meetings not working, and conveying of agile mindset are the **most frequently identified challenges** (Cervone, 2011).

Rather than opposing Lean and Agile**, the question is about how to make agility and the new ideas brought by Agile coexist with already existing components in Lean**. This is done by considering overall cohesion and using **a unified terminology in the relevant organizations. Several transactional (banks, insurances), digital, and start-up companies use Agile, alone or with Lean, so a clarification was necessary**. Now let us look at what Digital technology means in detail.

2.3 Digital Technologies Including Artificial Intelligence

I have mentioned digital technology several times, but what is it about exactly? The transition from Analog to Digital is one of the great technical evolutions of the second half of the twentieth century thanks to computers. Analog technology was born with the coming of electricity and data processing (sound, images, and text) where the format is continuous, so to speak, similar to a wave, close to a natural representation of real-world phenomena because they could reproduce them from a recorded signal (audio or video for example).

The Digital evolution meant the conversion of information into discrete binary signals (as numbers, 0 s and 1 s), also called bits, for "**bi**nary digi**ts**". This was a Breakthrough Improvement with regards to **complex mathematical calculations** and also with regards to **storage** (the same amount of information taking much less space) and **transmission, too because information** can reliably travel long distances without any significant loss in quality, as opposed to analog technology.

I therefore for "Digital" suggest the following definition:

> Digital technology is the set of **all the means** that use data in a **digital format (numbers)**, such as **material means** (hardware: computers, servers, etc.), **software** (including algorithms, etc.), **networks** (means of communication and connecting equipment, such as the Internet, Wi-Fi networks, 4G/5G, etc.). It is made up of and also includes the *process*[26] to:
>
> - **Convert information**: analog to digital and vice versa,
> - ***Process***[27] **data**: transmit, detect, control, use, analyze, and convert data (generate, predict),
> - **Store information**: using databases, servers, and the cloud (which is defined below).

Note that concerning basic vocabulary, there is a subtlety in English between the words "digitization" and "digitalization". They are two central constructs used concerning digital technology and the innovation *process*.[28] However, the definitions of these two important constructs are often implied and, in practice, there is **no clear consensus on the conceptualization** (Achtenhagen & Achtenhagen, 2019):

- **Digitization** is the **creation of digital artifacts** through technical processes of conversion, representation, and enhancement. This definition clearly demarcates digitization as a technical *process*,[29] which results in conversion, enhancement, or representation of analog or physical objects or real-world phenomena into digital artifacts. By digital artifact, we refer to the creation of something artificial.
- **Digitalization** is a socioeconomic ***process***[30] **of adoption and application of digital artifacts**.

Digitization is a more U.S.-centric information management construct and digitalization is a more European-centric general management construct (Achtenhagen & Achtenhagen, 2019).

This subtlety being made, let us return to the question of digital transformation or evolution, which concerns all organizations whatever their activity, whether it is **transactional** or **industrial**. There is no standardized classification for methods and tools, so I suggest making a separation. On one hand, the basic components and associated elements that concern only data, but are used in transactional and

---

[26] A more precise definition is found in the Glossary (See No. 2).

[27] A more precise definition is found in the Glossary (See No. 2).

[28] A more precise definition is found in the Glossary (See No. 2).

[29] A more precise definition is found in the Glossary (See No. 2).

[30] A more precise definition is found in the Glossary (See No. 2).

industrial activities alike. On the other, the available physical solutions with their data physical solutions, which we will look at in the section on Industry 4.0 (Vinardi, 2019).

2.3.1 Data Core and Induced Components

As was previously mentioned, the digital evolution has had an impact on storage size and, therefore, the capacity to handle large volumes of data. Basic components came from this evolution (Vinardi, 2019):

- **Big Data**: is a large volume of data (either structured or not), which cannot be handled by a human, nor by a traditional machine. These data are extracted and used with specific applications (data analytics).
- **Cloud**: is the dematerialized data storage on delocalized servers instead of local hard drives. That also often includes other related services.. There are different types of cloud:
  - Public Cloud: Resources are provided over the internet and are shared among many users. The market is dominated by a few major players, with 65% of the market share held by Amazon Web Services (32%), Microsoft Azure (23%), and Google Cloud (10%) (Haranas, 2023).
  - Private Cloud: Resources are dedicated to a single organization. These resources can be managed internally by the organization itself or externally by a third party.
  - Hybrid Cloud: Combines elements of both public and private clouds.
  - Multi-Cloud: This approach involves using cloud services from multiple providers. It enables service diversification and reduces reliance on a single provider.
- **AI**: is the mathematical approach that enables complex calculations. I will specifically look at various forms of AI in Sect. 2.3.2 because it deserves special attention when talking about performance.

As a preliminary, let us note that the Cloud needs Big data, but the opposite is not true. Big Data, for its part, needs AI to extract and analyze data, and AI needs the volumes that Big Data provides (Vinardi, 2019).

Besides the basic components, there are also secondary components common to all activities. They ensure trust and maximize performance:

- **Network**: before the digital age, there already was a phone network but its evolution and new generations of mobile technology and 5G for example, greatly increased the throughput and reduced the latency, i.e., the time needed for data exchange.

- **Cybersecurity**: is the set of laws, frameworks, methods, and devices that ensure the integrity of the components, of the software, and of the network. It gives trust in digital technologies.
- **Horizontal integration of data**: the fast and effective sharing of data with customers and suppliers between various departments of the organization (from design to after-sales services)
- **Vertical integration of data**: the fast and effective sharing of data between all the teams in the company from Top to Front Line Leaders.

The difference between the cloud and data centers are:

- **A data center is a physical facility** that has computer systems and components, including servers, storage systems, and networking equipment. Managing a data center requires dedicated resources.
- **The cloud is a virtual model** for delivering IT services, providing access to computing resources such as servers, storage, and applications over the internet.

The cloud relies on data centers and is managed by service providers.

2.3.2 *Focus on Artificial Intelligence*

The Term AI was used for the first time at the Dartmouth Conference in the summer of 1956. Researchers were brought together to explore the potential of machines to simulate human intelligence. It was a foundational event for the emergence of this new scientific discipline of AI. In the beginning, the term related to marketing and was almost a voluntary falsehood, resorting to a sensational expression to obtain research funding from the US public authorities (Barraud, 2020).

Another point about the word intelligence is that in English, it can refer to the management of data, information processing, what the human brain does, and this translates rather as "*Artificial Cleverness*".

Lastly, **several experts, such as Luc Julia, consider that the acronym AI is better suited to mean "*Augmented Intelligence*".** This would suit better for a hand-held tool, appropriately, which is generally the case. When it is not appropriately handled, regulation would be needed (Julia, 2019).

Following wording considerations, here is a suggested definition of AI (Vinardi, 2019) as well as a schematic representation in Fig. 2.17.

> It is a discipline based on the use of **advanced algorithmic techniques** that use very large quantities of data to optimize, anticipate, and predict by identifying correlations between multiple parameters.

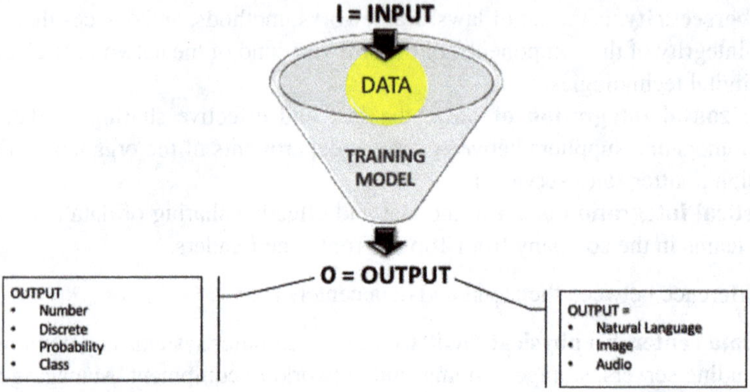

Fig. 2.17 AI transforming Input into Output using a training model

AI is based on a **training model** that integrates **algorithms and problem solving** (a basic part of Lean seen above). Complex problems are broken down into smaller solvable problems whose suggested solutions are recombined to answer the initial (complex) problem.

Figure 2.18 highlights three AI types:

- **Machine learning** (ML) is a subfield of artificial intelligence that gives computers the ability to learn without explicitly being programmed (Brown, 2024) (Fig. 2.19). A Framework is provided in Fig. 2.20

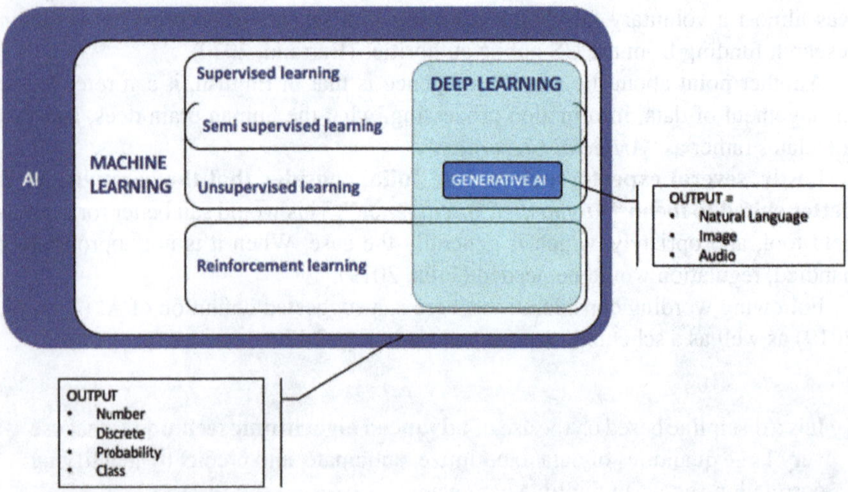

Fig. 2.18 Different AI types

Fig. 2.19 Example using of Dall-E 3 in the free version of Chat GPT

Examples of applications are chatbots (conversation robots) used in customer services, e-mail filtering, Google search engine, analysis of medical documents, etc. Moreover, there are types of learning tasks:

- **Supervised:** supervision from the developer. It is a task-driven approach that uses labeled training data to make classification or regression.
- **Unsupervised**: no direct control from the developer. It is a data-driven *process*[31] that analyzes unlabeled datasets to make clusters or associations. The accuracy of output is lower than for the supervised mode.
- **Semi-supervised:** a hybrid of both the supervised and unsupervised approach. It starts with the classification and then uses clustering.
- **Reinforcement:** the model finds the solution with a feedback loop. It is an environment-driven approach.

• <u>Deep Learning (DL)</u> is a **subfield of Machine Learning.** Whereas ML algorithms are linear, DL algorithms are stacked on top of one another (with multilayers organized in Artificial Neural Network) in a hierarchy of increasing complexity and abstraction.

Examples of applications are autonomous car, analysis of customer behavior (Amazon), forecasts of inventories and sales, predictive maintenance of equipment, etc.

DL is highly dependent to (Sarker, 2021):

- **Data amount:** The reason is that when the data volume is small, deep learning algorithms often perform poorly (Fig. 2.21).
- **Hardware:** DL relies more on high-performance machines with Graphics Processing Units (GPU).

[31] A more precise definition is found in the Glossary (See No. 2).

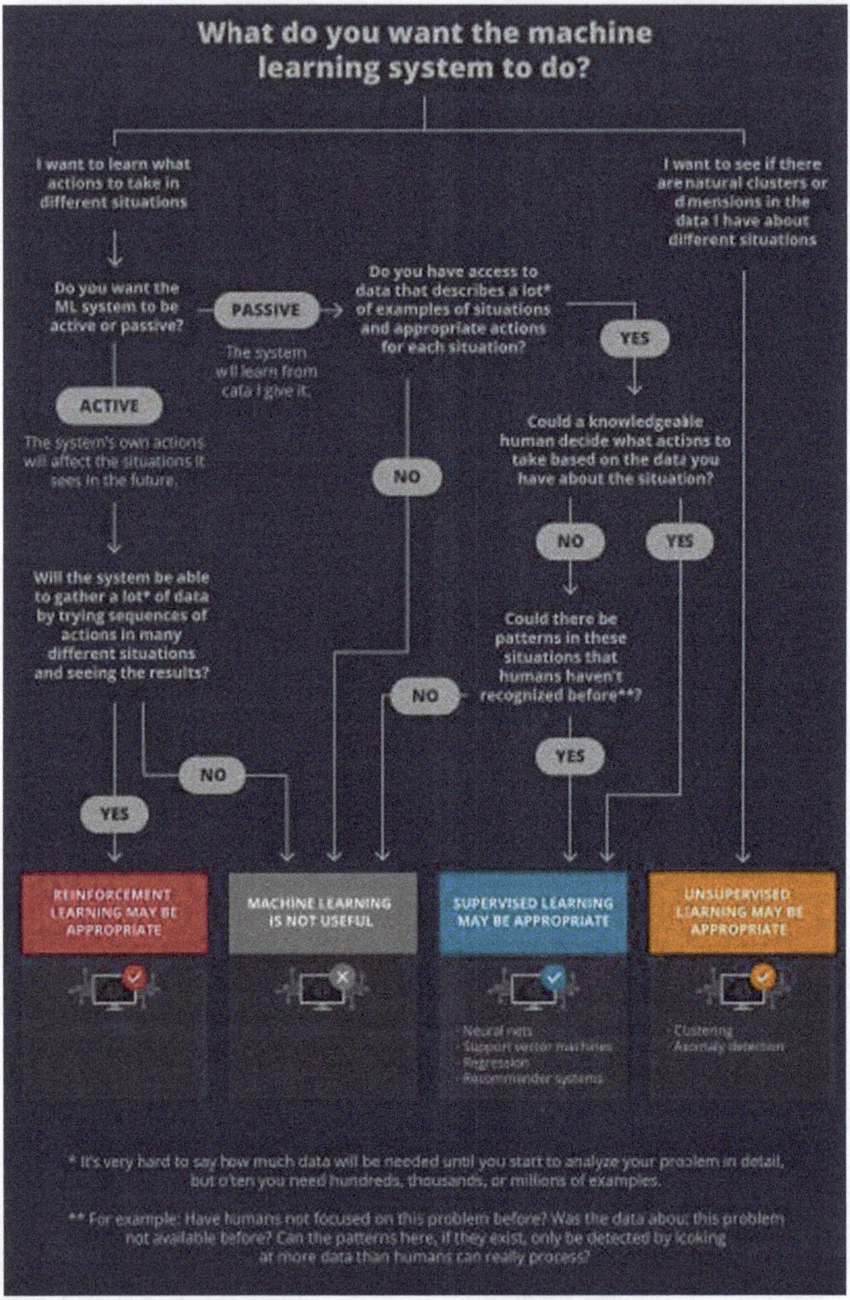

Fig. 2.20 Framework for Predicting where Machine Learning is useful (*Source* Malone et al., 2020)

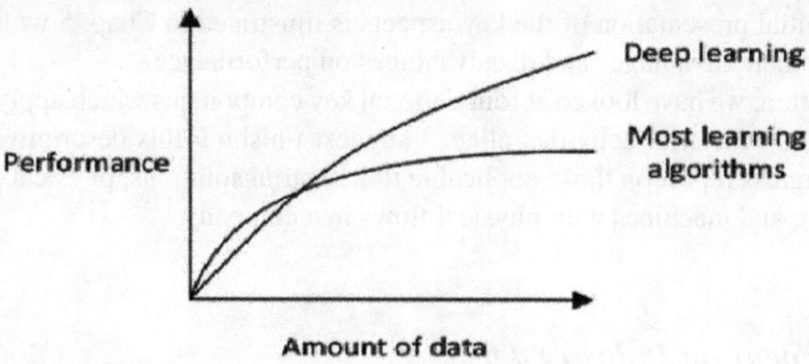

Fig. 2.21 Performance of Deep Learning (DL) and other Machine Learning (ML) algorithms (*Source* Sarker, 2021)

- **Feature Engineering *Process*:**[32] It is a *process*[33] of extracting features (characteristics, properties, and attributes) from raw data using. DL attempts to extract high-level characteristics directly from data.
- **Model Training and Execution time**: In general, due to a large number of parameters, DL model training *process*[34] takes longer than for ML.
- **Black-box Acceptation:** It's difficult to explain how a deep learning result was obtained. Whereas the ML algorithms, particularly, provide explicit logic rules for making decisions that are easily interpretable for humans.

• **Generative AI also** called "Gen AI" is a **subfield of Deep Learning with Unsupervised Learning.** It generates new content (output is Natural Language, Audio, Images). A prompt, i.e. a computer instruction in the form of a question is used to scope the requirement or problem to be solved. This **facilitation is a real evolution in terms of use compared to DL**.

This kind of AI made itself known by its application to generate text with ChatGPT (GPT for Generative Pretrained Transformer) starting from text, and its Dall-E equivalent to create images, both created by OpenAI. Note that DeepSeek, a Chinese alternative, is being released as I am writing Chap. 5. I will come back to the potential impact.

Examples of applications: summaries and documentation translations and associated offerings, the creation of motifs or single images, generation of musical content.

To illustrate GenAI, I created for this book with my (free) Chat GPT account which includes Dall-E 3. The 4 prompts (that I purposely wrote badly) and the transcribed results are summarized in Fig. 2.19. The expected level of precision is given together with the unsatisfied demands, despite repeating the requests (in red). Prompting should be as precise, short and complete as possible, etc. and needs patience to obtain the results! We shall look at use cases and limitations in Sect. 5.2.5.

---

[32] A more precise definition is found in the Glossary (See No. 2).

[33] A more precise definition is found in the Glossary (See No. 2).

[34] A more precise definition is found in the Glossary (See No. 2).

This initial presentation of the key aspects is illustrated in Chap. 5 with detailed use cases, their advantages and disadvantages on performance.

In addition, we have looked at foundational key components which apply to transactional and industrial activities alike. I suggest finishing this descriptive chapter with a progress report on those applicable to industrial solutions, physical products, equipment, and machines with physical flows in a company.

2.3.3 Focus on Industry 4.0

Manufacturing industries started using computing in the 1970s, and the first use of robots dates back to even before.

In 2011, the concept of **Industry 4.0 (I4.0**) was first used in Germany in the framework of the government program called "2020 High-Tech Strategic Planning Action Plan" (Vinardi, 2019).

Currently, even if the term I4.0 is one of the most used and seems to gain momentum, depending on countries and continents, there are other terms, as listed in Table 2.2 which is non-exhaustive.

The number "4.0" refers to three preceding industrial "revolutions" which are found in Fig. 2.22. This phase is supported by the increasing availability of groups of digital and connectivity technologies to reorganize the value chain (Vinardi, 2019).

I4.0, therefore, describes the imminent changes the industry faces and all the digital solutions involved (Vinardi, 2019).

In representing the four major phases it should remembered that Industry did not start with the first steam engines. From Latin, "Industria" may be translated as "skill or dexterity to do something", and there is no doubt that people did not wait until the eighteenth century to be industrious. We shall look at this subject again in the next chapter.

Table 2.2 Summary of the terms used (*Source* Vinardi, 2019)

Country	Terminology
Germany	Industry 4.0
Rest of Europe	Industry 4.0/Factory of the future/ "Usine du future"ou "Industrie du future" (France)
USA and English native countries	Smart factory/Smart manufacturing/Digital Manufacturing/ Industrial internet of Things/Advanced Manufacturing
Switzerland	"Industrie 2025"
China	Made in China 2025/Internet +

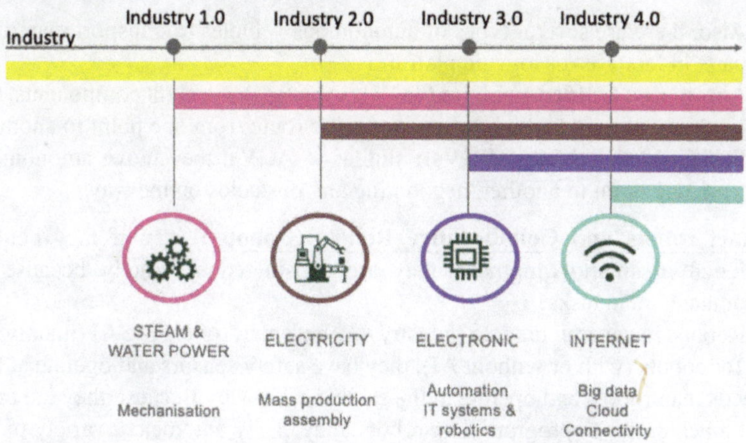

Fig. 2.22 From Industry to Industry 4.0, four Industrial (R)Evolutions

Contrary to the majority of representations, in Fig. 2.22, I suggest a cumulative approach of the various industrial contributions. This way of considering progress combines the tested methods with the latest available evolution.

Without repeating the basic and secondary components that I referred to in Sect. 2.3.1, it is important to briefly describe the key solutions most used by the industry (Vinardi, 2019):

- **IoT:** The Internet of Things is an interconnected network of physical devices (computers, sensors, and machines placed in other equipment or products) and software applications, which work together to connect objects. They collect digital data at a reasonable cost and act on shared production parameters. They are therefore directly connected to the Internet. They include Radio Frequency Identification (RFID), QR codes, etc.
- **Smart equipment** is the means or a networks of means (equipment, machines) that use connections, sensors, and actuators to solve complex problems and to make decisions alone. Here are three particularly useful use cases for industries that are highly equipped with machines:
  - **Smart *process* control**:[35] parameter control, analysis of their interactions and adjustment to optimize quality (services or products including energy consumption).
  - **Smart inspection**: measurement and visual control of manufactured parts to identify and eliminate defects and act on smart *process*[36] control.
  - **Smart predictive maintenance**: the conditional and real-time monitoring of key parameters to anticipate the best time for maintenance actions.

[35] A more precise definition is found in the Glossary (See No. 2).

[36] A more precise definition is found in the Glossary (See No. 2).

Also, there are several types of autonomous vehicles to transport materials or parts in factories or logistic platforms:

- **Automated Guided Vehicles (AGV):** vehicles that freight components, tools, products, etc., and which follow a specific route from one point to another.
- **Self-Guided Vehicles (SGVs):** similar to AGVs, they move autonomously from one point to another, negotiating any obstacles on the way.

- **Smart robots and Collaborative Robots (Cobots):** even if they could be included in smart equipment, they are considered separately because they automate human tasks.

  Robots, in general, predate Industry 4.0, but smart robots use AI (mainly ML). As for cobots (with or without AI), they have safety sensors and operate at lower speeds than robots and operate in the same workspace.. Because they are mobile and much easier to program than robots, they easily automate a variety of tasks and have better man-machine interaction. We will look at this again in Chap. 5.
- **Drones**: they are flying robots also called Unmanned Aerial Vehicle (UAV) or Unmanned Aircraft System (UAS). They operate without a human pilot onboard and they are controlled either autonomously by computers or remotely by a human operator on the ground. There are a variety of shapes and sizes, from small handheld devices to larger ones.
- **Exosqueleton:** similarly to the previous category, some classify these as smart equipment. I suggest separating them insofar as in industry at least, they relieve humans in painful tasks which cannot be done otherwise. Some of them exist for the upper or lower part of the body.
- **Virtual Reality (VR) and Augmented Reality (AR):**
  - VR: recreates the whole environment virtually. They are used for training (given to operators, managers, technicians, or customers), maintenance optimization, information sharing, etc.
  - AR: here, the environment is real, and virtual information is overlayed using digital equipment (such as glasses or a tablet, for example). Tasks can be guided by operators or technicians, for example, or experts can interact remotely to guide a person without them being physically by their side.

- **3D Printing, also called Additive Manufacturing**: is printing in three dimensions using specific equipment. In particular, prototypes can quickly be created for the benefit of innovation or to rapidly make a single part (such as for spare parts) or a small series. Currently, the operational limitations are mainly the speed of execution and the tolerance, but these aspects are gradually addressed to open up possibilities.
- **Advanced Simulations:** are simulation tools to digitally show optical, physical, or chemical experiments. Based on Machine Learning or Deep Learning, it confirms, predicts or suggests different outcomes.
- **Digital Twins**: they are digital replicas of, for example, parts, machines, buildings, factories (or even cities). Digital twin is a virtual representation of a physical object or system, capturing its behavior, performance, and life-cycle in real-time.

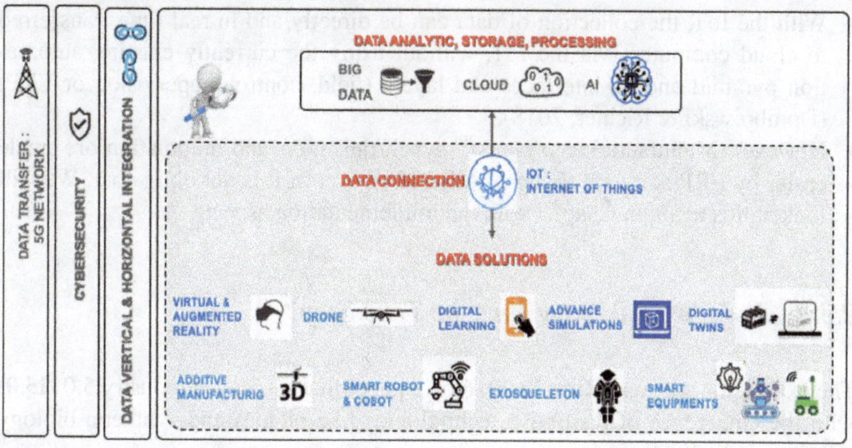

Fig. 2.23 Industry 4.0 framework

Digitalization can be made using a scanner for the small parts or using drones, for example. Once the twin is created, it is used to share information or to train in close to real conditions, to test and produce prototypes quickly in a virtual manner, to implement new lines of production or a new *process*[37] of manufacturing, new functionalities, by reducing costs or risk-taking, or to facilitate the preparation of major operations (or in a risk environment) or large modifications of the industrial tools. This technology is often combined with VR.

- **Digital Learning** is a virtual representation (digital replica) of a physical object or system, capturing its behavior, performance, and life-cycle in real-time. It is part of a set of new tools that often come from the gaming world. With the help of digital technologies, interactive and fun training modules are adapted to suit specific individual needs.

  These tools can also be accompanied by VR and AR on a computer or smartphone.

  It should be noted that a training tool using AR and VR is showcased by **Metaverse** (the term comes from "meta" and "universe"). It is a collective and shared virtual space aiming to reproduce the real world. The metaverse is not strictly speaking a technological building bricks of I4.0, but it is built on AR, VR, and AI (mainly on unsupervised DL).

Figure 2.23 summarizes the key components of the I4.0 Framework.

To finish this summary let me precise that I did not intentionnaly mention Enterprise Resource Planning (ERP) which is an information system to manage key flows of the functions of the value chain of a company in a centralized and unified way.

- ERP is part of Industry 3.0 even if currently it evolves with Software-as-a-Service (SaaS) which is a cloud application.

[37] A more precise definition is found in the Glossary (See No. 2).

- With the IoT, the collection of data can be directly and in real time transferred to cloud computers via the IoT, without using the currently existing automation pyramid and its interfaces and layers (field, control, supervision or ERP) (Dombrowski & Richter, 2018).
- However, Manufacturing *process*[38] synchronization and automation are made easier by ERP software (Feng & Ali, 2024), even if it is not obligatory. We will look at this again in Chap. 5 with the implementation aspects.

2.3.4 Industry 5.0: New Buzz or True Step?

The expression was **used first in 2016** on a paper that associates Industry 5.0 **(I5.0)** with the emergence of disruptive technologies like bionics and synthetic biology (Coelho et al., 2023).

In 2021, at the ten-year mark of the introduction of I 4.0, the European Commission formally called for the Fifth Industrial Revolution: I5.0. The main differences highlighted are the following ones (Xu et al., 2021; Coelho et al., 2023):

- I4.0 is considered to be technology-driven, whereas Industry 5.0 is **value-driven, such as** human-centricity, ecology, and social benefits
- I4.0 is considered a technology-driven revolution to achieve higher efficiency and productivity and, as a high-tech strategy of the government, to enhance Germany's competitiveness in a global market.
- The introduction of I5.0 is based on the observation or assumption that I4.0 focuses less on the original principles of social fairness and sustainability but more on digitalization and AI-driven technologies for increasing the efficiency and flexibility of production.

Other authors also mention a new revolution from I4.0's technology-centric approach to a **human-**centric model (Fani et al., 2024), with a switch from Mass Production to Mass Personalization (Reddy et al., 2022). Most of the other works associate the expression I5.0 with a fifth industrial revolution, **which will do what I4.0 did not achieve** (Coelho et al., 2023).

The objective is also to move Industry 4.0 technologies toward two areas: **sustainability and resilience** (Caggiano et al., 2023):

- **Sustainability** for **efficient use of natural resources as energy and material** (Caggiano et al., 2023) or for the development of circular *processes*[39] that reuse, repurpose, and recycle natural resources, reduce waste and environmental impact, and ultimately lead to a circular economy with better resource efficiency and effectiveness (Xu et al. 2021).
- **Resilience** is the **capability of any system to recover from unexpected drastic events and find a new equilibrium state that allows it to survive**. The system's

[38] A more precise definition is found in the Glossary (See No. 2).

[39] A more precise definition is found in the Glossary (See No. 2).

resilience involves investing in AI technologies (and predictive approach) for managing risk and contingency planning, implementing smart infrastructure, and advanced communication technologies (Caggiano et al., 2023). It would also refer to the need to develop a higher degree of robustness in industrial production, arming it better against disruptions and ensuring it can provide and support critical infrastructure in times of crisis (Xu et al., 2021).

However, there is no emerging consensus in the various published articles. For authors, **it is the same key enabling technologies as for Industry 4.0** (Reddy et al., 2022; Caggiano et al., 2023).

Sometimes **Edge computing** (EC) is highlighted as a difference (EC: computer processing is carried out where the user or the data source in situ, or at least close). It would enable to minimize communication overhead, guarantee applications are protective in remote area. Moreover, it would filter data by minimizing the volume of data sent to a centralized server or by promoting Cobots to improve safety and performance (Reddy et al., 2022).

The use of Cobots to improve safety and performance is a particular focus of attention for a few others.

To conclude this quick tour of the different points of view, others have put forward six key technologies (Xu et al., 2021):

- Individualized human–machine interaction technologies that interconnect and combine the strengths of humans and machines.
- Bio-inspired technologies and smart materials that allow materials with embedded sensors and enhanced features while being recyclable.
- Digital Twins and simulation to model entire systems.
- Data transmission, storage, and analysis technologies that can handle data and system interoperability.
- AI to detect, for example, causalities in complex, dynamic systems, leading to actionable intelligence.
- Technologies for energy efficiency, renewables, storage, and autonomy.

Some comments are needed here.

- The first is about the first occurrence of the term I5.0. Whereas researchers started to use it in 2016, in 2024, it is still not widespread in companies. This may be due to several factors:
  - Whereas I4.0 is a set of digital technologies, its use and goal in deployment are specific to each organization. Even if we frequently refer to digital deployment to reduce costs, some companies already use cobots, for example, for health and safety (this will be seen in Chap. 5).
  - Organizations that have deployed Lean, as described in Sect. 2.1, have already made their employees a priority.

- Environmental actions, in particular in Europe, are already included in the ***Strategic Objectives***[40] of companies, in addition to being constrained by the European CSRD (Corporate Sustainability Reporting Directive) and European standards on durability information (European Sustainability Reporting Standards or ESRS) which have been gradually coming into force since January 1st, 2024 (Based on "**double materiality**", companies need to identify "outside-in" and "inside-out" impacts, risks and opportunities").
- **The main elements for I4.0 are often the same as those for I5.0.** Moreover, apart from the prospective technologies that will be seen in Chap. 9, there are no new technologies other than those described in Sect. 2.3. Publications of Xu et al. (2021) mention Bio-inspired technologies, for example, but it is important to say that these are not new and are not, strictly speaking, digital.

• Other comments concern the relevance of refocusing (if necessary), as suggested by I5.0. This should not be an issue. Even if the ideas are not new, it will be beneficial if academics and businesses agree to position or reposition humans and the planet in the center of the digital contribution, as a fundamental challenge.

Consequently, whereas the content and the goals of I5.0 completely make sense for I4.0, there is the question of whether the shape must change. Some authors such as Xu et al. 2021 also raise the question even if they study I5.0: **a word of caution is that a proliferation of buzz words**, such as Industry 4.0 +, Industry 4.5, and even Industry 6.0 and Industry 7.0 in a not-too-distant future, may dawn upon us. These buzz words may be inviting for paper-writing or grant applications; they are not conducive to making any business decision and facing technological challenges.

Similarly, as we will see in the next chapter, the question is also if the term "industrial revolution" is used appropriately.

References

Achtenhagen, C., & Achtenhagen, L. (2019). The impact of digital technologies on vocational education and training needs. *Education Training, 61*(2), 222–233. https://doi.org/10.1108/et-05-2018-0119

Barraud, B. (2020). *L'intelligence de l'intelligence artificielle*, hal-02327501v3, extract from L'intelligence artificielle—Dans toutes ses dimensions (L'Harmattan, 2019)

Barrand, J., & Moïchine, S. (2024). *Histoire de l'agilité dans les organisations.* Harvard Business Review France. https://www.hbrfrance.fr/organisation/histoire-de-lagilite-dans-les-organisations-60415

Berthou, V. (2022). *Les deux racines conceptuelles de l'agilité*, 3 novembre 2022. *Cahiers Costech*, No. 5. https://doi.org/10.34746/cahierscostech151

Brown, S. (2024). *Machine learning, explained.* MIT Management Sloan School. https://mitsloan.mit.edu/ideas-made-to-matter/machine-learning-explained

[40] A more precise definition is found in the Glossary (See No. 5).

Caggiano, M., Semeraro, C., & Dassisti, M. (2023). A metamodel for designing assessment models to support transition of production systems towards Industry 5.0. *Computers in Industry, 152*, 104008. ISSN 0166-3615. https://doi.org/10.1016/j.compind.2023.104008

Cervone, H. F. (2011). Understanding agile project management methods using Scrum. *OCLC Systems & Services: International Digital Library Perspectives, 27*(1), 18–22. https://doi.org/10.1108/10650751111106528

Coelho, P., Bessa, C., Landeck, J., & Silva, C. (2023). Industry 5.0: The arising of a concept. *Procedia Computer Science, 217*, 1137–1144. ISSN 1877-0509. https://doi.org/10.1016/j.procs.2022.12.312

Dombrowski, U., & Richter, T. (2018). The Lean production system 4.0 framework—Enhancing Lean methods by Industrie 4.0. In *IFIP international conference on advances in production management systems (APMS)* (pp. 410–416). Seoul, South Korea. https://doi.org/10.1007/978-3-319-99707-0_51 (hal-02177896)

Fani, V., Bucci, I., Rossi, M., & Bandinelli, R. (2024). Lean and industry 4.0 principles toward industry 5.0: a conceptual framework and empirical insights from fashion industry. *Journal of Manufacturing Technology Management, 35*(9), 122–141 (Emerald Publishing Limited 1741-038X). https://doi.org/10.1108/JMTM-11-2023-0509

Feng, C., & Ali, D. A. (2024). Levering digital transformation and ERP for enhanced operational efficiency in manufacturing enterprises. *Journal of Law and Sustainable Development*. https://doi.org/10.24857/rgsa.v18n3-101

Furlan, A., Grandinetti, R., & De Toni, A. F. (2023). Managing the Lean-Agile Paradox in complex environments. *Systems, 11*, 258. https://doi.org/10.3390/systems11050258

Found, P., Lahy, A., Williams, S., Hu, Q., & Mason, R. (2018). Towards a theory of operational excellence. *Total Quality Management & Business Excellence, 29*(9–10), 1012–1024. https://doi.org/10.1080/14783363.2018.1486544

Haranas, M. (2023). *AWS, Microsoft, Google's Cloud Market Share Q1 2023*. CRN. https://crn.com/news/cloud/aws-microsoft-google-s-cloud-market-share-q1-2023

Godfrain, J. (2006). Le viaduc de Millau Tour Eiffel du Larzac. *Le journal de l'école de Paris du management, 60*(4), 7 12. https://doi.org/10.3917/jepam.060.0007

Julia, L. (2019). *L'intelligence Artificielle n'existe pas* (1st ed.).

Kischelewski, B., & Richter, J. (2024). Implementing large-scale Agile—An analysis of challenges and success factors. In *28th European conference on information systems (ECIS)* (p. 176). Marrakech, Morocco. https://hal.science/hal-04603145v1

Malone, T. W., McGovern, P. J., Rus, D. A., Viterbi, E., & Laubacher, R. (2020). Artificial intelligence and the future of work. MIT work of the future. *Research Brief, 17*. https://workofthefuture-taskforce.mit.edu/wp-content/uploads/2020/12/2020-Research-Brief-Malone-Rus-Laubacher2.pdf

Năftănăilă, I. (2008). *Critical analysis of the Scrum Project management methodology* (pp. 436–440). Academy of Economic Studies Bucharest, Management Faculty.

Maddikunta, P. K. R., Pham, Q.-V., Prabadevi, B., Deepa, N., Dev, K., Gadekallu, T. R., Ruby, R., & Liyanage, M. (2022). Industry 5.0: A survey on enabling technologies and potential applications. *Journal of Industrial Information Integration, 26*, 100257. ISSN 2452-414X. https://doi.org/10.1016/j.jii.2021.100257

Sarker, I. H. (2021). Deep learning: A comprehensive overview on techniques, taxonomy, applications and research directions. *SN Computer Science, 2*, 420. https://doi.org/10.1007/s42979-021-00815-1

Sliger, M. (2011). *Agile project management with Scrum*. Global Congress 2011—North America, Dallas.

Vinardi, C. (2013). *Le Lean : atouts, impacts et limites* (Vuibert ed.). Paris.

Vinardi, C. (2019). Thesis on industrial engineering. In *Les défis du Lean à l'ère de la mondialisation et de l'industrie 4.0*. Université de Technologie de Compiègne, Français. NNT:2019COMP2500 (tel-03201462)

Vinardi, C. (2023) *Business strategy with Hoshin Kanri.* SpringerBriefs in Applied Sciences and Technology. https://doi.org/10.1007/978-3-031-20963-5_1

Xu, X., Lu, Y., Vogel-Heuser, B., & Wang, L. (2021). Industry 4.0 and Industry 5.0—Inception, conception and perception. *Journal of Manufacturing Systems, 61*, 530–535. ISSN 0278-6125. https://doi.org/10.1016/j.jmsy.2021.10.006

Chapter 3
A Little Bit of History

Abstract This chapter reflects on history to understand the development of the Lean System and Digital Technologies, examining the key stages and timelines involved. This retrospective analysis will help address whether these changes represent a revolution or a transformation. Additionally, it will explore the role of constraints in scenarios of change and innovation and identify the ideal conditions for adopting new managerial and technological solutions. The chapter concludes with initial lessons learned from history regarding two promising and relevant solutions: Artificial Intelligence (AI) and the Metaverse.

In itself, going back in time doesn't help to look at the future but understanding history is fundamental to human evolution, as sometimes even history repeats itself. In any event, there is no crystal ball, but the analysis of the past and present is useful for learning, and setting perspectives too. Just like using the rearview mirrors when driving, looking back is useful to anticipate parts of the problems and risks ahead.

This historical retrospection also helps to answer the hard questions set at the end of the previous chapter.

? Concerning I4.0, should we be speaking more of industrial revolutions or evolutions? Is it appropriate to talk about a digital revolution? Similarly, should we talk about a digital evolution or transformation?

The word revolution comes from the Latin "revolvere" ("re" for back and "volvere" for "roll"). It can mean a "**change of direction with respect to oneself**". In practice, it applies to situations where **rules are suddenly shaken.** The word transformation comes from the Latin "transformare" ("trans" for through or beyond and "formare" for "form"). It can be translated as "**changing form".** The action can be **slow or fast (breakthrough).** The usage of the words revolutions and transformation are addressed after the summary of key dates.

I suggest a retrospection on the Lean and Digital domains. Table 3.1 does not claim to be exhaustive, nor does it list all the discoveries that have had a long-standing impact on society and industry, such as plastic, for example. But it has important

C. Vinardi, *Sustainable Performance in the Digital Age*,
https://doi.org/10.1007/978-3-032-01252-4_3

dates, right back to 300 BC. Indeed, both Lean and Digital had their origins at the time. Here are some initial comments:

- The Digital column also includes dates in the development of "Quantum Intelligence" (QI). This will be seen in the prospective part in Chap. 9.2. Whereas AI is based on mathematics, QI is based on (quantum) physics and is part of future technologies, whether or not combined with AI.
- The Lean column mentions *Karakuri*, which is a type of low-cost automation that we will look at in Sect. 5.1.3.1.

3.1 Overview of Key Dates and Comments

Table 3.1 Key dates in the Lean and Digital/Quantum domains

	LEAN events	DIGITAL and quantum events
300BC		**Algorithm:** Best-known mathematical algorithms (Euclid)
100 to 400BC	**Problem solving:** WhoWhatWhereWhenHowWhy (In Latin Quis? Quid? Ubi? Quibus auxilius? Cur? Quomodo? Quando?) used by Roman investigators	
	Standards: First documents to produce vases in a repeatable manner	
820		**Formalization of algorithmics** by Al-Khwarizmi who explored the solution to equations. The word algorithm comes from his name
807	**Automation:** A water clock that moved as the vase was filled was a gift by the King of Persia Haroun-Al-Rachid to Emperor Charlemagne	
1200	**Karakuri automation**: Automation without an energy source was first defined in Japan	
1500–1550	**Standard line production**: A "line-based" production system, where hulls of boats move along a line and various parts are assembled according to a schedule. This is set up in the Venice Arsenal, using work standards	
1623		**Calculator:** first instance of a calculating machine by Schickard
1765–1775	**First industrial machines**: The steam engine by the Scotsman James Watt and the powered loom in the United Kingdom	

(continued)

Table 3.1 (continued)

	LEAN events	DIGITAL and quantum events
1780	**SMED:** The French Army used a system of interchangeable components for fast changes as described in the SMED	
1822	**Autonomous production unit using machines:** Springfield Arsenal in the USA set up autonomous production cells with automation. (precursor of *Jidoka*).	
1830s	**First uses of newly discovered electricity**	
1900–1905		**Additive manufacturing:** Its precursor, the photosculture, appeared in 1860 invented by François Willème **Quantum physics principles** devised by A. Einstein with the theory of relativity and the radiation of black bodies, which when heated radiate with energy is released in package ("quantum" of energy) – by M. Planck
1910–1915	**Continuous flow:** Made popular in the assembly lines of H. Ford	**First robot** is an "Electric dog" in 1915, having sensors designed by Hammond and Miessner
1930–1935	**Takt time**: In Germany, aircraft fuselage was produced in pace with demand	**Transistors** invented by Norman B., are the basis of Analog to Digital integrated circuits with the Alan Turing "machine" in 1936
1947		**Premier transistor fonctionnel** inventé par les physiciens John Bardeen, Walter Brattain et William Shockley **AI:** Alan Turing publie son premier article sur l' "Intelligent machinery"
1950–1960 **1956–1957**	**Principles of the Toyota production system** began without being properly devised. Japanese companies including Toyota, suffered from a lack of materials and human resources, so made only a small series at a time focusing on production quality **Robot automation:** The first robots used by car manufacturers to automate some tasks in the 1960s	**Computer**: the first scientific computers appeared in 1952 with IBM and its early period with the military and government **AI:** The term was used for the first time in 1956. The first AI-generated music: "The Illiac Suite" in 1957

(continued)

Table 3.1 (continued)

	LEAN events	DIGITAL and quantum events
1960s		**Virtual reality/metaverse:** first immersive concept CAVE (Cave Automatic Virtual Environment), and films on the virtual universe **Quantum physics applications:** first inventions (laser)
1969		**APARNET:** The first precursory Internet network in the US
1970–1980	**Toyota:** business performance inspired the whole industrial world, especially during the oil crisis	**AI Winter no. 1**: "Winter" refers to has period where the enthusiasm cooled down drastically due to failures and slow results
The early 1980s	**Formalization of the Toyota production system** **6SIGMA**: Started at Motorola in 1986	**MINITEL:** French video-text device having a screen and a telephone connection, the first successful application of online services for the general public **Additive manufacturing**: first patent
1985–1995	**LEAN system** appeared in 1988 in an article by J. Krafcik ("Triumph of the Lean Production System"), officialized in 1996 with "The Machine that Changed the World" by J.P. Womack, D.T. Jones and D. Roos of MIT	**IoT:** First connected objects **AI Winter no. 2** due to overhyped projects, and economic downturns
The early 1990s	**TPS & HOSHIN KANRI:** first instances outside Japan **6SIGMA**: used and enhanced by GE **AGILE:** the term used for the first time in IT teams in the USA	**Additive manufacturing:** first 3D printer built **World-Wide Web** created and accessible to the general public
2000s	**LEAN, 6SIGMA, OPEX, AGILE:** widespread use of the terms ….that companies try to replicate and to adapt with more or less success	**Quantum calculator**: first calculator built **Deep learning:** The first version with various layers in 2006
2010		**AI acceleration** with Siri in 2011 and Big Data
2011		**Industry 4.0:** introduction of the term in Germany
2012		**ALEXNET:** Deep Learning marked a breakthrough in image recognition

(continued)

Table 3.1 (continued)

	LEAN events	DIGITAL and quantum events
2016		**Augmented & virtual reality** (AR and VR): first VR visors and first general public AR application with the game Pokemon Go
2022		**ChatGPT:** OpenAI launched ChatGPT based on Deep Learning (GenAI)

Table 3.1 can be read horizontally or vertically.

- Starting with the vertical reading for Lean, several points stand out:
  - Most **of the parts of the Lean system took many centuries to take shape**, mainly through trade and political exchanges with Asia, Europe, and Africa.
  - The very concept of automation existed well before any so-called industrial revolution.
  - Whereas Lean began to spread in the '90 s, despite not needing anything more than an investment in human resources, **the implementation of its full potential** is far from established even in 2025 (this will be seen in Sect. 5.1.2).
- Concerning the Digital/Quantum column:
  - The implementation of both digital and quantum technologies **was not built in a short time but took several decades,** similar to Lean.
  - The development of AI has taken quite some time since it began in 1956, even with the help of the Internet in the 2000s.
  - QI is based on the principles of **quantum physics which were understood back in the early 1900s,** and early applications such as laser, date from the 1960s.
  - The main components of I4.0 are presented in Sect. 2.3.3 from a historical perspective. These have been around for several decades (Vinardi, 2019). A few dated entries are in the table for IoT, Additive Manufacturing, VR, and metaverses, to which we can also add:

    Automated Guided Vehicle (AGV): the first ones date back to the 80 s.
    BAR codes: appeared in the 1950s, and became widespread in the 1970s.
    RFID tags: were created in 1960 but deployed in 1990.
    QR codes: used in Japan in the 1990s by the Denso car company.
    Cloud: came with the Internet, with a general use in the 2010s.

  - I cannot resist adding a final example, which was not included in Table 3.1 because it is rather anecdotal, yet it symbolizes the development time in digital technology well: **Videoconferences with both text and images are quite recent, but the first one was given in December 1968 by the American engineer Doug Engelbart!**

Reading Table 3.1 horizontally, other points stand out:

- Before the early 1900s, very little happened in the Digital arena, but performance levers, such as those consolidated later in Lean, were already apparent and were in use.
- Whereas the "electronic revolution" is associated with the 1970s in most academic articles, still **forty years of growth were needed for electronic applications to be widespread**.
- The 1970s were a flourishing time for both the fields of Digital and Lean. Indeed, companies became interested in Toyota during the first oil crisis. It provided the opportunity for Toyota to gear up on digital technologies and robotics **(a second significant exchange between Lean / Digital).**
- In the 2000s, digital technological bricks were strengthened, and Lean started its acceleration using (non-technological) bricks that had already shown their performance.
- Lastly, **Digital solutions emerged or were accelerated by overcoming technical and technological constraints** (for example, electronics leveraged on transistors and printed circuits, and digital technology leveraged on the Internet).

Lean was helped by other types of constraints such as economic, social, and societal. Here are some examples:

- 1500–1550: the Arsenal of Venice had to find a way to build a fleet in record time (100 ships in two months where other naval yards produced 1 ship per month) while minimizing costs and resources because the war with the Ottomans threatened the Venice province.
- 1822: the Arsenal of Springfield developed small autonomous production cells, which they automated to make various types of weapons quickly and easily because the area was in the grip of conflict, but men were more valuable on the battlefield than building weapons.
- 1930–1935: the production of aircraft fuselages kept pace with demand because Germany needed to build aircraft, but at least cost and especially, in the least space so as not to raise suspicions from the Allies.
- 1945: at the end of the war, Japan had to rebuild itself but lacked all kinds of human and material resources. The country, which was not known for its production quality before the war, also had to change its reputation if it wanted to regain its place.
- 1970–1980: the oil crisis forced companies to re-examine their performance model, hence their interest in Japan and Toyota in particular.

The reading of Table 3 being now made allows us to return to the relevance of the terminologies "revolutions and transformations" that I discussed at the beginning of the chapter.

3.2 Revolution or Transformation?

The word revolution often applies to large industrial periods and also to the digital world alike. When it refers to industrial periods, scientific publications state that (Vinardi, 2019):

- The 1st revolution is said to come from water and steam driven mechanization, transitioning the move from manual to machine production in the second half of the eighteenth century.
- The 2nd revolution is said to have come from mass production on electric assembly lines at the end of the nineteenth century.
- The 3rd revolution is said to have come from automation using electronics and Communication and Information Technologies (CIT) in the 1970s.
- The 4th revolution came from the large-scale development of the Internet in the 2000s and the new era in digital technology (and the power of data management that came with it).

Each of these transitions was slow, over a few years, so the word revolution can be put into question. For historians, industrial revolutions have rather a long *process*,[1] which notably did not take place at the same time everywhere and happened differently according to the country (Vinardi, 2019).

Even if the term "revolution" is often used, it doesn't seem to correspond to what happened, either in the industrial or digital context. For some, that particular revolution hasn't happened yet insofar as it hasn't yet produced tangible results (mainly because usage is still small). For the most pessimistic, the subject is also criticized because Cyber Physics Systems are part of "engineers' dreams" and are supported by governments that do not have even the basic knowledge of how to get the best in production performance. The interest of Germans in Industry 4.0 could be explained by the 250 million euros provisioned to explore its potential (Vinardi, 2019).

If we put aside the notion of Industrial Revolutions, referring to any "Industry X.0" is therefore not an issue.

As regards to "transformation" which is a change of form, whether slow or rapid, at least the term seems fitting, concerning digital or industrial transformations, or even societal evolution, since the term "Society X.0" is also sometimes used.

The way to engage these transformations needs to be defined. Let us look again at aspects highlighted in the historical summary in Sect. 3.1 with the role constraints play in changes and innovation (whatever the type of activity: scientific, organizational, managerial, industrial, etc.).

[1] A more precise definition is found in the Glossary (See No. 2).

3.2.1 Do We Need Constraints to Change and Innovate?

The role of constraints, whatever their nature, is addressed in both Lean and Digital methods. That seems to be what drives change. Let us have another look at the key stages of lasting change in a simplified way, as in Fig. 3.1:

$$\text{Awareness} \rightarrow \text{Way of thinking/behaving} \rightarrow \text{New habits} \rightarrow \text{Culture}$$

The grieving curve shows how to give up a way of thinking or a behavior pattern. It is applied concurrently until the step where new behavior sets in.

Awareness is the first step for change; it must be factual AND emotional to engage in change. The emotional part, either negative or positive, is the trigger that gives the impulse. The factual part is also an acknowledgment: the rationality to drive change. **Constraints are facts that generate emotion.**

After the awareness step, constraints **generate ideas and, therefore the corresponding new behavior and new actions** that follow. New behavior must be relevant to the teams to carry out the next step concerning habits. Indeed, training needed to have habits often generates some measure of pain. The compensation will be the relevance and motivation. New practices are anchored with determination and are soon effortless because they will be part of the organization's culture (Vinardi, 2019).

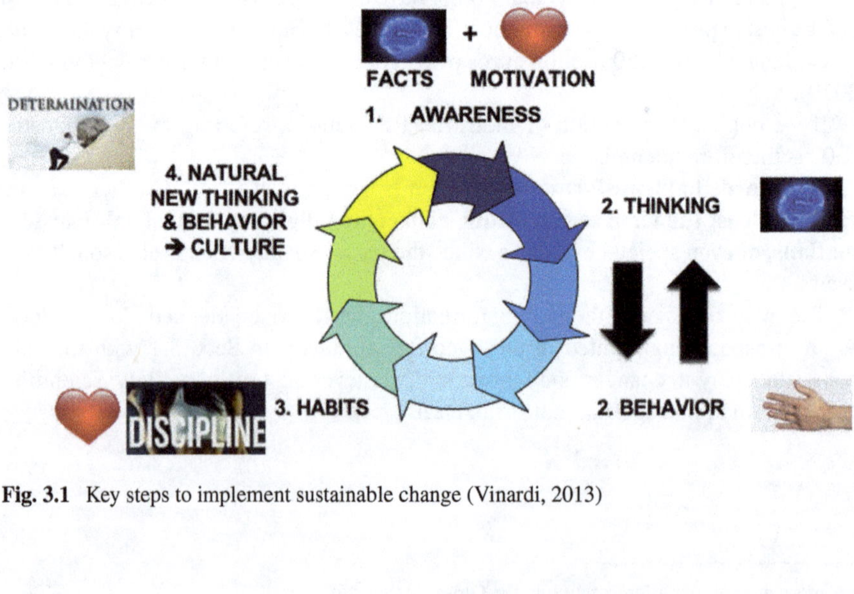

Fig. 3.1 Key steps to implement sustainable change (Vinardi, 2013)

The change is to be defined, understood, and supported. The main components are as follows:

- Purpose: why
- Action: what
- Direction: where
- Way or "choreography": how
- Resources and connection: with whom
- Scale effect: how much
- Capacity: limitations in change management
- Development: training
- Flexibility: ability to adapt.

The more companies clarify and work on these points, the more they will be empowered to change. These companies are generally found in the 1st quartile regarding profitability (Vinardi, 2019).

Now that the key steps of change are clarified, let us look again at the role constraints in the *process*[2] involved in the creation, innovation, and therefore decision making.

- **Creativity thrives under the right amount of pressure**. This is the **"Paradox of Choice":** our brains struggle to *process2* all the possibilities, often **resulting in indecisions.** Almost everyone faces a paradox of choice every single day of their life. **Having a choice is not a bad thing, but having too many choices can be a bad thing** (Schwartz, 2015).

  Decision fatigue is not a myth; being tired of making decisions is incredibly common and surprisingly draining. In order to **reduce it, we create habits, we take breaks, and we can limit our choices.**
- **Constraints can force decision**, narrowing your focus through a funnel. Moreover, the right sort of constraints can, in fact, "stoke the fire of innovation." Academic research confirms day-to-day experience (May, 2013).

  Sometimes it seems there is no obvious solution, but situations can make sense, as in the case of the Arsenal of Venice or the rebuilding of Japan and its industry, which led Toyota to reinvent itself. Even if we often have a choice, the budget constraint of our organizations is always there. However, that is not the only constraint to be "challenged," and in any case, it does not necessarily yield the expected creativity. This is debatable (May, 2013).

? How then can constraints create positive tension?

- **With a well-scoped and defined problem** (which focuses on the right teams) **and/or challenging (and relevant) goals.**

[2] A more precise definition is found in the Glossary (See No. 2).

Example: In the kingdom of France in 1665, military expenditure drained finances that were already strained by a severe economic crisis. It became **critical to create jobs but also to minimize imports** (Strategic Objective). With this aim, Louis XIV resolutely innovated by **developing exceptional industries** (Strategic Action). At the time, France imported twice as much as it exported, and every country preciously kept its specialities secret. In the face of these difficulties, luxury goods were developed across royal manufacturers using technical and scientific innovation and new forms of know-how and craft. However, its success rested on less respectable methods: industrial espionage, systematic head-hunting, and, if need be, pure and simple abduction of people for their skills.

The wager on the luxury sector paid off twice, albeit using some dirty work, by filling the state coffers while contributing to making the King and his kingdom shine. Versailles imposed French luxury in Europe, and more than two centuries later, France remains an international emblem of the luxury business, from haute couture and cosmetics to high-class wines. A **sophisticated industry was created from nothing** to conquer new markets due to the ambition of Louis XIV and the *vision*[3] of his minister, Colbert (Arte, 2020).

- **Seeing the opportunity in a constraint.** Embrace voluntary constraints on your freedom of choice instead of rebelling against them.

Example 1: Let's take the case of Nutella. Before the war, an Italian pastrycook was renowned since the 1920s for his block of chocolate called Gianduja. His name was Pietro Ferrero, a pastrycook from Alba, a small town of the north of Italy - the place where the Ferrero Group is still located today. At the time, the paste was hard to cut. In 1946, a **cocoa shortage** pushed the pastrycook to review his recipes to continue to do business. In that area, **many hazelnut trees were not harvested much,** and he had the idea **to mix as much hazelnut as possible, with "just enough cocoa"**, but it was still a hard paste to cut. In 1949, according to the legend, a heat wave melted the loaf of chocolate. Children ate the paste by spreading it out onto bread, and it made them an enjoyable snack. **The Ferrero brothers took on the idea** of making their chocolate loaf into a paste spread, which still has a huge success today.

Example 2: At the end of the 1960s in Japan, there was a heavy demand for leather. At the time, the country lacked all kinds of animals, including cattle, to meet demand. So the Japanese tried to create an **alternative that looked like leather. Artificial leather** was invented in the 1970s by the researcher M. Okamoto by combining advanced skills of the chemical industry and this of the traditional weaving industry.

Taichi Ohno, one of the founders of the Toyota System (TPS), used to say:

[3] A more precise definition is found in the Glossary (See No. 4).

> *"Need is the mother of creativity."*

He reiterated in fact, the motions of constraint which science has since greatly supported. To this could be added a quotation attributed to Galileo.

It is an important condition for change and innovation but not the only one. Two other points must be mentioned, the first is the **role of doubt**, which can be presented with a quote by Galileo:

> *"Doubt is the father of invention."*

It is completely in line with the role of scientists looking to check data to verify WHETHER they are right (leaving room for doubt), whereas a person who does not doubt will look for data to verify THAT they are right. **This is called confirmation bias** (Georget et al., 2024). When inventing, creating, and changing, **curiosity** is needed **to call into question existing ideas and solutions (requiring an open state of mind). One other key point is intuition.** In research in this field, intuition is defined as the rapid and automatic recognition of patterns or relationships between elements in our environment. Often relevant, it can also deceive us, by our emotions in particular (Gobet, 2019; Georget et al., 2024).

Intuition involves long-term memory. Gradually, **as learning progresses, there is a shift from conscious to non-conscious processing**. With beginners (whether in medicine, chess, or elsewhere), problem resolution is conscious at first. Work is necessary to develop intuition to acquire reflexes. How? By choosing a field we enjoy preferably (the role of emotion), with expertise, thought will become faster and is almost completely intuitive in the end. **There are links between creativity and intuition** (Gobet, 2019).

As soon as you have a vast amount of knowledge in a field of expertise, this knowledge is encoded in the form of patterns, and these can emerge at any time according to the constraints of the creative work. This opening up the possibility of making discoveries in this domain. Is anyone able to think outside the box? **Prudence is needed, namely when the field of innovation requires solid expertise; in this case, all those who make true innovations necessarily have great knowledge in that field** (Gobet, 2019).

It is important to encourage **divergent and wandering thoughts** to use intuition creatively.

For that, the brain needs to be at rest and free from restraining tasks (such as screen, activities…). To awaken creativity, calm is needed: in the shower, on walks in the forest, listening to music… these situations are often mentioned by the creative. (Bohler, 2024).

In addition, 80% of the Nobel Prize winners questioned said that ideas appeared "in a flash" (Georget et al. 2024), that seems to confirm it is necessary to let the brain rest to invent better.

The winning ingredients for change and innovation seem to be: **to have a real need, to doubt, to be curious, to use its intuition, and to have divergent thoughts.Being able to create the necessary emotions and the discipline to change one's habits in the long term** (by aligning emotions, thinking, and behavior) seems to be necessary to complete the changes.

3.3 Necessary Conditions for Mass Adoption

Innovating and being creative means offering a new service or a new product, and also means **finding a way to reinvent itself** to do so, i.e., by changing the organization or the Management mode, by changing the *process*[4] and corresponding standards or changing the skills. Adopting new ways of managing by optimizing the ways of working implies creating or benefiting from favorable conditions for large-scale adoption.

The following aspects can be found frequently in literature and experience and create massively adopted favorable conditions in personal and professional domains alike:

- **Common needs**: match with real needs of a majority of people.
- **Benefits:** provide a benefit for the users (safety, working condition, services…).
- **Costs:** be affordable.
- **Protection**: be secured (with cybersecurity for Digital) and respectful of local regulations to protect people and society (legal framework).

These conditions are necessary but not sufficient because the adoption of a new approach or a new technology, for example, must also **be relevant to the organization** (in terms of interest, strategic coherence, etc.) for all employees. The organization first needs to gain **maturity to make use of these conditions.** We will see this in greater detail in the implementation of the two approaches in the following chapters.

? Nevertheless, are there guidelines for making choices, in particular, of technological ones?

[4] A more precise definition is found in the Glossary (See No. 2).

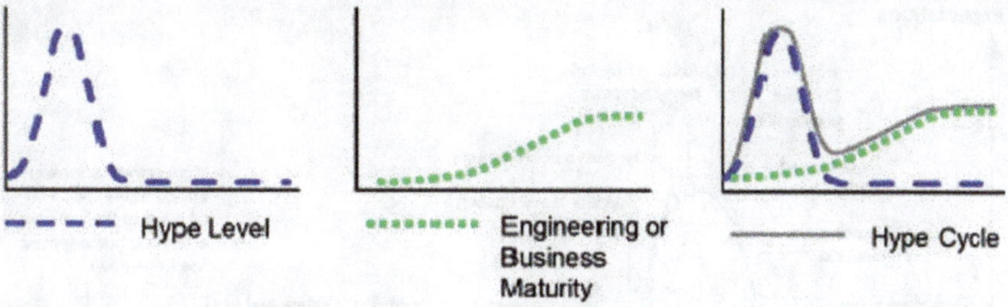

Fig. 3.2 The two curves that form the hype cycle (*Source* Steinert et al. 2010)

Some organizations offer expertise aimed at helping companies choose innovation. For example, **Gartner has its Hype Cycle to chart the maturity and adoption of technology** (and also the systemic methods, similar to Lean). There are few models as well known and used as this. That is why I suggest a short introduction.

According to Gartner, the model is supposed to help avoid adopting technologies too late or too soon, giving up too soon, or hanging on too late while adopting. Gartner's hype cycle is one of the most prominent and influential consultant models for advising large companies on their technology strategy and investment decisions (Steinert et al., 2010). Two curves form the Hype Cycle as it is represented by Fig. 3.2. Gartner's model is made of five steps and all in all, it can take 3 to 5 years to reach the last one as represented in Fig. 3.3. It is also accepted that some technologies disappear before the end of the find steps.

Theoretically, the model has not been mathematically defined properly; neither the dependent nor the independent variable can be operationalized or quantified, as its definitions remain vague. Empirically, it is not easy to compare hype cycles over time; links with previous and/or following years are not provided, and the titles of the hype cycles can change over time. Individual technologies do not necessarily behave according to the hype cycles model assumptions (Steinert et al. 2010).

Since 2010, it is expected that Gartner has included a mathematical formulation to collect, analyse and report the evolution of technologies consistently and in a structured way as it was one of the weaknesses reported. However, as the form has not evolved, the second suggestion made by Steinert et al. is still relevant: a better contemplation and integration of existing research of the two underlying phenomena, hype curves and technology s-curves (with a discontinuity between maturity and adoption) could provide a more stable theory foundation.

The pathway of innovation is more complex than it appears. Gartner is not responsible for not having a crystal ball, but a lot can be learned from Gartner's database and method of analysis, in particular in *benchmarking*.[5] However, it is **recommended to take a step back when using predictions,** particularly when the Hype Cycle model is used without a strong scientific standard.

[5] A more precise definition is found in the Glossary (See No. 1).

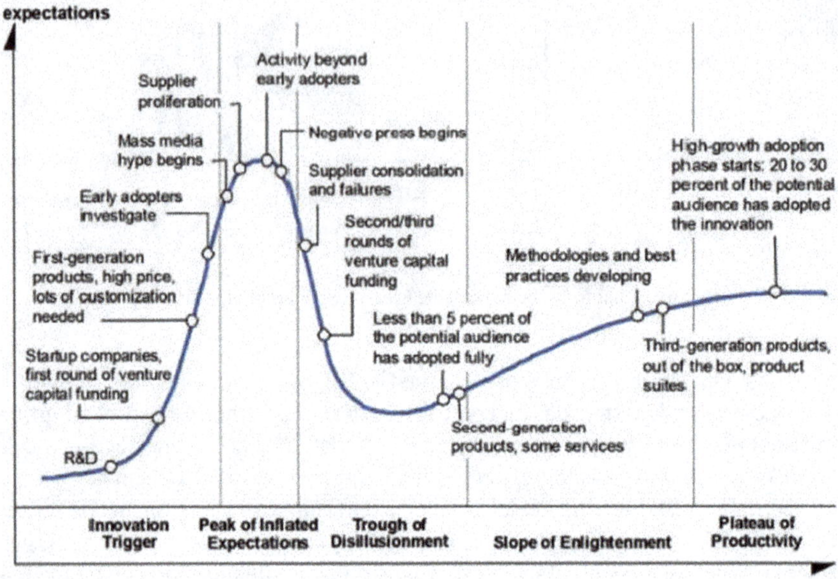

Fig. 3.3 Hype cycle and its stage indicators (*Source* Steinert et al. 2010)

3.4 Lessons from AI and Metaverse Adoption and Maturity

Let us take two notable cases, one from AI which is an obvious digital transformation, and that of Metaverse which was announced in 2021 as the revolution for the future.

3.4.1 Lessons from AI

First, let us note a few dates: ML (machine learning) appeared in 1997 but was officially created in 1956, DL (Deep Learning) in 2006, GenAI (Generative AI based on mainly unsupervised DL) in 2021 and ChatGPT which was founded on GenAI, at the end of 2022 for a proposed use in 2023 (it is not a technological breakthrough, but a usage breakthrough).

In the middle of the 1950s, waves of concern followed one after another, each time with the same prophecy: humans would eventually be replaced by machines. However, these predictions did not come true.

One date that is not on the historical list deserves a mention. On November 7th, 2019, a communiqué published in Nature stated that "*Neural network can discover physical concepts without any prior knowledge*". However, heliocentrism is known through the works of Nicolas Copernicus in 1533. So what is new?

AI discovered that the Earth spins around the Sun thanks to a network of neurons known as "SciNet", starting with experimental data, therefore from simple objective observations given by humans. SciNet understood that the Sun had to be in the center of the solar system depending on the movements of the stars and Mars as seen from Earth. Could AI, as such, discover new laws of physics in the future? It would be more fitting to say that these hypothetical discoveries are made **by man using the most sophisticated tools** because these AI tools are themselves made by man**.** (Barraud, 2020). Besides, let us have another look at Augmented Intelligence mentioned by Luc Julia in particular. The history of the techniques is patterned by **successive threshold crossing: ratchets of irreversibility.** For example, the invention of the printing press by Gutenberg led to the disappearance of copyist monks of the Middle Ages, and digital files and *streaming* meant that a fair number of CD players were no longer used. There is no doubt that **artificial intelligence does/will constitute a new - and very powerful - ratchet of irreversibility** (Barraud, 2020). In what time frame and how? It is difficult to answer.

Indeed, AI has already had two "winters": the first in the 1970s happened for lack of results and because it made too many errors, and the second was at the end of the 1980s/beginning of 1990s, mainly due to overhyped projects and economic downturns. **According to MIT, in the 1970s, some predicted that Gen AI would be generalized twenty years later, that is to say in the 1990s. Yet, we are there still very far….**

AI developed when the capacity (and speed) of computers increased, large amounts of data could be accessed by Big Data, and finally, when storage costs and computer prices decreased.

You might think that news of "major AI breakthroughs" would do nothing but help ML's adoption. **But ML made AI step out from the second winter with measurable success that hyped up AI in the first place**.

After all, improving measurable performance is supervised machine learning. Most practical use cases of ML—designed to improve the efficiencies of existing business operations—innovate in fairly straightforward ways. The purpose of **ML is to issue actionable predictions**, which is why it's sometimes also called **predictive analytics**. For example, by predicting which customers are most likely to cancel, a company can provide those customers incentives to stick around. Practical ML use cases like those that deliver the greatest impact on existing business operations, and the advanced data science methods that such projects apply boil down to ML and only ML. However, "AI" content and drivers suffer from an unrelenting, incurable case of vagueness. (Siegel, 2023).

In 2025, most people including professionals, cannot tell the difference between ML, DL, and GenAI. Often, it should be said, this is due to a lack of curiosity of Leaders but also because the popular media are not skilled educators. On this subject, it is not enough to "push" information. There are courses on offer, but besides these, it would be enough just to surf a few hours on scientific platforms reading articles in the public domain to acquire the relevant fundamentals. But that is often not done, and many people confuse AI with "Artificial GENERAL Intelligence" (AGI), which is supposed to reproduce human behavior one day. (This, we will see in the prospective Sect. 9.2). Finally, some people confuse AI with ChatGPT (the text generator), whereas it is only a small part of its potential. However, it is what the general public has access to, I grant you.

The opening up of ChaptGPT to the public is far from philanthropic, but few know the real reason: **the more users and requests on ChatGPT, the more its algorithm self-learns.** Not to mention that ChatGPT is pre-trained on an immense corpus of 500 billion text files and has a variety of training types. Continuous training based on requests from over 180 million active monthly users is a **huge advantage in gaining the market lead** (Tran, 2024).

In this particular market as in others, "when it is free", in general, "you are the product!"

Over the last decade, 250 billion dollars were invested in AI start-ups (first in the United States then in China, according to Stanford Institute's AI index (Tran, 2024). Return on investment is expected by the markets or even by states just as it is in any economic model. That was the case in the past for AI. **When there is disappointment or the wait is too long, the chopper can fall again at any time.**

Yet many companies and organizations have taken up but a tiny fraction of the potential of ML available, and yet it is more accessible than DL. Will research on GenAI and the use of ChatGPT or Dall-E (its equivalent for image generation) produce the results the market expects in the years to come? Especially as the general public is not yet completely onboard. The recent rise to prominence of GenAI and the debates sparked by ChatGPT **specifically are the latest manifestation of both technological resistance and hype** (Willet & Na, 2024).

? Even considering many AI start-ups and researchers work on important subjects for humanity, will that be enough to avoid a third winter?

> *"**I predict we will see the third AI Winter within the next five years**... When I graduated with my Ph.D. in AI and ML in '91, AI was literally a bad word. No company would consider hiring somebody who was in AI." – Usama Fayyad, June 23, 2022, speaking at Machine Learning Week* (Siegel, 2023).

The stakes by the military may still "justify" investments. Indeed, the aim of war is not so much to win territories but to win **populations**. For a long time, that **was done by invading territory.** Today, populations are won remotely, thanks to *Data*, the "soft power" (Barraud, 2020). But in that case, we would not consider the same research and civil implications anymore, in the short run.

The adoption of AI needs improvements in its training models, its feedback on training, and the generated applications (as for GenAI). Apart from financial considerations and the conditions for massive adoption addressed in Sect. 3.3, the success of artificial intelligence also depends on the **trust given by the general**

public. This trust supposes transparency and traceability—as ethical and political requirements—whenever decisions are taken by algorithms to **avoid them being unexplainable**.

3.4.2 *Lessons from the Metaverse*

Quoting worldwide market figures in 2023 and projections for the Metaverse:

- $83.9 bn in 2023 up to 1300.9 bn by 2030 (Source www. marektsandmarket.com).
- $82 bn in 2023 up to 820 bn by 2030 (Source www.grandviewresearch.com).

A few years ago, 70% of start-ups in Silicon Valley were positioned on the Metaverse, whereas they are currently 80% on AI.

In 2022, the worldwide market was estimated at $800 bn for 2024, and we were going to spend an hour per day on it in 2026, according to an article in a well-known French economic newspaper (Meunier et al., 2022).

? So what happened?

In 2021, the idea of "the metaverse" reached **peak hype** with Mark Zuckerberg's announcement that Facebook was changing its name to Meta—and would spend up to $10 billion a year on metaverse development—reflecting the fact that it was becoming a metaverse-first company (Wu and Higgins, 2023).

However, already in 2022, several factors for the deployment of technology were missing or lacking. Taking the criteria given in Sect. 3.3, it can be seen **that technology was not accessible in terms of costs, and the need was still to be clarified.** Other crucial points were also missing (Meunier et al., 2022):

- **the legal status of avatars** was not yet defined (copy of the penal code?),
- the **risk of addiction was already mentioned,**
- **Web 3 technology** (where users have control over data using *blockchain*[6] technology, which the current state of Web 2 cannot provide) was not available.

The speculative bubble of *cryptoassets*[7] (mistakenly called cryptocurrency) was already predicted (the bankruptcy of FTX was a strong warning), yet exchanges in the metaverse are based on these. Lastly, the metaverse is based on unsupervised DL technology presented above, which was not yet—nor by far—completely mature in 2022. Neither the technology nor the infrastructure existed to enable the development of new immersive virtual worlds at scale (Dwivedi et al., 2022).

---

[6] A more precise definition is found in the Glossary (See No. 8).

[7] A more precise definition is found in the Glossary (See No. 8).

In 2023, Metaverse-related expenses were dragging down Meta's earnings, headset adoption was lagging expectations, user growth had stalled, and the promising shine of being "the next big thing" was starting to wear off the entire category. Meta's big pivot looked more like a costly misstep than a stroke of strategic genius (Wu & Higgins, 2023).

Back in 2021, it could already be seen that even without a crystal ball, the deployment of the metaverse would still take several years. In 2023 some still waited for short-term profit in this technology: manufacturers anticipate that the industrial metaverse could lead to a 12% gain in labor productivity, which might help address the ongoing labor shortages (Coykendall et al. 2024) The surveyed industrial leaders will certainly need to resort to other solutions to face their challenges in human resources in the short run.

Immersive technology will continue to develop, in particular with gaming, education, training, the development and sharing of expertise, and even perhaps remote management. At which pace and depth will our activities be concerned? Nobody can say yet.

References

Arte, T. V. (2020). *L'invention Du Luxe à La Française—Colbert*. https://archive.org/details/LuxeColbert

Barraud, B. (2020), *L'intelligence de l'intelligence artificielle*, hal-02327501v3, extract from L'intelligence artificielle—Dans toutes ses dimensions (L'Harmattan 2019)

Bohler, S. (2024). *Comment devenir créatif?* Cerveau & Psycho No. 169

Coykendall, J., Hardin, K., & Morehouse, J. (2023). Deloitte Insights: 2024 manufacturing industry outlook (2023). Deloitte Research Center for Energy & Industrials. https://www2.deloitte.com/us/en/insights/industry/manufacturing/manufacturing-industry-outlook.html

Dwivedi, Y. K., Hughes, L., Baabdullah, A. M., Ribeiro-Navarrete, S., Giannakis, M., Al-Debei, M. M., Dennehy, D., Metri, B., Buhalis, D., Cheung, C. M. K., Conboy, K., Doyle, R., Dubey, R., Dutot, V., Reto, F., Goyal, D. P., Gustafsson, A., Hinsch, C., Jebabli, I., et al. (2022). Metaverse beyond the hype: Multidisciplinary perspectives on emerging challenges, opportunities, and agenda for research, practice and policy. *International Journal of Information Management, 66*, 102542. ISSN 0268-4012. https://doi.org/10.1016/j.ijinfomgt.2022.102542

Georget, P., & Jaubert, E. (2024). *L'intuition face à la science: s'y fier ou s'en méfier?* Podcast. https://www.polytechnique-insights.com

Gobet, F. (2019). *L'intuition ça se travaille*. Cerveau & Psycho No. 112, Juillet 2019.

May, M. E. (2013). How intelligent constraints drive creativity. *Harvard Business Review*. https://hbr.org/2013/01/how-intelligent-constraints-dr

Meunier, L., Vairet, F., Wong, C., & Lemarchand, J. (2022). *Les sept questions qui fachent sur le métavers. Journal Les Echos du Lundi 30*. https://start.lesechos.fr/innovations-startups/tech-futur/exemplaire-gratuit-de-notre-journal-metavers-the-new-place-to-be-1410104

Siegel, E. (2023). The AI Hype cycle is distracting companies. *Harvard Business Review*. https://hbr.org/2023/06/the-ai-hype-cycle-is-distracting-companies

Staudt Willet, K. B., & Na, H. (2024). Generative AI generating buzz: Volume, engagement, and content of initial reactions to ChatGPT in discussions across education-related subreddits. *Online Learning, 28*(2), 1–24. https://doi.org/10.24059/olj.v28i2.4434

Steinert, M., & Leifer, L. J. (2010). Scrutinizing Gartner's hype cycle approach. In *Picmet 2010 Technology Management for Global Economic Growth*, 1–13.

Schwartz, B. (2015). *The paradox of choice. Why more is less* (Stephen Joseph ed.).

Tran, S. (2024). Stratégie. *La guerre des IA aura-t-elle lieu ?* Harvard Business Review France

Vinardi, C. (2013). *Le Lean : atouts, impacts et limites* (Vuibert ed.).

Vinardi, C. (2019). Ph.D., Thesis on industrial engineering. In *Les défis du Lean à l'ère de la mondialisation et de l'industrie 4.0* (Université de Technologie de Compiègne, Français, 2019) NNT: 2019COMP2500, tel-03201462.

Wu, A., & Higgins, M. (2023). *Metaverse wars*. Harvard Business School Teaching Note 724-400, October 2023.

Chapter 4
From Business Constraints to Sustainable Performance

Abstract Professional organizations have always been subject to a range of constraints and objectives that remain, for the most part, valid even today. These are the original and intrinsic performance criteria of commercial negotiation. After recalling these and their impacts, a summary of the new areas of constraints to be considered for the future will be proposed. Finally, their translation into additional performance criteria to be taken into account will make it possible to define all the filters that will be proposed to assess the sustainable deployment of Lean and Digital in the following chapters.

I have discussed the concept of "organization" several times, encompassing the activities of companies and organizations in both the private and public sectors as they relate to my subject matter. However, when discussing constraints and objectives, it is necessary to clarify what is common and what is not. Let's begin with the shared aspects.

All organizations have customers, even if they are called "patients" in the medical field, "members" for mutual companies, or "the general public/citizens" for administration and associations. All organizations have resources—both human and material—and they transform inputs (such as data, components, parts, documents, etc.) into outputs (including parts, products, advice, services, documents, etc.) and they also have *processes*.[1]

The divergences between the different forms of organizations appear mainly in the **governance** and the **financial** part, particularly with share capital and financial flows:

- **An administration** pays its employees and delivers outputs, most of which are not commercial (when a sum of money is requested, it goes back to the community). Its purpose is not to maximize its profit, and it does not distribute profits to its members. Nevertheless, the state borrows from markets and, for the most part, has to behave like a company by balancing revenues and expenditures.

[1] A more precise definition is found in the Glossary (See No. 2).

C. Vinardi, *Sustainable Performance in the Digital Age*,
https://doi.org/10.1007/978-3-032-01252-4_4

- **An association** operates on the same principles as an administration, but its members can be salaried or volunteers. It can also offer paid or free services and products but does not distribute profits to its members like the administration.

  It must balance its accounts (income and expenditure) according to the subsidies it receives. Governance operates on a democratic model (one member, one vote).
- **A capitalistic company (known as "classical," capital-driven)** remunerates its employees and owners (shareholders) and sells its output on the market to maximize its profit (in the more or less long term).
- **A mutual company** sells its outputs on the market to serve its member customers above all with a limited search for profit (service is supposed to prevail over the search for profit). It has no share capital (the funds come from the members) or shareholders (as it belongs to its member customers).

  It delivers services to meet members' expectations (as policyholders) while involving them in their governance (as insurers). Its model of governance is democratic (one person, one vote). It does not remunerate its members and sets aside the profit results (Seran, 2021).
- **A cooperative** is a form of business that shares some characteristics with a mutual company: it has no shareholders and is a group created by voluntary people with democratic governance. However, it has social capital and aims to maximize its profit while maintaining a strong social and cultural commitment to common projects.

The word "mutuality" expresses a need for solidarity, fraternity, and equality because it has its roots in the harshness of 19th-century European social history. It is characterized by the statute of partnerships and claims to be part of the corporate and social economy (Sammut, 2011). In Europe, the mutualist model is found in all the countries of the union, and the cooperative model is found in all countries for the traditional sectors (agriculture, wine-making), as well as in some industrial companies and banks. Mutual insurance companies and cooperatives are, therefore, not rare.

Capitalistic Companies and cooperatives are different from other categories by their goal, their finality, which is to generate and maximize profit. This implies constraints associated with profitability. Capital-intensive companies also have the specificity of remunerating their owners and have a specific governance model. We will choose this category to examine the most complex field of constraints (with the shareholder part in particular). **However, what is specified below will resonate for the other organization types mentioned above.**

4.1 General Business Constraints and Related Performance Indicators

4.1.1 General Business Constraints

Let's start with the external or extrinsic field constraints of organizations and, in particular, capitalistic ones. It should be considered that they generally operate in a **competitive market,** and so they must **be competitive** because **their customers do not face them alone.** The ability to offer a product or service that appeals to customers by standing out from the competition is key. My grandmother ran her small business as a "real businesswoman," and she had the "good sense" to say: "*You have to do what you must do so that customers want to come/return to buy from you rather than go elsewhere.*" In the language of the business, this means **meeting and anticipating customer needs (innovation)** and also often by reassuring them, in particular ensuring **continuity during the life of the product or service with after-sales service.**

This does not stop them from often competing with other categories, such as so-called international or multinational companies. These companies have been around ever since humans built ways to travel long distances, and they have grown enormously during the last century. "Multinationals" account for two-thirds of world trade; they represent half of world GDP and a third of manufacturing production. It is estimated that there are over 80,000 (Vinardi, 2016). Some deliver to customers all over the world with **location-related specifics** (distances, logistics, delays, languages, currencies, customs duties, legislation, availability of human and material resources, etc.). Suppliers are often also multinationals, and as a result, the transformation of inputs into outputs within international companies leads to the **management of competition-related complexity in a globalized market and with multiple players.** International companies have a diversity of national *cultures* with a powerful impact on values, behaviors, and management methods.

Culture is "collective programming" that is linked to the history of the group, **its values** (the importance given to things), its **beliefs** (the way things work), and its **habits and traditions**. There are **three levels of culture in a company**: the *culture* of the nation (countries and regions according to size), the *culture* of the function and the profession, and the *culture* of the company (Vinardi, 2016). We will return to this point because it is an intrinsic and extrinsic factor. We will return to this point because it is an intrinsic and extrinsic factor.

In Sect. 4.3, we will discuss three other extrinsic constraints that have recently emerged and are directly related to the subject of this book: constraints related to respect for the planet and society, the maturity of digital technologies, and the security of associated systems (equipment and data). These have begun to impact companies, even more so in the future.

We have already mentioned customers, suppliers, and the competitiveness they create (local, national, or global), which are company extrinsic factors. They lead to an intrinsic constraint: the **ability to meet and anticipate customers' needs**

during the life of the product or service. We have also mentioned another intrinsic constraint: the purpose of the company and its choice of organizational model. But there are others to take into account. Here are the main ones (Vinardi, 2019):

- Shareholders (if there are any, and their origin).
- Culture at the national and corporate level, including specificities relating to:
  - age and history,
  - sector(s) of activity,
  - size of the organization,
  - position (rank) in the value chain (Supply Chain).
- The kind of production.

? Let's take the case of shareholding. What is its impact on performance?

The first important point is that even if money and profit have always been at the heart of companies' concerns, **their importance in evaluating performance has increased with the financialization of markets**. Questions of governance and shareholding have been raised since the end of the nineteenth century, with the **rise of capital companies in Western capitalism**. After the 1970s, the gradual movement towards the liberalization of financial structures both at the international level (with financial globalization) and at the domestic level (with the deregulation of the securities markets and banking structures) changed the perspective (Darcillon & Rebérioux, 2020).

- **Family and independent shareholding tend to consider the long-term** and the **company's sustainability over time.**
- Even more so, **"Financial" shareholding** is when it is set up with a takeover, such as a leveraged buy-out or LBO (thanks to a bank loan that the acquired company itself helps to repay). In this case, shareholders will favor **short-term financial returns (over a few years maximum).**

Remember that the company's governance includes shareholder representation and defines the **strategy**. The type of shareholding influences the company's culture, which brings me to the second question about the field of cultural constraints.

? How can culture influence the performance of organizations?

In terms of influence on individuals, **the culture of the nation/region is the most influential,** followed by the culture of the profession/function and **finally the culture of the company**. However, when we talk about culture, it should be noted that we rely on trends and averages (Vinardi, 2019). This being reaffirmed, experience and scientific research converge to shed light on the relationship between culture and performance.

- The first impact concerns the performance of **multicultural teams** (with at least two different national cultures) that make up international companies.

  They are indeed a concentration of diversity. Beyond the mix of monocultural teams (those with the same nationality) with differences in jobs but also gender, personality, age, religion, social origin, etc., they include national differences, possibly with different languages. Figures 4.1 and 4.2 illustrate the two main impacts on performance.

  Multicultural teams **are slower to express a higher performance potential than monocultural teams** (Vinardi, 2016). Indeed, any team will go through various stages, particularly the storming step where (normal) friction appears. In this stage, they often underperform, but when the norming step starts, they can overtake the performance of classical teams. To do this, they will have to move from diversity (creating a team of different people) to inclusion (making the group perform by making each different people count).
- Differences only produce richness when combined to create interaction. This explains why **multicultural teams either outperform** their monocultural counterparts (when they pass the storming step) or **underperform** when they get stuck at this stage, as illustrated in Fig. 4.2.
- An international company is **often a set of multicultural teams, so the effect should be considered over all the teams.**

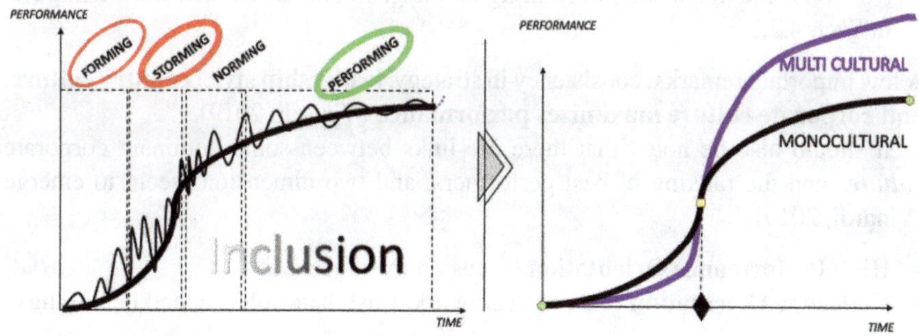

Fig. 4.1 Longer to develop a multicultural team…but better performance (*Source* Vinardi, 2016, 2019)

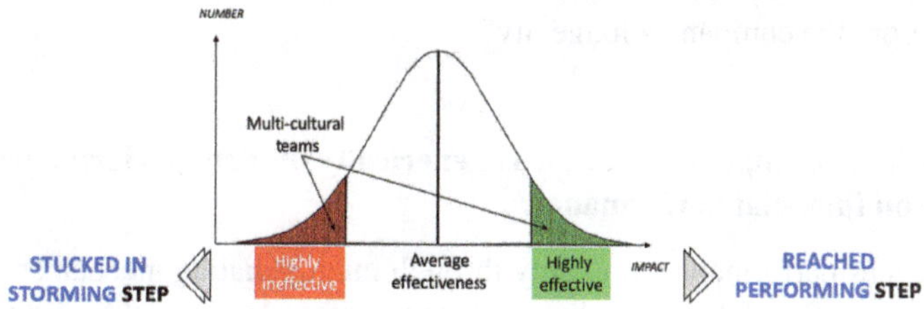

Fig. 4.2 No average effectiveness for multicultural teams (*Source* adapted from Adler, 2008)

- Apart from performance relating to multicultural factors, the management style and leadership must also be effective when dealing with people of different nationalities or regions. It was highlighted that **multinational companies must adapt their managerial practices to the national cultures in which they operate to achieve the best level of performance.**

  Differences must be **recognized, respected, and reconciled** by leaders who will have to adapt and create an inclusive approach. This goes against a management style that American business schools have predominantly disseminated: there is no **universal way to manage effectively.** (Vinardi, 2019).

It is possible to use the cultural dimension to share differences. They make it possible to characterize a nation or a region (they also apply to other *cultures,* such as those of organizations and professions). According to experts, they can vary, for example, with Power distance (Low to High), Application of rules (Low to High), Time Management (Sequential or Synchronous), Time Orientation (Short or Long Term), etc. Clarification is necessary to fully understand the cultural trends contained in the different dimensions (Vinardi, 2016):

- The **positioning on the dimensions is expressed using an average.** Because of the dispersion, the results do not necessarily reflect all the members of a community, especially in the extremes.
- **Time and generations can change the results**: the scale of dimensions can move from one side to the other depending on the generations. We will see the impacts in Sect. 4.3.3.

A few important remarks: consistency in strategy, **leadership style, country culture, and corporate culture maximizes performance** (Vinardi, 2019).

It should also be noted that there are links between some dominant corporate *cultures* and the ranking of best performers, and two dimensions seem to emerge (Vinardi, 2019):

- **High Performance Orientation** (focus on the objective)
- **Collective Orientation** (valuing teamwork, trust, benevolence, and belonging).

Other elements impact the corporate culture, and we will analyze them: age, size, activity, and rank in the Supply Chain.

? Is there a trend in terms of performance when it comes to age and therefore the company's longevity?

Most studies are empirical and focus generically on firm performance; only a few focus on financial performance.

Age influences performance, probably through intermediating mechanisms such as routinization, accumulated reputation and organizational rigidity (Coad et al., 2018).

However, as with size and sector, it is difficult to isolate this parameter from others that influence performance. Although firm age has attracted the attention of many scholars, **results have often been mixed** (Rossi, 2016).

Let's summarize the key takeaways from the academic field. **If we focus on financial performance, a few principles of enduring success seem to emerge:**

- Exploit before you explore,
- Diversify your business portfolio,
- Remember and learn from your mistakes,
- Be conservative about change.

In addition, younger firms seem more successful at converting employment growth into sales, profits, and productivity growth. Meanwhile, older firms look to do better at **converting sales growth into growth of profits and productivity** (Rossi, 2016). Other research suggests that, **as firms grow older, their profitability seems to decline** (Loderer & Waelchli, 2010). But many observed relationships between age and performance are U-shaped or non- Linear (Coad et al., 2018). The non-linear connection is also mentioned for **family-owned companies** with a **convex relationship between age and profitability.** It would suggest that **young companies should have higher performance than old companies, but after a certain age,** the direction of the relationship changes, and **older firms outperform younger firms** (Haykir & Çelik, 2018).

If we focus on global performance (other than financial), a **"liability of newness"** emerges. It is a way to describe how **young organizations face higher risks of failure**. Young firms must **promote social interactions within their organizations and with external organizations** to sustain the additional learning costs involved in new roles and tasks (Rossi, 2016).

In addition, two periods of an organizational life cycle would impact the risk for failures:

- **Death risks are low in the early phase** (after the start-up phase) because **decision-makers monitor performance.** Meanwhile, organizations often live on a stock of initial resources.
- **In a later phase**, initial monitoring will end, and organizations will be **subject to the usual risks of failure.**

"Born global" firms seem to experience faster international sales growth **than their older counterparts.** They interpret this finding as evidence that younger firms can better develop export capabilities because they can learn how to succeed in uncertain environments. It illustrates the concept of **"learning advantages of newness,"** which confirms the **usefulness of knowledge-based and learning views** for understanding international expansion issues (Rossi, 2016).

As firms get older, they **develop learning impediments** that **hamper their ability to grow in new environments successfully.** The **relative flexibility of newer firms allows them to rapidly learn the competencies necessary to pursue continued growth in foreign markets** (Rossi, 2016).

However, two different *vision*[2] exist regarding the organization evolution of firms: the first one suggests that **old and large organizations become increasingly dominant** over their environment, and the second one suggests that as **organizations age, they become less able to respond to new challenges** (Rossi, 2016).

? How do the size of a company and its constraints specifically influence its performance?

In the nineteenth century, industrial companies emerged as SMEs before large companies appeared. In history, and still today, **SMEs** (including micro-enterprises) **represent the bulk of companies:** more than 90% of total firms and more than 70% of paid employees worldwide (www.investopedia.com).

It is difficult to say whether SMEs are more or less efficient than large companies because a comparison would have identical parameters, and there is no study of this type. **In addition, team size, sector, and managerial skill**s seem to be **interrelated aspects of growth and performance** (Orser et al., 2000).

In Chap. 5, we will also see that company size can have advantages and disadvantages in deploying Lean and Digital approaches.

It is worth mentioning one category of SMEs that has seen the day since the advent of the internet: start-ups. They are small structures that are characterized by three main elements (Vinardi, 2019):

- Their **age** (a few years of existence).
- The **bet they make on innovation** and how it is received by **potential markets.**
- Innovation in **new technologies** (the Internet in particular).

It differs from young and small "traditional" companies mainly in that it operates in the field of **new technologies and also in that it is looking for its market.** It does not aim **to be the most profitable immediately but to be the first in a market**, explaining the large number of failures and bankruptcies.

In Silicon Valley, efficient operations are not a priority but often an afterthought; culturally, the focus is on product innovation and growth. So the ***culture* of a start-up is not one of maximum performance** but has a few cultural dimensions, some of which have been presented previously (Vinardi, 2019):

- **Low Power Distance:** no formal organizational charts, a concentric character with two or three levels at most around the founder and leader who is the only authority, low distinction of tasks.
- **Low Performance Orientation**: the objective is to find the market.
- **Short-Term Orientation**: to quickly find a market, investments, and, therefore, the fundraising to obtain them are the sinews of war.

Weak performance orientation does not mean that the search for performance for start-ups does not exist, as Luc Julia, who founded three and sold two, points out.

[2] A more precise definition is found in the Glossary (See No. 4).

> *"Even when you develop a start-up, money is still crucial, and you must avoid waste. Therefore, the basic **but mandatory drivers to implement first for performance are mainly Lean and a focused mindset**."*
>
> *From an interview of Luc Julia with Carine Vinardi*

Start-ups are young companies, and their failure rate is high (9 out of 10?), which aligns with the above remarks on the age and the very nature of the start-up, which has not yet found its cruising speed in performance.

? So, what is the impact of a company's activity on their performance prospects?

First of all, let's go back to the main types. Service companies (commercial or liberal) have always existed alongside craft companies (sales of mainly specific manual know-how to carry out manufacturing/repair, transformation). From the nineteenth century, industrial companies emerged and dominated the economic landscape until the democratization of information systems and computing. The emergence of the internet has allowed the rise of large transactional enterprises (services) to the point of enabling the advent of so-called digital companies (whose output, service, is providing digital solution(s)).

Even if craft and industrial companies can also offer services, the main flow, known as primary flow, is physical, where products drive the flow of information (secondary flows) and guide the organization.

Waste analysis is more straightforward when there is a physical flow because the process[3] and a significant part of the waste can be seen. This does not mean the solutions to eliminate this waste are simple or more straightforward just because they can be seen. Indeed, an industrial company needs more or less equipment to produce its product and has induced costs (Capital Expenses (CAPEX) or Operational Expenses (OPEX)). These can be more or less difficult to move or change and can be a significant part of costs. In the 1970s, W. E. Deming, who greatly inspired Japanese leaders, considered that **10 to 40% of production costs in the United States** were **waste-related** (Vinardi, 2013). It isn't easy to know how this percentage has evolved.

In the case of transactional companies, the main flows are not physical but are information flows (services offered). These main flows guide secondary flows, which are also information flows. As mentioned, there is no physical flow or very few to guide the information flow in these companies, so human organization and the associated *processes*[4] guide the information flow without ever meeting. **Diagnosis is much more complex because waste is not or hardly visible** (piles of files when they still exist or in waiting when staff do not try to hide them). These companies

[3] A more precise definition is found in the Glossary (See No. 2).

[4] A more precise definition is found in the Glossary (See No. 2).

may have the impression that they are efficient and not wasteful. Empirically, it has been shown that the **rate of waste increases when there is no physical flow**: about ten years ago, it was not uncommon to see **more than 90% of waste in some transactional businesses** (Vinardi, 2013).

This difficulty is stated and commented on by Éric Ciampi, who advises this type of company, and Stéphanie Perchet, who has continually evolved in these environments.

> *"For **Transactional Business**, it can **be difficult to understand what part of the savings could have been made** without investing in digital."*
>
> From an interview of Eric Ciampi with Carine Vinardi

> *"From my experience, in many transactional businesses, **Lean or Agile principles are not deeply embedded.** This is often because **top leaders either fail to recognize inefficiencies** and the associated impact on internal performance or **feel they have 'enough money,'** choosing to hire additional staff to address problems rather than tackling the root causes."*
>
> From an interview of Stéphanie Perchet with Carine Vinardi

Apart from expensive products sold beyond the market logic (as may be the case, for example, in the luxury industry or sometimes in aeronautics), an organization's ability to **maximize performance** will most often depend on its **ability to eliminate or limit everything that weighs on them, particularly waste**.

Finally, it is important to understand that transactional and industrial companies are not subject to **traditional fields of constraint to the same degree.** Indeed, all companies have human resources (organization, people, Leaders, skills), material resources (Raw materials, Material Assets: machines, equipment, buildings, IT, etc.), financial resources, and lastly, *processes* [5] to coordinate the whole.

However, in the case of industrial companies, material resources can generate safety risks, and *processes* [6] include physical parts that can bear a risk. Constraints on people's and equipment safety will, therefore, weigh more heavily on industrial companies than on transactional ones.

In addition, material resources can be very expensive and are a significant part of industrial companies' costs. Their ability to make the right investment choices and to optimize the operation of these resources with human resources is a much stronger constraint in industry. Conversely, the ability to integrate and develop the best human resources will be crucial and an even greater constraint for transactional companies.

---

[5] A more precise definition is found in the Glossary (See No. 2).

[6] A more precise definition is found in the Glossary (See No. 2).

One last point that influences company culture is its place in the Supply Chain because it impacts its role and its activity.

? Does a company's rank in the supply chain affect its performance?

As a reminder, when you deliver directly to a customer, you are a Tier 1 supplier. When you deliver a Tier 1 supplier, you are a Tier 2 supplier, etc., up to the last Tier N, which is the one who supplies the raw material.

Ranking in the Supply Chain often implies a change in company size, especially in industry, but not only. The **ability to negotiate** is essential for economic performance, in particular, more so than its rank in the Supply Chain:

- If there is a trade balance, negotiation will probably be balanced too.
- If there is no trade balance:
  - Either the company represents too much or not enough for its customers. If its product or know-how is not unique, it will not necessarily be in a favorable position to negotiate acceptable prices to ensure the financial performance of its organization.
  - Or the company operates in a niche market (with a unique product or service), regardless of your size or market weight, in which case it is often possible to negotiate favorably.

? Finally, does the production mode have an impact on performance?

As companies tend to change production mode and do not necessarily apply the same for all products and services, isolating this factor is even more complicated than before. However, performance levers depend on different modes and can be examined. Let's take the main ones:

- **Make To Stock (MTS)**: the product or service is not new; the standard for making it is known. A new service or product is produced when it is consumed.

  It is mainly used in industry when the product is not expensive or is very often requested (High Runners product). It is simple to organize, and managing **finished products is key to** customer and internal **performance**.
- **Make To Order (MTO)**: the product or service is not new; the standard for making it is known. The production of a new product or service is made **to order only.** It is used in transactional or industrial activity (to avoid stocks of finished products). This mode of production is more complex to organize than the previous one because there are no stocks to absorb variations in demand. The stock of **raw materials and the ability to produce (skills, resources) are key elements of** customer and internal performance.

- **Design To Order (DTO)**: the product or service is new; the standard for making it is unknown. It is used in transactional or industrial activity (to avoid stocks and because the product or service can be **customized to order**). It is by far the most complex production method. Since there is no stock, **the availability of raw materials (including data), design, and production capacity makes the difference and impacts performance**. To deliver on time and when working at a finite capacity, you have to be able **to anticipate the time needed to create something new and eliminate unknowns**. In industry, it is rarely used for the entire complex product for performance, but for the "customizable" part and at the end of production.

4.1.2 Connected Performance Indicators

Now that the main traditional areas of constraints have been clarified, the question of measuring their impact is also raised.

Let's recall how organizations, particularly capitalistic companies, evaluate/are evaluated for their performance. As I explained in this book introduction, performance is assessed on three main components: **effectiveness** (the comparison with an objective), **efficiency** (the comparison with resources for transforming inputs into outputs), and *benchmarking*[7] (the comparison with other similar organizations). This result is expressed as a Key Performance Indicator (**KPI**) (Ionescu et al., 2024).

Performance is assessed on the company as a whole, with its customers (and potentially its suppliers), its employees, and shareholders (if applicable).

? So, what are the main "traditional" KPIs?

The goal is not to make an exhaustive list but to identify the main lasting ones. Let's begin with the starting point of a company's flow: customers.

- FOR CUSTOMERS: there are at least five KPIs that are common to companies regardless of their activity.
  - **Customer Lead Time (LT):** A leading indicator of the time between customer demand and customer delivery of the product or service. The customer has expectations of LT, and the company must ensure it meets them. There may be slight differences between the ideal LT, the LT validated (with the supplier), the required LT, etc.

    The suggested formula for LT = Delivery Date or Time – Date or Time of the Request.
  - **Customer satisfaction**: A lagging indicator, which is the degree of customer appreciation concerning the products or services offered. It is the ratio of the

[7] A more precise definition is found in the Glossary (See No. 1).

number of satisfied customers (and/or very satisfied customers, depending on the requirement) to the total number of customers.

The suggested formula is = (Number of (very) satisfied customers/Total number of customers) × 100.

- **Satisfaction/problem solving:** lagging indicator of the problem solving effectiveness, especially relevant for the after-sales part.

  The suggested formula is = (Number of customers (very) satisfied with the resolution/Number of customers with a problem) × 100.

- **Retention rate**: a lagging and leading indicator of a company's ability to retain its customers over a given period.

  The suggested formula is = ((Number of customers at the end of the period – Number of new customers)/Number of customers at the beginning of the period) × 100.

- **Net Promoter Score (NPS):** A leading indicator of the likelihood that a customer will recommend the company to other customers; this score is a reliable indicator of long-term loyalty. Customers express their willingness to recommend the company on a score of 0 to 10. Scores of 9 or 10 mean they are considered promoters, 0 to 6 detractors, and 7 or 8 neutral.

  The suggested formula is = Percentage of all promoters (Number of promoters/ Total number of customers surveyed x100) – Percentage of all detractors (Number of detractors/Total number of customers surveyed x100).

It should be noted that very small businesses or craftsmen do not all use KPIs. Still, some have a qualitative appreciation of the concepts mentioned below, which is also true for the indicators below.

It should also be noted that these indicators can be applied to assess suppliers and the quality of their relationships. In some industries, such as automotive, which has a mature supply chain in terms of data and performance measurement, it **has been empirically shown that there is a relationship between customer performance and supplier performance**: the best companies in terms of customer satisfaction seem to be the best at satisfying their suppliers (and getting the best from them, starting with innovation ideas).

- FOR EMPLOYEES: all human resources who use material resources, define and execute the company *processes*.[8] At least four KPIs measure their satisfaction. The first two are must-haves and are the minimum satisfaction expected by employees who come to work.

  - **Frequency rate:** A lagging indicator of the frequency of accidents.

    The suggested formula is = Nb of accidents (with or without lost days)/ millions of working hours.

[8] A more precise definition is found in the Glossary (See No. 2).

- **Severity rate**: A lagging indicator of the severity of accidents.

  The suggested formula is = (Nb of Lost Workdays/Total Hours Worked) × 1000.

These two indicators are important for the company, which must ensure that employees' health is not deteriorated by their work. As mentioned, the **constraints are tighter in industrial companies** that have equipment, machines, and vehicles because they are at greater risk.

- **Rate of unplanned absenteeism**: unplanned leave because of illness, accidents or commuting, decisions, and personal events. This absence is the most disruptive because it is unpredictable. Its level is often related to the employee's social protection of the country. The rate is higher when this leave is paid. It is a lagging indicator.

  The suggested formula is = (Number of hours' leave in a given period/Number of theoretical working hours over the period) × 100.
- **Turnover rate**: the proportion of employees who have left the company in a given period. This rate should be normalized against the job market in the region concerned. It is a lagging indicator.

  The suggested formula is = (Number of leavers/Average number of employees over the period) × 100.
- **Engagement rate**: the level of engagement of employees. It is a lagging indicator.

  It is expressed as a percentage of employees who consider themselves (very) committed based on a survey whose content may vary. A simplified and more ad-hoc evaluation can be done with a visual icon such as smileys using digital applications (Vinardi, 2016). Employee engagement refers to employees' **emotional and intellectual commitment toward their work roles and organizational objectives.** A significant body of research consistently demonstrates a **positive correlation** between **engagement and organizational performance metrics** such as **productivity, profitability, customer satisfaction, and employee retention.** The employees should be **stimulated and motivated not only by salary** but by **training, health, social, and other additional programs that will raise their interest, trust, and commitment to the company** (Jakšić et al., 2013; Bin, 2016; Deepalashmi et al., 2024).

  Lastly, note that engagement can be measured **every year with a more detailed survey) and also more regularly using an immediate format**.

  Part of dissatisfaction can also be expressed by unanticipated absenteeism and employees leaving, and may also be included in the engagement rate.

- FOR THE COMPANY: in this case, performance is systemic as it measures overall work, the association of human resources, materials, and associated *processes*[9] with its own performance indicators. I suggest just three indicators for current activity in "everyday life" (existing products or services) and three indicators for future activity (future products or services):
  - **On Quality Delivery (OQD) rate:** It is a quality lagging indicator that measures the ability to produce a product or service right the first time (in the format and at the expected level of quality/compliance) by looking at the input (by suppliers) and the output (to customers). To this, can be added the OQD rate, which is an internal leading indicator at each stage in the company.

    The suggested formula is = (Number of Compliant Goods or Services/Total Goods or Services Delivered) × 100/
  - **On Time Delivery (OTD) rate:** It is a delivery lagging indicator that measures the ability to produce and deliver a product or service at the right time by examining the input (by suppliers) and the output (to customers). The OQD rate, which is an internal leading indicator at each stage in the company, can be added to this.

    The suggested formula is = (Number of goods or services delivered on the agreed date/Number of total goods or services delivered) × 100.
  - **Equipment Effectiveness (EE) rate:** It is an equipment lagging indicator for equipment or machines, but can also be applied to an interdependent human organization to evaluate overall performance. It is a function of the OQD and OTD rates and other organizational factors such as speed and stoppages (whether planned or not). In short, it provides information on capacity utilization.

    The suggested formula is = (Time to produce products or services at the expected speed/Total time used) × 100.

To measure the ability to anticipate future revenues and, therefore, innovation, there are at least three key indicators:

- **R&D or Innovation rate**: This is a lagging indicator of R&D expenditure or, more broadly, expenditure on innovation (including in production) in relation to company revenues. It is a ***benchmarking***[10] indicator that does not, however, presume effectiveness or complete efficiency.

  The suggested formula is = (R&D or Innovation expenses + capital expenses (CAPEX) + operational expenses (OPEX), internal and external to the company/ Revenues for the period) × 100

- **Time To Market**: It is a leading indicator of the time between the decision to design a new product or service and its launch on the market.

  The suggested formula is = Launch Date – Decision Date.

[9] A more precise definition is found in the Glossary (See No. 2).

[10] A more precise definition is found in the Glossary (See No. 1).

– **Technology/Customer Maturity rate**: this is a leading indicator. This is a primary indicator for new products or services to ensure that market maturity, called Business Readiness Level (BRL), is aligned with the technology maturity of the new product or service, called Technology Readiness Level (TRL). There are generally 9 Readiness Levels, 9 being the maximum maturity level (the customer or the market is ready, the technology is proven and ready to be brought to market). I suggest a ratio greater or equal to 1.

  The suggested formula is = TRL/BRL

Other indicators include the number of patents (against revenue-producing patents) or indicators of investment profitability. However, the three just mentioned are fundamental.

It is important to add that the above indicators also influence employee satisfaction. Indeed, we are **all the more motivated and committed when we work in a high-performance environment**, i.e., with the appropriate means and the right compliant and reliable data, as and when they are needed (as measured by the previous indicators), and we are reassured that it will produce benefits for the future.

When it comes to speed, however, **working conditions and respect for human physiology and psychology** will need to be considered **for its increase to be sustainable.** (We will come back to this in Sect. 4.4).

Many indicators are expressed as a percentage, making observing lost percentage points possible. **These points have a cost** (which can be calculated) that impacts the company's economic performance, and **many are related to waste that can be reduced.** Similarly, the resources for achieving an outcome (measured by its KPI) also have a cost. It amounts to just the right resources and skills needed to minimize costs (Not the strict minimum but the right amount). By seeking this balance, we ask ourselves the right questions in terms of skills, choice of equipment, and choice of organization. Internal performance is often the first driver of financial performance, which is mainly expressed by the ability to create wealth, profitability (the ability to generate money to finance its operations), and available liquidity.

So, let us end with two entirely financial KPIs to complete the measurement of company performance:

– **Economic profit** is a lagging indicator of a company's ability to create wealth, i.e., its financial performance.

  The suggested formula is = Net Operating Profit After Tax – (Invested Capital x Weighted Average Cost of Capital)

  • **Net Operating Profit After Tax,** also called Earnings Before Interests and Taxes (EBIT), measures the profit generated by core operations, excluding Taxes and the cost of capital. It is directly impacted by internal performance.

    The suggested formula is = Revenues – (direct costs + operating expenses (OPEX) + depreciation)

- **Invested Capital** is a measure of capital invested in the business, including equity and debt (with assets like buildings, equipment, and other investments necessary for operations).
- **Weighted Average Cost of Capital** measures the weighted average of equity and debt costs, considering the proportion of each in the company's capital structure.

– **Working capital:** It is a leading indicator of the difference between a company's current assets (items a business owns that can be turned into cash in the next 12 months) and current liabilities (costs and expenses the business incurs in that period. For example, short Term obligation but also expenses connected to operating cycle: inventories, operating costs and wages ...).

The suggested formula is = current assets – current liabilities.

- SHAREHOLDERS: are the owners and investors of the company, whether natural or legal persons. The size of their participation depends mainly on the number of shares owned. They **invest on the expectation of financial return in the more or less long term**. They **participate in the strategy through the governance of the company**. There are two essential KPIs, and even more so when the company is listed on the stock exchange because the markets examine them:

  – **Dividends per share**: share of profits redistributed to shareholders. Dividends are linked to the Economic Profit, i.e., the company's ability to generate revenue and profit, and to the dividend policy (a standard that defines the percentage and criteria).

    The suggested formula is = total amount of dividends paid/number of distributed shares.

  – **Capital gains per share:** gains made when shares are sold if the sale price exceeds the purchase price.

We have quickly reviewed the leading traditional performance indicators that have been around for a long time to assess the results specific to each category. These **KPIs are assessed dynamically and also smoothed over time to understand their evolution.**

One of the keys to performance is for a company to find the right balance to transform inputs (skills, data, materials, components, liquidity) into outputs (services, documents, products, and, therefore, turnover) and maximize these indicators. So, the transformation produces areas of constraint (in legislation, funding, recruitment, commercial agreements, organization, operation, etc.) that are more or less strong depending on the region of the world. Also, here is Fig. 4.3, a summary representation of the different stress fields where the above KPIs will apply. We will develop this representation with the fields we will examine later.

Lessons from past crises should contribute to innovative, adaptable solutions to future challenges and disruptions.

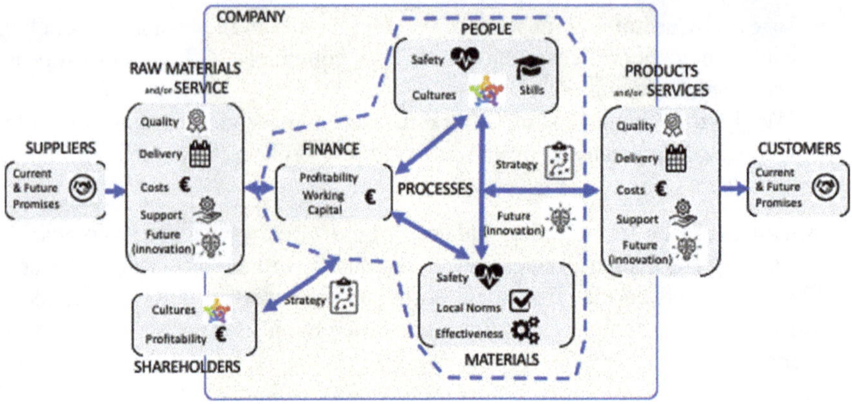

Fig. 4.3 Simplified representation of fields of constraints

4.2 Lessons Learned

4.2.1 The Complexity in Satisfying Customers

Customer satisfaction has rarely been neglected in history, especially where there is competition. It is both the fuel and the company's destination. However, the customer can be less of a priority when the company (even a capitalistic one) has a (near) monopoly, for example. When there is no competition, the objective of satisfying its customers can indeed be somewhat less imperative. This can also be the case for other forms of non-capitalistic organizations.

Since the 1990s, neuroscience **gradually helped organizations understand their customers**. Indeed, functional magnetic resonance imaging (fMRI) machines have become more accessible and enabled to measure brain activity.

The Science Behind Neuromarketing (Term introduced in the early 2000s) **helps marketers understand how people react based on biology and not conscious choices so that they can predict consumer behavior.** Unconscious choices can drive buying decisions, and the technology can help to understand the customers better and deliver a more favorable experience. **There is a truth to neuromarketing that can't be replicated by traditional marketing research** tactics like focus groups (Kirk, 2024).Here are five ways businesses use neuromarketing to improve their marketing efforts and drive sales: Testing Ads, Improving Packaging Design, Enhancing Website and App Design, Informing Rebranding, and Optimizing Conversion Rates. As a famous example, let's talk about IKEA. The company has designed its stores to showcase everything it sells before a consumer can leave the store, thus increasing the likelihood of a purchase. The layout was developed using neuromarketing research. It's easy to see why red is the favored logo color of so many iconic brands, including Coca-Cola, Target, McDonald's and Netflix (Kirk, 2024).

Customer satisfaction is not often debated, but **understanding your customers can be complex** and 'requires science,' as we have seen. Once you have identified the needs (whether they are expressed or not) and made the promises, you must be able to deliver. **Implementation complexity is a real issue to keep the promises made to customers.**

In a competitive market, customers choose according to their requirements and the company's ability to meet them. Complexity **arises from the frequency of demand, as well as from the variety, novelty, and the (up or down) trend of the** products or services requested.

Some products or services are requested often (the so-called "High Runners"), while others are requested more rarely (the so-called "Low Runners"). For High Runners, the company must anticipate, for Low Runners, the company must find the capacity to respond.

To tackle variety and novelty, the company will offer an ever-increasing number of products, services, and innovations. All this has a cost.

When faced with increasing or decreasing demand, the company will anticipate and respond to trends by increasing or decreasing resources at the right time because under- or over-resourcing is costly.

In addition to the complexity and the effort to find a balance in response, the company must face **the uncertainty of demand** in markets and its competition, which has become increasingly globalized since the last century. Also, customers often do not communicate directly with the entire company but through sales or marketing departments via B to C (Business to Customer) and with other companies via B to B (Business to Business). **Intermediate filters are difficult to overcome,** especially if trust is an issue.

Lastly, the company has a particular complexity inherent to its culture or its mode of production, as we saw at the beginning of this chapter. It has only finite human and material resources to face instant demands that may seem infinite if they are not aligned with the company's capacity. We will take up this point again in Sect. 4.2.3.

4.2.2 The Hidden and Unfulfilled Potential of Employee Satisfaction

Unlike customer satisfaction, **employee satisfaction was often neglected** until the beginning of the last century, almost everywhere in the world. Unfortunately, this is still the case today but in fewer and fewer countries. The reasons are found at different levels:

- **Employees who fought/fight for a decent salary are less focused (in their priorities) on their working conditions.** In some extreme cases (such as after wars), there is so much to do that the focus is on the "Must Have" and not on what is "Nice to Have."

- **When the workforce exceeds demand, someone is always willing to accept unadvantageous working conditions.** Some employers, therefore, do not worry about employee satisfaction because, in any case, they have no choice.
- The recruitment, retention, and motivational benefits of employee satisfaction are most valuable in countries where firms face fewer hiring and firing constraints (Edmans et al., 2014).
- **Managers are sometimes far from their operational teams.** They are not aware of the actions that could be taken. In the same way, many managers never started their professional life in the field because of their academic background or other personal choices. Naturally, they would not see the effect of working conditions on performance.
- Lastly, **for a long time, the effect of employee satisfaction (through safety, commitment, and organization, for example) on performance was never studied.** And when it had been established, managers did not always understand it, particularly because the effects of actions undertaken are not immediately seen.

As far as safety is concerned, and the minimum expected by employees, few studies look at this parameter. However, experience shows that **industrial companies that are the best in terms of safety (maintaining safety performance at the best level over time) are also the best at overall internal performance.**

Some studies highlight **proactive safety risk management and its contribution to overall resilience and performance** in industrial environments (Renaud & Berland, 2007). Safety performance indicators (such as the frequency and severity of accidents) can also **assess organizational maturity, particularly collective discipline.** Companies that excel in these areas tend to do better on other *Strategic Objectives*,[11] showing a link between security and overall success.

Even if we have highlighted two lagging KPIs, Frequency and Severity rate, which are essential to evaluating safety performance output, it is necessary **to have the best working conditions** (those that ensure the best performance) upstream and proactively, which is what best performers do. This is done by **looking at human physiology and psychology** at workstations and, therefore, in human–machine interactions by applying Ergonomics. Measurement of the difficulty of workstations is required by law in some countries (such as France), and some companies chart the ergonomics of workstations.

Ergonomics is not a new scientific discipline but a proven one that aims to design and arrange workplaces, products, and systems to ensure they fit and adapt to the people who use them. It is based on physical elements with anthropometry (body sizes and shapes), biomechanics (muscles and movement), environmental (noise, light, temperature), as well as cognitive elements with applied psychology (autonomy, skill, learning, errors, memory, reasoning, …) and social psychology (group dynamics, communication).

Science has also confirmed what we all know by experience every day (in sports, music, gardening, etc.). The best task performance is reached when our body is in the

[11] A more precise definition is found in the Glossary (See No. 5).

best **posture using the most appropriate means.** In larger companies, dedicated ergonomists work on different categories of workstations. Ergonomists can be called ad hoc as part of external services. Also, it is advisable to give ergonomic training to leaders and support functions who define and optimize workstations (to identify problems and propose improvements) and set working hours from which cost prices are determined (this is to establish realistic assumptions).

One of the main mistakes in the deployment of Lean was precisely to have the task times set by people who did not have the training and skills to do so or who were under pressure to dismiss ergonomic principles. I personally experienced this pressure at an automotive supplier where I had to fight back at the risk of losing my job. This has sometimes led some companies to default to "dry" minimum times as standards without considering anthropometric or environmental constraints.

When the intention is to use minimum times as a reference to identify waste, it could make sense. **Still, when they are used to calculate production costs, it is doubly pernicious**: the target is not achievable except in very exceptional circumstances (with young athletic men, for example), and it discourages and demotivates staff. This problem is far from having disappeared, and when it exists, it negatively impacts engagement.

As mentioned above, we have established a relationship between **engagement and performance** (Deepa Ashmi et al., 2024). **Employees worldwide appreciate the interest shown in them through their working conditions.**

What is striking is that companies often focus more on disengagement (reacting when there is a problem) than on engagement (being proactive in creating engagement).

There are often pragmatic reasons for disengagement. **Contradictions between words and deeds insidiously erode motivation. Incoherence is the "public enemy n°1" of commitment** (Roy, 2024). Companies that promote well-being or benevolence but tolerate toxic management fall into **paradoxical injunction** (also called "*double-blind*"). A person receiving contradictory messages from another one in authority experiences frustration and stress and, above all, a complete loss of trust (Coté, 2024).

In recent years, promoting the feeling of usefulness has become fashionable, especially with the "raison d'être" of companies. This is supposed to federate employees. Quite apart from the inconsistencies, it is often a mistake to consider employees' contributions from a philosophical or societal perspective. Indeed, employees want to feel useful, but that doesn't necessarily mean they want to change the world or find deep meaning in every daily task. A good Leader must ensure that each team member feels useful by **relating individual actions to *Strategic Objectives*[12]** (which is what the *Hoshin kanri*[13] approach does) and **by identifying and valuing "the trick" the talent of each** (Roy, 2024).

[12] A more precise definition is found in the Glossary (See No. 5).

[13] A more precise definition is found in the Glossary (See No. 3).

It is not about organizations being perfect, but they must be **irreproachable in general**. Employees want to engage in an environment that **values their contribution, helps them progress, and, most importantly, is consistent** (Roy, 2024). People would outperform even the most sophisticated technology for two main reasons: **potential and appreciation,** and the **value of human capital (in trillions of dollars) is significantly higher than that of physical assets across the world**, even in nations such as China and India (Pearlman et al., 2024).

Lastly, we discussed multicultural teams, their exceptional performance, and the challenges they face in being effective while navigating their differences and transitioning from diversity to inclusion. Beyond multiculturalism, team configuration exemplifies the significance of social ties in fostering commitment and enhancing performance. To become effective and efficient, teams must progress through several stages (forming, storming, norming, and performing) to build **trust, transparency, and constructive exchanges** (Vinardi, 2016).

Also, the **bond or social contract is based on reciprocity**. At the beginning of the last century, the French ethnologist M. Mauss studied the origin and impact of **"gifts" and "counter-gifts," which is a form of the social contract** (i.e., belonging to society) for market economies (Mauss, 1925).In our societies, money has primarily contributed to technoscientific, industrial, and urban development and has participated uniquely in the rationalization of the world. To do this, money has replaced barter. However, anthropologists and sociologists point out that this general monetarization is not as "revolutionary" as is claimed: the **dynamics of gift, debt, and counter-gift** continue to be the **basis of our most substantial social relations for a long time**. Numerical **indicators never tell us everything that is done around a goal** (Drulhe, 2011; Lasida, 2013).

The relationship created by the gift is not that of pure commodity exchange; the value of the link is not to be confused with the value of exchange. When you give, there is no obligation to return. **The relationship woven through the gift is of the order of call and not the order of obligation.** The "relationship" at the center of each of these notions and practices allows us to link gifts and markets and give them the role of social mediators (Lasida, 2013).The current and future satisfaction of customers and employees is a fundamental performance lever and also helps companies optimize *processes*[14] and internal organization.

4.2.3 *Sustainable Performing Organization: Resilient, Visionary but Grounded in Reality*

By examining the constraints that apply to companies today and in the past, we see that they do not affect all businesses in the same way. The impact varies based on the company's nature. Industrial companies, in particular, **face unique challenges**

[14] A more precise definition is found in the Glossary (See No. 2).

due to their reliance on material resources and the critical importance of workplace safety, making their constraints especially stringent. This is important when accounting for environmental and digital constraints which we will discuss in the next chapter.

Concerning purely customer innovation (on products or services) that prepares for the future, we addressed the **need for the "right timing."** It is important to **choose the start** time when the **market is mature and ready for it** (BRL, mentioned above) and the launch time (TRL, also mentioned above). **Companies that want a "head start" do not hesitate to develop technologies with low TRL**. These are often bets on the future.

> *"For each Engineering & Innovation project, we evaluate opportunities to effectively implement* ***solutions with low TRL levels to make our approach more pioneering."***
>
> From an interview of Richard Vinardi with Carine Vinardi

Going back to the timeline in Sect. 3.1, we mentioned that the emergence of digital solutions that we know today dates back to before the 2000s. **For these solutions to be deployed, they must be technologically, economically, and socially mature to find their market.** Internal company innovations (related to skills, organization, or equipment, for example) must be **done upstream, well before the desired results**, to satisfy the customer, employees, the company, and shareholders in the future.

We also talked about **how constraints are both problems that help organizations move forward and opportunities.** This is only true if problems are anticipated, resolved, and turned into opportunities, too (see Sect. 3.3).

The case of Toyota is a significant source of learning. The company was able to rebuild itself after the war and convert an enormous field of constraints with a lack of human resources and skills, material resources (raw materials, components, equipment, energy), a lack of money, and finally, a lack of experience to implement the new strategy (competing with the Americans on the automotive market). The context and the future were very uncertain because only a progression in quality could allow the company to recover. Toyota was able to invent a model with effective problem-solving. From this situation (one of the basics of Lean explained in Sect. 2.1.1) and to establish itself in the long term as a high-performance and innovative manufacturer. On a small-series production, the company had to learn to avoid all material and human waste quickly (but we often forget this). **Waste elimination was necessary because it could not afford the luxury of creating or having it.** It was also necessary not to waste time defining, executing, and adapting the strategy. In this context, the Lean and *Hoshin kanri* [15] system, described in Chap. 2, were gradually created.

Lastly, Toyota's performance was also driven mainly by its ability to create an ecosystem around the company with a **privileged network of suppliers**. Like other

[15] A more precise definition is found in the Glossary (See No. 3).

companies in Japan, Toyota initiated special partnerships with its suppliers through *Kereitsu* that still exist today. This Japanese term means "Business network" or "Grouping of Enterprise"; there are different types, but the principle is always the same and centers around a powerful **relationship with suppliers and customers** (Boers, 2020):

- "**Mutual sustenance**": the business success would trickle down to the parts suppliers
- **Economic protection** is offered by a larger firm with global production capacity.
- **Advice and business support are routinely offered** to its suppliers, helping improve their profitability and, in turn, securing the relationship.
- **Direct involvement in the design *process*** ,[16] participation directly in product-development meetings, fostering closer interpersonal relationships between producer and supplier (particularly true for Toyota)
- **Bonded together by creating economies of scale** for all members of the conglomerate and by **reducing production costs for those members.**
- **High level of trust between all stakeholders** allows transparency and quick information sharing.

Beyond **boosting the ability to innovate,** the **integration of Toyota's Supply Chain** (for example) has contributed to the financial performance of the companies involved and allowed them to **react quickly to changes.**

? Are positive contributions of upstream Supply Chain integration on performance seen outside of the Japanese *Kereitsu*?

First of all, supply chain management is defined as the management of the flow of goods, services, and information that includes all value-added *processes* [17] in turning raw materials or data into final products or services (Tarigan & Siagian, 2021). Several studies carried out in emerging countries and/or countries beset by strong and frequent instabilities show that **strong links in the Supply Chain at all levels with suppliers and customers are necessary,** even more so in very uncertain contexts (Shukor et al., 2021).

Even if it is not the only parameter, **Supply chain integrations** were shown to have a positive impact at different levels:

- on the firm's Supply Chain **agility and organizational flexibility** (Shukor et al., 2021),
- on better **sharing information about the market demands and supplies**, thus enabling them to **be more responsive and agile** (Piprani et al., 2020),
- on the **coordination mechanism between the functional areas,** thus resulting in **improved cross-department communication** and **improved business performance** (Piprani et al., 2020),

[16] A more precise definition is found in the Glossary (See No. 2).

[17] A more precise definition is found in the Glossary (See No. 2).

- on the **operational capabilities of a firm** (Saragih et al., 2020);
- on **cutting down the influence of supply chain disruptions** caused by uncertainties (Piprani et al., 2020),
- and therefore on **operational, financial, and business performance** (Piprani et al., 2020; Yu et al., 2021),
- **the level of trust** (Alshurideh et al., 2022).

The trust created from the Supply Chain integration seems to **promote transparency in exchanges and agility and responsiveness in case of problems. Therefore, it** has a significant impact on many aspects of company performance. This helps to **better adapt to changes in uncertainty.** As Daniel Hager points out, the supply chain is also one of the main performance levers mentioned by manufacturers.

> *"One of the main drivers of performance is to create a coherent value chain* ***from the supplier and the supply up to the customer."***
>
> From an interview of Daniel Hager with Carine Vinardi

? Apart from integration, how can we efficiently organize physical flows and information flows between suppliers, the company and its customers?

To achieve the best balance between strategic and operational planning, companies must have *process*[18] and decision-making entities to reach a consensus between Sales, Marketing, Operations, and Finance objectives. So, the S&OP *process* [19] (Sales & Operations Planning) matches the sales plan with the company's medium-term capabilities and the service and financial goals to produce a single achievable plan. This focuses on critical factors while identifying the risks.

S&OP is recognized as a relevant and powerful element in promoting **inter-functional integration** (Tarigan & Siagian, 2021). The **strategic planning** set by the company is needed to maintain a balance in the company's supply chain flow to achieve efficiency, effectiveness and adaptability (Tarigan & Siagian, 2021). Note that it is done when *Hoshin kanri* [20] is deployed (Vinardi, 2023). On the other hand, some **S&OP enablers** are related to Information Quality, such as fore**cast accuracy, information flow management, and cross-functional integration** (Sagawa & Nagano, 2021; Tarigan & Siagian, 2021).There are several challenges to overcome in implementing the *S&OP process*[21] (Tarigan & Siagian, 2021):

---

[18] A more precise definition is found in the Glossary (See No. 2).

[19] A more precise definition is found in the Glossary (See No. 2).

[20] A more precise definition is found in the Glossary (See No. 3).

[21] A more precise definition is found in the Glossary (See No. 2).

- **S&OP aims to align all stakeholders** on a single, synchronized data set on a single plane. However, uncertainties and randomness are present at every level, and the single plane may prevent sensitivity analysis and alternative scenarios.
- **S&OP helps with regular exchanges** between functions and enables communication between all the players. However, this requires **a common and shared language, extrinsic transparency,** and flexibility that the organization may not have. In addition, it requires the participation and representation of functions that are stakeholders but cannot be sufficiently involved.
- **It requires up-to-date and reliable data**. However, it faces the problem of nonexistent or unreliable critical data. Too often, companies do not try to build or clean that essential data.
- **A monthly cycle is needed to anticipate internal and supplier needs**, with numerous waiting phases for or validating data. However, it is often difficult to fit all the tasks of the cycle into one month.
- Lastly, anticipating strategic investments requires the long-term support of Top leaders. However, leaders often need to understand the power and importance of S&OP to get involved.

As Daniel Hager points out in full transparency, these difficulties are not uncommon today, even with a lot of determination.

> *"Transparent **data sharing and lack of fluidity between departments** is still a problem. I was often called upon to be the referee between Operations and Marketing, and what was shocking was that they didn't understand each other. They are all smart people, but each had their own goal."*
>
> From an interview of Daniel Hager with Carine Vinardi

When there is little or no trust between the players, despite the S&OP *process*,[22] the Forrester effect (named after the person who defined it) can appear. It amplifies the signal from the customer, producing disproportionate reactions (each player in the Supply Chain plans to produce more or less the demand because they do not trust their customer's data). This can lead to interruptions in delivery despite the capacity of each player being sufficient or on the contrary, over-inventories and related obsolescence.

Mastering the Supply Chain is a primary factor in business performance. Its efficiency and agility contribute to coping with changes arising from an uncertain context and maintaining the company's performance results.

Adapting to changes is fundamental for durability and for overcoming hazards. As Charles Darwin pointed out in one of his quotes, not one of the most famous but the most "straight to the point":

[22] A more precise definition is found in the Glossary (See No. 2).

> *"The **most important factor in survival** is neither intelligence nor strength but **adaptability**."*

Adapting does not assume that the situation is well-lived. We discussed the concept of **resilience** in Sect. 3.3.4 (the capability of any system to recover from unexpected drastic events and find a new equilibrium). We also talked about having a **real need, doubting to be curious**, using your intuition, and acting out divergent thinking to provoke change and to innovate. To complete the cycle of change, you would also need to be able to **create the emotions** and discipline to change habits in the long term (by aligning emotions, thinking, and behavior). **Companies should recruit and develop leaders who can support all stages of change to complete transformations.** Adapting to change is the first short-term step to be implemented to survive without assuming that the changes are well experienced. Resilience suggests the ability to transform oneself over time to **rebuild a new balance that makes you stronger.**

? What makes an organization resilient?

Studies into how organizations become resilient and operate in high-hazard environments sprung from the research conducted on High Reliability Organizations in the 1990s (Klockner, 2017). Since COVID-19, which was traumatic at a global level, studies analyzing the relationship between the resilience of professional organizations and their performance have increased exponentially at the academic level and within the largest strategy firms. Most studies indicate that **organizational resilience influences business performance** (Beuren et al., 2022; Garrido-Moreno et al., 2024).

To be resilient, an **organization must be a learning one**, turning outwardly towards others and the outside world to anticipate and manage future threats better. This implies having **leaders who are resilient** and able **to support their teams in the changes** and build organizations and *processes*,[23] that are resilient themselves.

? What makes a resilient leader?

First, studies (especially medical ones) show **that a lack of resilience is associated with health problems or even burnout** (Walsh et al., 2024).

Therefore, a resilient leader is preferable at all company levels, not only at the top but also in Front-**Line** Managers and especially Middle Managers.

There is a need to **operate holistically across all organizational levels,** and **middle managers play a critical role** as they link system unity and are a channel

[23] A more precise definition is found in the Glossary (See No. 2).

for information exchange. (Klockner, 2017). **A resilient Leader is** (Klockner, 2017; Lombardi et al., 2021; Levey, 2019; Garrido-Moreno et al., 2024; Brassey et al., 2024):

- **Visionary**, setting a compass or North Star to help people move in a common direction.
- **Mindful** (present-focused consciousness), not ruminating about the past and worrying about the future. Note that global interest in mindfulness has expanded exponentially since the 1960s when the first Western researchers carried treasured teachings from different mindfulness traditions of SE Asia.
- **Secure base,** capable of building a **psychologically safe community**, not just a workforce.
- **Capable of adapting leadership style** (use the right and different leadership approach depending on context)
- **Collaborative and Social** (create performing networks and partnerships)
- **"Gardner and Learner"** engaged in continuous preparation (i.e., gardening), being in the system, protecting the organization from the temptation of changing everything while at the same time feeding the capacity to learn as the crisis unfolds, being out of the system. The capacity to move from Inside-Out and Outside-In is key.
- **Role Model** (to provide consistency) demonstrating resilience and adaptability and serving as role models for others so that they become resilient leaders.

Training programs, the sharing of experience, and teamwork, for example, can help develop resilience (Levey, 2019; Walsh, 2025). **Recruiting resilient leaders** through their experience is also a way to integrate this skill: they will be able to help the rest of the organization.

Very few companies have put the development of resilient organizations and leaders at the top of their agenda, despite their interest and the many publications. **When a crisis is over, the tendency is to return to "business as usual".** As with the development of employee satisfaction, the development of resilient leaders and *process*[24] takes time and does not pay off immediately… at least not before the next crisis!

Reality is sometimes different, with the profiles of the actual leaders in place far from the resilient leaders. Top leaders who are selected and succeed, are "sometimes" far from what we described earlier: Narcissism, Overconfidence, Low Emotional Quotient (EQ).

? Why do we persist in selecting for leadership traits that hamper organizational progress—and leave the right potential leaders in the wrong roles?

An organization's **"rules of the game" seem to define it**. If you succeed at playing within the current rules of the game, you're going to get further, but then

[24] A more precise definition is found in the Glossary (See No. 2).

you're going to make things worse. **And if you don't, you might never be selected** (Chamorro-Premuzic et al., 2024). Change the rules of the game means:

- **De-emphasize things like credentials, expertise, and experience**, especially when they apply to something people have done before but are **not so relevant for the future.**
- **Less focus on hard skills** and more **focus on the right soft skills.**
- **Less focus on individualistic, selfish, egotistical, and career-enhancing traits** and dispositions and more emphasis on those that **make other people better.** Fundamentally, leaders need to make other people better, including by helping them collaborate effectively (Chamorro-Premuzic et al., 2024).

Even when the need is known, change is not necessarily initiated. This brings us back to the cycle of change seen in Fig. 3.1. **Factual awareness is one thing**, but it takes an **"emotional click" to commit.** Building resilience is not a new idea, and fortunately, organizations don't wait for a string of crises in this new world that is portrayed as increasingly uncertain. We will see this in Sect. 4.2.4. First, let's finish with company shareholding, especially in a capitalistic context.

4.2.4 Just Maximizing Returns for Shareholders is not Enough

At the beginning of the nineteenth century, the family institution guaranteed trust. Companies were often managed by people who were like merchants and who had mortgaged their family home over many years, that capital being received and transmitted by inheritance. It, therefore, meant being a "prudent" manager. Family names also showed long-term involvement and were intended to guarantee the company's sustainability to the extent of also being the company name. However, "inheritance" capitalism began to decline at the beginning of the twentieth century due to investments in mass production exceeding the capacity of families or putting them at too significant an economic risk. Many began to diversify their assets into financial holdings and move away from industrial activities. (Gomez, 2020).

In the 1930s, a new legal structure emerged as the dominant one, which shifted liability away from individual shareholders and directors regarding their personal assets. This change weakened the direct relationship between decision-makers and the company. Additionally, as shareholders became anonymous and their liability was limited to their financial contributions, **the responsibility for liability increasingly fell on the companies themselves**. (Gomez, 2020).

Starting in the 1980s, the term "return of shareholders" became common to describe the changes taking place in corporate governance and the influence shareholders were beginning to exert. By 2015, 80% of the compensation packages for the 500 highest-paid CEOs in the United States consisted of stock market-related items, such as stock options and free share allocations. This component was virtually absent at the beginning of the 1980s. In France, by 2018, this percentage for the eighty largest publicly traded companies was approximately 40%. Similarly, in the United States, investment funds and pension funds now own about two-thirds of the shares in listed companies, a stark contrast to their almost nonexistent presence in the 1950s (Darcillon & Rebérioux, 2020).

From the 2000s onwards, there was a turnaround. The recurrence of stock market crises (2000, 2008, 2011, etc.) and financial scandals (Enron), as well as the growing urgency for real ecological and social transition, led private **and public players to have more balanced governance** (Darcillon & Rebérioux, 2020). Companies' governance and executive committees have acquired social and environmental requirements, with a mission and a raison d'être (**once upheld by the owner family**) (Gomez, 2020).

Finally, **new governance standards** emerged a few years ago and gave ever greater responsibility to the latter.

4.2.5 What is Really New About VUCA and BANI?

The acronym VUCA was coined by the US Army War College in the late 1990s. When the USSR collapsed, the Cold War binary faction made way for multilateralism. It became difficult for the US Army War College to describe this new world, and VUCA terminology appeared. It is a military-derived term to describe the "multipolar new world order" after the end of the Cold War era. It was later applied to the business environment to describe **chaotic, unstable, and rapidly changing work** settings (Taşkan et al., 2022).

In recent years, organizations have faced significant and unexpected events, such as financial crises, the COVID-19 pandemic, climate change, and war, with a large impact on the world at several economic and societal levels. Therefore, the acronym VUCA has been frequently used by scholars and practitioners to try to understand such environmental dynamics (Taşkan et al., 2022). The four letters of the acronym VUCA stand for the following characteristics (Taşkan et al., 2022; Levey, 2019):

- **Volatility** as many things change rapidly.
- **Uncertainty** can be defined as a situation characterized by a lack of knowledge.
- **Complexity** can be explained by many interconnected parts, making it difficult to understand the reasons and factors in a problem.
- **Ambiguity** can be defined as situations where the causes and questions "who, what, where, when, how, and why" behind the causes and events are difficult to identify.

It should be noted that another acronym, less widely used, was also created in 2020: BANI. The futurologist J. Cascio introduced it after the very anxiety-provoking situation created by COVID-19 while the VUCA was also spreading. What does this acronym mean?

- **Brittle**: The increasing fragility of systems and institutions highlights their lack of robustness in the face of sudden disruptions.
- **Anxious:** Constant uncertainty breeds collective anxiety at the level of individuals and organizations.
- **Non-Linear**: causes and effects are no longer linear; events follow unpredictable trajectories.
- **Incomprehensible:** the complexity becomes such that some situations escape human comprehension (Cascio, 2022).

The lack of robustness is stated in "Brittle" and was not in VUCA. It refers to situations that do not seem new. "Anxious conveys emotion (more than fact) generated by uncertainty, a notion found in "Non-Linear. "Lastly, "Incomprehensible" describes a negative bias of a situation that cannot be understood.

I suggest looking at VUCA. Changes in needs and contexts are often unpredictable, and problems generate typical instability. Volatility has, therefore, always been part of our professional lives. Complexity is part of the daily life of companies, mainly through the need to satisfy customers while considering the intrinsic constraints of a company, which we have mentioned. **The world has always been complicated to understand because of its multiple cultures and possible interpretations.** In 77, Pliny the Elder, in his work Naturalis Historia, wrote:

> *"Only one thing is certain, and that is that nothing is certain."*

Almost two millennia ahead of his time, this was already wholly VUCA. However, if we are honest, there is a significant evolution between the world of yesterday, that of Pliny the Elder, and today: the **globalization** of information systems and flows:

- **We are more than ever virtually interconnected** by information systems and the media.
- **We are more than ever physically interconnected** through transport, whose costs have become affordable.

As a result, we are increasingly interdependent: what happens locally can generate positive or negative consequences globally. In reality, it is not so much the characteristics of the VUCA that define our time but the **connectivity and the associated speed** that generate a new wave we must learn to surf. Within companies, this means knowing how to quickly share information with all who need to be involved to make the best decisions together as soon as possible at the point of impact. Heterogeneous groups require **much more effort and collaboration than in the past for wise decision-making.** It is far from easy, as we have pointed out in previous testimonies, but it makes even more of a difference because the need is increasing.

A global study on VUCA Leadership Preparedness assessed leaders' readiness in these four key areas and found that (Levey, 2019):

- **Less than two-thirds of leaders rated themselves as "highly confident" or "very confident"** in their ability to meet the four VUCA challenges.
- When human resources professionals were asked to assess the VUCA preparedness levels of their organizations' leaders, **about one-third viewed their organizational leaders as incapable of meeting** the challenges of VUCA times.
- As reflected in bottom-line financial performance, the **top 20% of organizations performing well financially are three times more likely to have "VUCA-capable" leaders** than the bottom 20%.

Leaders who are "VUCA-capable" are, in fact, the resilient leaders that we have talked about previously.

4.3 Future Constraints to Be Included and Connected Performance Indicators

4.3.1 General Trends

Some futurists talk about the 2040s or 2050s, but most experts project trends to 2030. It will always be wise to be cautious because thirty years ago, many forecasters talked about China as the future "factory of the world", which would supply "cheap" products.

? What are the major trends for 2030, 2040 or even 2050?

Without the ability or the pretentiousness to make a complete world tour of trends, I have tried condensing what is expressed on different continents (with obviously different biases) using four recognized studies summarized in Table 4.1.

As a general trend, the topics that come up repeatedly concern **climate issues, social and societal issues, technological issues, the implementation of standards, growth, and potential crises.**

- **The four selected studies all examine the impact of technology and digital solutions**, as well as **the associated security and standards**. This trend does not seem debatable.
- **Social and societal issues** are significant in Europe, Japan, and North America.
- **Climate issues and our ability to face them** are expressed prominently in Europe and Japan.
- **Future growth** is expressed mainly only in Asia, where it is a particular concern.

Considering future crises (considered inevitable) has become essential with the global crisis due to COVID-19. The other points are not new, and the future should see **the acceleration of general trends that have already been present** in the past twenty years.

Table 4.1 Selection of future trends (*Sources* Le Goulven et al., 2024; Singh, 2023; Takeda et al., 2022; MRI, 2022; Aggarwal & Bharadwaj, 2024)

	EUROPE from INSEAD	GROWTH	**PLANET**	**SOCIAL**	REGULATION	**TECH.**	CRISIS
TRENDS	Climate Change		■				
	Rise of Emerging Technologies					■	
UNEXPECTED TRENDS	Lack of adequate understanding of Emerging Technologies among government regulators to regulate them effectively.				■	■	
	Civil society increasingly distrusts leaders and institutions' ability to tell the truth.			■			■
EXPECTED TRENDS	The three main interdependent capitals to preserve are the social, the human, and the natural.		■	■			
	NORTH AMERICA from Forbes	GROWTH	PLANET	**SOCIAL**	REGULATION	**TECH.**	
TRENDS	Economic powerhouses of the future – Baby-bust economies : China, India, Indonesia, Brazil, Mexico, Saudi Arabia, and Nigeria could emerge as the ultimate and top 15 economic powerhouses by 2050.	■					
	The Hyper-Connected World					■	
	Dawn of Technology Singularity					■	
	Industry 5.0 - The Symbiotic Era			■		■	
	Increasing urbanization rates in small and intermediate-sized cities, particularly in Africa and Asia.			■			
EXPECTED TRENDS	The Energy Transition		■				
	Innovation to zero: A zero-concept world with zero crimes, zero diseases, zero hunger, zero deaths, carbon-neutral buildings, cities, and net-zero nations.		■	■		■	
	JAPAN from Mitsubishi Research Institute	GROWTH	**PLANET**	**SOCIAL**	REGULATION	**TECH.**	CRISIS
TRENDS	Impact of Climate Change		■				
	Digital Economy					■	
	Changing Life through New Technologies					■	
	Data explosion					■	
	Natural disasters will happen.		■				■
	Changing Role of Government			■	■		
	International Order without Hegemony			■			
	Increasing time for hobbies, travel, and social participation			■			
UNEXPECTED TRENDS	Inequalities remain unsolved and are amplified			■			
	The digital economy is divided among major powers, narrowing its growth opportunity.					■	
EXPECTED TRENDS	Interaction between behavior change and technological innovation for a smooth transition to carbon neutrality.		■			■	
	A Circular Economy with Less Carbon Emissions		■				
	Decarbonization of combustible energy sources, which can be easily stockpiled and transported for both day-to-day normalcy and extraordinary emergencies.		■				■
	ASIA from the World Economic Forum	**GROWTH**	PLANET	SOCIAL	REGULATION	**TECH.**	CRISIS
TRENDS	Asia's influence across the global economy and international affairs is growing. By 2031, ASEAN exports are expected to surge by nearly 90% compared to the overall global trade growth of less than 30%.	■					
	India is emerging as a leader in the development of digital public infrastructure.	■				■	
	China continues to redefine global standards and push the boundaries of innovative global industries such as electric vehicles, renewable energy, and the digital economy.		■			■	
	General trade interconnectivity in Asia					■	
	Supply chain resilience is at the top of mind for corporate decision makers.					■	■

In addition, our modern organizations were built over time around three main pillars: **materialism** (we mainly exchanged material goods), **the verticality of organizations** (the printing press enabled communication from one to all, from the leader or the expert to the "masses") and with **infinite resources** (in a linear system, once we used something, we generally didn't consider anything else but discarding it). These elements have recently been turned upside down.

- The Internet has become a primary force in the economic system by promoting the emergence of **increasingly intangible goods.** The dematerialization of purchases, the creation of new intangible offerings, and the weakening of social ties are some examples of this. **A world characterized by immateriality is developing and gradually replacing materialism, without us knowing the extent of this substitution.**
- The Internet communication has allowed **everyone to connect with everyone else.** The expert or the leader is therefore more easily questioned because everyone can express an opinion on everything, even if it is wrong or potentially harmful.

Human **organizations tend to have a more horizontal structure,** whether in family or professional settings.

- Finally, we must recognize that the Earth's resources are finite, we live in a world with **limited resources which will lead to reevaluating the linear cycle of production** (Barrand & Moïchine, 2024).

? At the level of a company seen as a global entity, will current performance levers for companies still align with new general trends?

For companies in a VUCA context and with the future trends we have specified, the traditional criteria and constraints they must meet **should be quite relevant, or even make even more sense/be relevant:**

- For industrial companies, managing material resources and security will be even more critical where there are high expectations for team engagement.
- It will also be crucial to innovate at the right time, when the market matures (and especially comfortable with technology), to understand the constraints to transform them into a driver of innovation.
- Eliminating waste will still be a priority where in addition to traditional waste management, there will be those related to the environment.
- The ability to implement a resilient organization with resilient leaders and *process*,[25] will be decisive. This includes:
  - Skills management that includes soft skills and anticipates future needs.
  - Greater promotion of staff (we will confirm this in Sect. 4.3.3)
  - Efforts to collaborate faster and respond at the speed required by interconnected global business.
- Lastly, the capacity to integrate suppliers will also be decisive in building a resilient and efficient Supply Chain, having:
  - Relationships based on increased trust, and bonds with transparent information sharing and mutual support (a win–win approach).
  - A simple but effective S&OP *process*,[26] to make the right decisions quickly.

4.3.2 Customer Trends

Regarding customers and their experience, the effort should be on **agility and a high level of automation** in 2030. By that time, 67% of interactions between brands and consumers using digital devices (online, mobile, etc.) will be handled by digital tools rather than by humans as they are today. Therefore, the relationship between

[25] A more precise definition is found in the Glossary (See No. 2).

[26] A more precise definition is found in the Glossary (See No. 2).

humans and machines is expected to strengthen rapidly in the next decade. In addition, Companies will need **to have a good understanding of the readiness of their customers about adopting technology.** This has not always been the case. In France, for example, 78% of the companies surveyed say that consumers do not yet feel comfortable with technology in stores, while only 35% of consumers expressed this discomfort. It will therefore be essential to find the right **balance between empathetic experiences based on human exchanges and purely digital interfaces** (Raymond, 2023). The **capacity to find a balance by adapting to how new generations** interact is a subject already addressed by Nicolas Siegler of the MAIF and Bertile Burel of Wonderbox.

> *"Recently,* ***ChatBots have been disappointing as customers need to be supported by a human interaction anyway.*** *The Natural Language Model could help us make a better interface. New generations also* ***talk more than they write,*** *so* ***making voice instead of typed entries*** *could be a game changer."*
>
> From an interview of Nicolas Siegler with Carine Vinardi

> *"We* started fully digital *(@sales), and it is now at 25%. We have reached a balance where* ***physical shops are still necessary****."*
>
> From an interview of Bertile Burel with Carine Vinardi

Tracking customers throughout their journey to provide them with the best deal requires a **scientific understanding of their countless interactions.** As a result, companies will increasingly need a **global, customizable data strategy at scale,** along with **journey analytics capabilities that can adapt in real-time.** Customers want to feel recognized and understood across various channels, touchpoints, and contexts.

At the same time, they are cautious of how brands handle their personal data. **More than 65% of global respondents believe that the use of their personal data is "out of their control" and are concerned about how brands use their data**. While consumers continue to use technologies that expose their lives publicly, their expectations of companies are twofold: they want to be seen as individuals and expect their privacy to be protected. Therefore, when developing customer experiences, it is essential to find the right balance between **customer experience leveraged on the richness of collected (user) data and the level of consumer trust** (Raymond, 2023).

Additionally, the need for data protection can come from non-sovereign clouds. Indeed, the extraterritorial practice of some countries can create problems for cloud services provided by companies located in those countries. **Data governance** must be established to take into account the imperatives of data management while also prioritizing customer and employee protection, as highlighted by Nicolas Siegler.

> ***"Security is an important risk**: it is essential as we **manage the sensitive data of our customers**. We are waiting for the establishment of a national Cloud and have currently partnered with two companies. However, we **have strict data governance and upload limited data only after validation."***
>
> From an interview of Nicolas Siegler with Carine Vinardi

Finally, customers are also citizens, and we have observed **a growing awareness of social, societal, and environmental issues.** When customers are not focused only on prices, when the countries in which they operate have expectations and transparent communication, these factors can have an impact on their purchasing decisions.

Social networks increasingly promote **frugality, and a notable trend is emerging towards renting, repairing, and purchasing second-hand items.** This is also true in the luxury sector. This new way of consuming could address the "joy of consuming," economic constraints, and environmental actions (Courrier International, 2025).

? In this future context, will the existing performance levers still be relevant?

The priority will be towards customer satisfaction, regardless of the requests. Let's keep in mind the traditional constraints for customer satisfaction:

- **Facing global competition**. This is likely to persist, even if protectionist measures in some countries may alter the rules unless a significant event (climate or health, for example) occurs.
- **The need to find a balance by managing increased complexity is expected to grow** with a need for personalization, rapid response, and a balanced approach between human interaction and digital solutions.
- **Dealing with the uncertainty** of demand will not go away, and the need to react quickly will be even more critical.

4.3.3 People Trends

In terms of employees and the ability to satisfy them, four main elements must be considered for the future:

- **The future availability of human resources** around the world. Indeed, as mentioned, when resources are lacking, companies have a stronger constraint to satisfy them.

- **The aspirations of human resources who will be the workforce of the future.**
- **The capacity for smooth interactions between humans and technology.**
- **Data security** and, in particular, the protection of personal data.
- **The power of data, its impact, and how to reassure employees.**

When it comes to the availability of resources, we have seen that customer satisfaction is taken into account, mainly when resources are scarce, and there is a need to attract or retain the workforce. **However, several studies predict a future labour shortage in particular, from 2030 onwards.** For example, the Korn Ferry study (Binvel et al., 2019) was published before the Covid-19 pandemic, so the timeframe may shift beyond 2030. This study classified employees into three groups: A, Highly skilled workers (those who have completed post-secondary education, such as college or university, or have a high-level trade college qualification), B, Mid-skilled workers (those who have attained upper secondary education, such as high school, or have a low-level trade college qualification), C, for Low-skilled workers (those who have less than upper secondary education).

Figure 4.4 summarises labor shortages by region and type of activity, with their impact.

Here are some general remarks to highlight (Binvel et al., 2019):

- A **talent deficit of 85.2 million workers across the economies analyzed is expected to be** greater than the current population of Germany. The shortfall of Level A workers could equal 21% of the highly skilled workforce of the 20 countries in our study.
- This **global skills shortage could result in $8.452 trillion in unrealized annual revenue**, equivalent to the combined GDP of Germany and Japan.
- **The developed markets would be hardest hit** by **imminent talent shortages.**

Globally	Workers deficit (number)	Unrealized output (annual revenues)
Total	85.2M	$8.452 trillion

Per Region	Workers deficit (number)	Unrealized output (annual revenues)
The Americas	23.9M	$2.307 trillion
EMEA	14.3M	$1.906 trillion
Asia Pacific	47.0M	$4.238 trillion

Per Activity	Unrealized output (annual revenues)	Workers deficit (number)
F&BS	$1.313 trillion	10.7M
FMT	$449.70 Bn	4.3M
M	$607.14 Bn	7.9M

Financial & Business Services (F&BS)	Unrealized output (annual revenues)	Workers deficit (number)
Total	$1.313 trillion	10.7M
USA	$435.7.7Bn	
China	$147.17Bn	
Germany	$136.97Bn	
Japan	$113.67Bn	
UK	$90Bn	
Brazil	$70.2Bn	
Australia	$68.1Bn	
France	$60.8Bn	
UAE	$13.3Bn	
Russia	$16.3Bn	
Mexico	$9Bn	

Technology, Media & Telecommunications (FMT)	Unrealized output (annual revenues)	Workers deficit (number)
Total	$449.70 Bn	4.3M
USA	$162.25Bn	
Japan	$47.8Bn	
China	$44.45Bn	
Germany	$30.7Bn	
Brazil	$28.1Bn	
UK	$27.7Bn	
France	$16.4Bn	
Australia	$15.7Bn	
UAE	$13.3Bn	
Russia	$6.4Bn	
Mexico	$4.5Bn	

Manufacturing (M)	Unrealized output (annual revenues)	Workers deficit (number)
Total	$607.14 Bn	7.9M
Japan	$194.6 Bn	
Germany	$77.9 Bn	
USA	$73.0 Bn	
China	$71.4 Bn	
Brazil	$53.0 Bn	
Mexico	$47.2 Bn	
Australia	$26.6 Bn	
UK	$17.5 Bn	
France	$9.8 Bn	
Russia	$8.8 Bn	
UAE	$0.3 Bn	

The Americas	Unrealized output (annual revenues)	Workers deficit (number)
Total	$2.307 trillion	23.9M
USA	$1.748 trillion	

Europe Middle East & Africa (EMEA)	Unrealized output (annual revenues)	Workers deficit (number)
Total	$1.906 trillion	14.3M
Germany	$629.89 Bn	
France	$214.56 Bn	
UK	$90 Bn	

Asia Pacific	Unrealized output (annual revenues)	Workers deficit (number)
Total	$4.238 trillion	47.0M
China	$1.434 trillion	
Japan	$1.387 trillion	

Fig. 4.4 Summary of expected impact of workers deficit (*Source* Binvel et al., 2019)

- The study also reveals a sizable **mismatch between the supply of available workers and the business demand at country level.**
- **India is the only economy in the study with a talent surplus in all domains,** driven by a burgeoning working-age population.

Here are some more specific remarks (Binvel et al., 2019):

- **Digital deployment versus labor shortage:** While digital transformation often seems unstoppable, it could be about to hit a wall of labor shortage. There is a mismatch between technological advances, including automation and artificial intelligence (AI), and the skills and experience workers need to leverage these advanced tools. Technology could not deliver the promised productivity gains without the proper skilled workers.
- **Regions versus labor shortage:**
  - The Americas: The United States already faces a significant shortage of skilled workers, which is set to worsen with an aging population. However, this problem is not limited to developed countries. Brazil's worker deficit across all skill levels. Mexico is also heading toward skills shortages that could hinder its development, particularly in its promising financial and business services sector.
  - For Europe Middle East Asia: All EMEA markets are expected to experience Level A deficits. In countries with already low unemployment and booming manufacturing production, including the Czech Republic, Poland, Hungary, and Slovakia, a labor shortage has already accelerated automation and increased use of robotics—not to replace people, but because there aren't enough of them to fill the factories. However, all EMEA countries in our study can expect Level C labor surpluses by the same milestone. This suggests a large scope for **skills upgrading to alleviate the talent crunch.**
  - For Asia Pacific: The talent crunch as a percentage of the economy is most pronounced in Asia–Pacific. While some economies in this region deal with rapidly aging populations, others have rising working-age citizens. Hong Kong and Japan face particularly stark deficits, for instance; in contrast, India stands out as the only country in our study that can expect a talent surplus. Although China has ambitious growth plans, talent shortages could curb this nation's vast efforts to fulfill its potential.

- **Talent management versus labor shortage:**
  - Organizations could be prompted to relocate their headquarters and operational centers to places with a more plentiful talent supply. The right talent is an organization's greatest competitive advantage—and that talent is getting scarcer every day.
  - To deal with a skills mismatch when existing, some companies are already building their own talent pipeline by hiring straight from school or college.

Internal schools and universities can supplement this for large groups in particular. **Labor will be available globally, but it may not match sector needs at the time.**

While technology will reshape the future of work, organizations will be unable to leverage it without the right talent. Only through the partnership of people and technology can the full potential of both be realized.

If we want to examine the aspirations of this workforce, which is expected to be deficient in the future, we must understand the generations involved in terms of time horizons.

- The Baby Boom Generation, born between 1945 and 1965, will no longer be active in 2030.
- Generation X (**Gen X**), born between 1965 and 1980, to which I belong, will still be at work in 2030 but not in 2040.
- From 2030 and still a little early 2040, the youngest members of Generation Y (**Gen Y**) or **"Millennials"**, born between 1980 and 1995, will still be at work.
- From 2030, Generation Z (**Gen Z**) or "Digital Natives," born between 1995 and 2010, will be the working youth.
- Finally, Generation Alpha (**Gen Alpha**), born after 2010, will not enter the labor market until 2030 and will constitute the bulk of resources at work from 2040 onwards.
- As I write, the Generation Beta (**Gen Beta**), born between 2025 and 2040, has just been born. It will be at work for the former at the beginning of 2040, and it is not yet possible to consider their aspirations.

In Sect. 4.1.1, we saw that generations can change their position on the cultural dimensions that serve to understand individual expectations. In the future, human resources are expected to be lacking in many regions of the world. The **aspirations of these generations in the workplace will be crucial to ensure they are fully committed to improving company performance.**

? What can we learn from the aspirations of these different generations?

Gen X grew up during **significant social and technological change** where they had to learn to be **adaptable and resilient,** in the **uncertainty brought by multiple crises** affecting a gradually global economy. There are a few additional points worth highlighting (Knight, 2024):

- **Generally loyal** to their company (low absenteeism rate), they are also the first generation to embrace the **ethical needs of companies.**
- It is also the first **generation to value work-life balance** without always being able to assume it. Women's access to work and the work of both parents has also generated a **need for independence.**
- It is finally the first **technologically savvy generation**.

They are often referred to as the **"forgotten generation"** because companies did not consider their aspirations. Having evolved in a context of abundant human resources, this did not encourage the consideration of their aspirations (as we have pointed out). When the first studies linking engagement and performance within companies were published, Gen Y, the famous Millennials, were already at work and the spotlight was focused on this generation.

Gen Y, appears to be the first really been able to challenge the norms post-2010s when resources began to run out in some countries (De Witt, 2024; Landecy, 2024):

- This generation has grown up with the internet and is distinguished in particular by its relationship to work. Having seen its loyal elders lose their jobs after years spent in the same company, it is the first generation to assume that **they do not want to completely sacrifice their personal lives for their professional lives, as their elders did.** It should be noted that this point was already expressed in Japan in the survey on general trends (Sect. 4.3.1).
- Even more than Gen X, this generation has **the habit of change.** And like Gen X, it seems to attach great **importance to ethics and meaning at work.**

Gen Z is the most studied of all (De Witt, 2024; Landecy, 2024; Vinardi, 2016):

- They are the first generation to be **truly born with digital technology**, which is part of their daily lives. This implies a **"global" approach** to information and links, a connection to the world at the click of a button.
- **An ultra-connected generation** evolving horizontally in a network involves its members in several forms of collectives.
- While Gen X and Y have become accustomed to the changes, **Gen Z rapidly expects change**: for them, it would be the norm, not an exception.

Some similarities also with Gen Y:

- Gen Z **cares about mental health and work-life balance.** They value the human experience and recognize that life is more than work.
- **Gen Z thinks differently about loyalty**. They also grew up in workplaces that were not very loyal to their employees.
- **Gen Z looks for trust and authenticity**: Authenticity is about trust. Words and actions need to match. Transparency is also essential.

Finally, it is a generation committed and **sensitive to social and environmental causes** that rocked its early years. In addition, this generation is preparing and expecting to change companies and professions several times, **so the cultures of the latter will have less of an impact on them**. On the other hand, they will always be influenced by their national culture, and their connection to the world does not imply that young people want and aspire to live elsewhere than in their country. **All the experts, therefore, envisage changes taking into account national specificities.**

Lastly, Gen Alpha, of which not all members have yet been born, is the subject of early predictions. Some experts describe this generation as "the most educated, the most connected, and overall the wealthiest of all time." Others suggest they will have "a different relationship to knowledge, characterized by a more utilitarian and "opportunistic approach" (De Monicault, 2024). At the enterprise level, Generation Alpha will have grown up with the rise of AI and will likely have high expectations for a sustainable future, as for Generation Z:

- Hyperconnected, its members will be used to evolving in a digitalized world, their **expectations are likely to be high to find in companies, the same standards of innovation and speed that they know on a daily basis.** Their ultra-connections also tend to generate **more acute social and emotional needs** with a need **for a work environment that develops collective and belonging.**
- As for Gen Z, but more acutely, the expected working conditions should include **collaborative spaces, teleworking, autonomy, and individualized training plans.** Finally, Gen Alpha will tend to join a company according to its values, with a **preference for authenticity, ethical commitment, transparency, social responsibility (diversity, inclusion, decent working conditions, etc.), and respect for the planet.**

As summarized in Table 4.2, there are no breaks in expectations, but there are elements that either attenuate (slightly) or amplify (significantly).

When examining the professional sphere, the relationships between employees and management **can influence specific cultural dimensions within companies and the countries to which they belong** (Vinardi, 2016). Several **common trends** have been identified by some authors and experts that must be considered to improve the commitment (and therefore the performance) of these generations compared to previous ones (Vinardi, 2016, 2019; De Witt, 2024):

- **More collective orientation**: Focus on the team and, therefore, on the collective importance of teamwork and learning. Collaboration must be all the more effective if we have to move quickly.
- **More flexible time management**: Tendency to multitask, multitask, want multiple jobs at once, and not necessarily work only in an office but from home. Changing companies after six months when you are unsatisfied is no longer exceptional for a young person of this generation.
- **Lower hierarchical distance**: Tendency to **question hierarchical authority,** they need a **more egalitarian structure.** They would not necessarily see elders as experts, and they would expect to understand why something is done in a certain way. They would also not be afraid to challenge why things are done the way they are.
- **Shorter-term orientation:** Accustomed to "acting" on social networks with a click, Gen Z could ask for more "real-time" transformation.
- **Need for more substantial feedback**: More regular feedback and support are needed.
- **Need for more substantial meaning**: in a world where ecology, geopolitics, and the economy are fragile, the quest for meaning in what they undertake is even more critical, and the family is a safe investment.

Table 4.2 Summary of main expectations of Gen X, Y, Z and alpha

Gen	Company loyalty	Trust and authenticity focus	Ethic focus	Social and sustainability focus	Need for meaning at work	Change habits	Confort with tech/digital	Level of digital connection	Need for work/life balance
X	+++	+	+	+	+	+	+	+	+
Y	++	++	++	++	++	++	++	++	++
Z	+	+++	+++	+++	+++	+++	+++	+++	+++
Alpha	–	++++	+++	++++	++++	++++	++++	++++	++++

The workforce of 2050 seems to have adapted to change and is comfortable with the frequency and the speed that goes with it. At the same time, there appears to be a strong need for meaning, to put humans back at the center of addressing needs. As a result, companies will need to increasingly develop their capacity to manage change swiftly but by maintaining peace of mind. **More than ever, leaders must employ strong change management skills at the heart of business development** to engage teams in the necessary transformations without causing any crushing anxiety.

In an increasingly connected world, **seamless interactions between humans and technology will be necessary**. As technology becomes universal and employees are placed at the center of the system, interactions with equipment and the virtual world will also need to be facilitated.

Effective interaction is the primary requirement to ensure fluidity. If digital solutions are ineffective or insufficient, they will generate frustration by taking up employees' time. As Nicolas Siegler highlighted, this can already happen with Chatbots (virtual communication robots) or during the implementation of AI-based solutions, as noted by Bertile Burel.

> *"We faced an issue implementing a tool, using AI, to automatically formulate the best answer to our customers (whereas it was done manually before).* ***Ten percent of the time, the proposition was wrong, so our team had to read every proposition before sending it.*** *Therefore, we* ***spent more time after implementation than before."***
>
> From an interview of Bertile Burel with Carine Vinardi

In addition, interacting with physical digital solutions, such as cobots or virtual solutions (virtual robots or chatbots, for example), is not intuitive.

While some interactions are facilitated by them on a daily basis (using applications to make reservations, for example), others are less so.

Therefore, it will be essential to support this transition for employees by integrating the steps we saw in Sect. 3.3. During this support *process*,[27] we must also **give confidence to employees** by ensuring **the efficiency of the robots** to prevent **redoing any work**. We will discuss how to achieve this in the following chapters.

Lastly, other important aspects to consider to ensure employee satisfaction are data, its security, and its impact. Let's start with protection.

In business data, **privacy with personal data is a concern, some of which is sensitive.** As we have seen in Sect. 4.3.2, customers are concerned about how their data is used, even if they may be less cautious on social networks. These customers are also employees of various organizations, and they expect companies to protect their data. In Germany, for example, internal applications that collect personal data are highly restrictive. Data protection differs from one country or continent to another,

[27] A more precise definition is found in the Glossary (See No. 2).

even excluding countries that do not protect their citizens but use data collected for "surveillance" purposes. Conversely, under pressure from its member states, some of which had already passed protective laws, the European Union has legislated the European Accessibility Act (EAA) in 2019. The act will be strengthened in 2025 with requirements involving access to data for people with disabilities or verifying informed consent (ensuring information is given without coercion), for example. Overall, the EAA provides a framework for the collection and use of data. Organizations that fail to meet these requirements may face penalties, including fines.

Beyond a regulatory framework, organizations must protect **their employees' and customers' data from theft.** The protection of employees must include physical and psychological safety (security of individuals and property) as well as virtual safety (cybersecurity).

Companies have long considered security, but cybersecurity is more recent, with the advent of the Internet. As a result, **cybersecurity is developing rapidly and has become a critical concern.** As for customers, data governance may be necessary to address management requirements and employee protection.

Beyond protection, data is becoming the new power issue, just as information was before the digital age. In the past, whoever possessed the information chose when and how to transmit it. This information could be specific expectations of customers or employees, a way to complete a task faster than the standard, a sales or production schedule, etc.; in short, anything that could enable employees to make decisions different from their peers. **However, the less confident employees or teams feel, the less inclined they are to share information**: this is already what we are noticing within companies. Information sharing within industrial activities is more important, even though it is not always ideal. One of the reasons is that there is a physical flow; information (stocks, security, malfunctions, etc.) can be more easily visualized than in transactional activities without physical flow. In both cases, if employees compete, are insecure about their jobs, or are uncomfortable with their skills, they are less inclined to share information.

The difficulties of sharing information are substantial in the S&OP *process*[28] mentioned earlier. It requires the collaboration of Sales, Marketing, Supply Chain, Production, and R&D functions (for example) to be effective. However, as we have seen, the latter may have different interests and difficulties in sharing. In the digital age, as data becomes increasingly accessible through various interfaces, new challenges will have to be addressed:

- For data that is automatically collected, directly available, and transmitted to everyone, those who were not in trust will be even more destabilized, and compensatory behavior is to be anticipated.
- For data collected and transmitted through humans, it is a safe bet that if trust is not sufficient or insecurity is too strong, retention will continue (this is already what customers do when it comes to the collection of personal data).

[28] A more precise definition is found in the Glossary (See No. 2).

Power struggles in the workplace can become harmful when employees feel insecure. As a result, **fostering trust and psychological safety within organizations becomes increasingly important especially as employees lose control over their data.**

Lastly, it will be essential to anticipate the effects of automating data collection, processing, and analysis. These are the responsibility of people for all or part of their work, regardless of their level or function in the organization. For digitalization to be implemented smoothly, several challenges will have to be addressed as well:

- If people realize that by sharing data, they might lose their jobs, it's a safe bet that they will be less cooperative. This is already what we saw when physical robots began to replace people in manufacturing.
- The people who provide data without the risk of losing their jobs will nevertheless have to gain something in return, including better working conditions. This is what Marie-Ange Giuseppin points out.

> *"If you want to digitalize, you need data, so you need the operator to confirm work orders, track logistic flows, enter Quality statuses... and data acquisition needs time. So,* ***at the beginning of digitalization and for some functions, you clearly lose productivity.*** *At that time, it is important to define the* ***right balance for each function around the constraints & benefits of digitalization.*** *For example, production operators must provide data about a work order. As a counterpart, they are given digital instructions/checklists so that they can also be supported by digital tools."*
>
> From an interview of Marie-Ange Giuseppin with Carine Vinardi

- For all those, and we are all concerned, who will have part of their activity automated, **the residual part of our work will have to make sense and be compatible with our abilities and skills.** As we saw in Chap. 3, our brain cannot spend all of its time making decisions or convincing people; a part must be done automatically and not "thought out." Moreover, because of skills and preferences, some are more comfortable with analysis than decision or negotiation. This is also what Marie-Ange Giuseppin points out.

> *"The* ***more transactional data management and analysis is made automatically****, the* ***more what remains for employees is to negotiate or convince****. However, they cannot always do it. (It is energy-consuming, and skills are not always available.) It's the same for Management Committees; what could*

remain is the ability to make quick and accurate decisions based on uncertain information..."

From an interview of Marie-Ange Giuseppin with Carine Vinardi

In summary, the power of data in the context of digitalization should force companies **to accentuate their actions to create trust and a basis of security, give meaning, anticipate changes in work, discuss them with employees, balance positions, and ensure the adequacy of skills.**

? In this future context, will the performance levers that have existed to date still be relevant?

As we have discussed, employee satisfaction is a vital performance lever that has been taken into account in a heterogeneous way to date, except when human resources are lacking and they need to be seduced. However, the lack of resources is a very strong risk for the years to come. The coming years should, therefore, confirm and accelerate the constraints of the past:

- Taking into account physical and psychological safety and proactive risk management to make working conditions as "pleasant" and efficient as possible, with:
  - Increased consideration of ergonomics on all workstations.
  - Increased respect for physiological constraints to define working hours.
  - A need to create a psychological safety base to create trust and share data effectively.
- Take **into account traditional criteria of commitment** that are transgenerational or reinforced by the aspirations of Millennials, Gen Z, and Gen Alpha that we have mentioned. By way of a summary:
  - Connected **companies that have integrated digital solutions when necessary**, and that organize themselves to **react (very) quickly to events. The organization will, therefore, have to be designed accordingly**: rapid and transparent flow of information, efficiency in problem solving with a small number of hierarchical levels, and autonomous teams.
  - Companies that **"speak the truth" without inconsistency and who match their words to their actions**. There is a need for greater coherence between word and deed (authenticity). For example, companies that avoid expressing a need for diversity where their executive committees and boards are single-sex, or companies that avoid expressing a need for autonomy where their decision-making is slow and complicated.
  - Companies that **assume their social (with Diversity and Inclusion), societal, and environmental responsibilities** with a **stronger sense of ethics.**

- **Individualizable development and training plans**.
- An increased need for **Leaders who act as role models** and can generate **social ties (role of giving).**
- An unavoidable need to **enhance the value of each person's "trick" to maximize the available human resources.**

In terms of cultural dimensions:

- **More collective orientation** with a more substantial need to belong.
- **More flexible time management** with **a preserved work/life balance, telecommuting, and autonomy.**
- **Lower hierarchical distance** with a transparent and rapid flow of information.
- **Shorter-term and even "real-time" orientation**
- **Need for consensus and more substantial feedback**
- **Need for more substantial meaning** by linking collective actions to strategy.

4.3.4 Governance Trends

Finally, as far as shareholders are concerned, since the question **of profitability** will not wane, it will have to be **harmonized with sustainability issues of the development model. This includes resilience in the event of systemic shocks and the integration of digital technology, which is now essential.** In this context, both regulators and investors look for new models that enable **performance and sustainability** to coexist (Darcillon & Rebérioux, 2020).

Corporate **governance responsibility has already increased,** and **this movement is accelerating**. Corporate boards face growing pressure and scrutiny from investors, regulators, and markets due to new disclosure requirements and performance expectations around managing **climate, as well as nature-related risks and their impact on society.** Stewardship by investors is growing, with activist investor pressure on companies to improve ESG (environmental, social, and governance) performance at record levels (CISL, 2023).Let us then recall the trends that seem to be emerging for the future (CISL, 2023; Chiang & Syn, 2024):

- **Navigating the storm:** Adapting to disruption. Boards must adopt a proactive, adaptive, and resilient approach. Some Asia Pacific (APAC) boards are embracing AI and other disruptive technologies to make more informed, data-driven decisions and stay ahead of the curve.
- **The New Boardroom: evolving compositions and roles.**
  - The skills and attributes needed in the boardroom are evolving rapidly. While **diversity remains a priority**, the emphasis is shifting from pure representation to diversity of expertise, particularly in areas like **ESG and Digital (with AI).**
  - **As boards' responsibilities increase**, more and more specialized committees are emerging. Boards are also increasingly taking on **roles traditionally reserved for management, such as talent development and succession**

planning. The independence of directors is key while ensuring they provide effective guidance and oversight.

  - **Continuous learning, a key to self-renewal:** Emphasis on learning extends to individual directors, who are increasingly valued for their curiosity, agility, and ability to contribute effectively in varied contexts**.**

- **The human Factor: Soft skills, trust, and relationship building.** While technical competencies are essential, soft skills, such as emotional intelligence, active listening, and strategic communication, are increasingly pivotal in board leadership. This is especially important in complex environments like family-owned businesses where dynamics can be intricate, or in corporate boards' interactions with CEOs.
- **The balancing act between performance and regulatory compliance**

  In Asia Pacific, family-controlled firms dominate, and the lines between corporate and family governance are increasingly blurred. It will become crucial to balance the nuances of family ownership with the rigors of professional governance.

Company governance has already redistributed this responsibility with its CEOs and Top leaders. The latter still have a significant part of their remuneration linked to stock market elements. However, in recent years, this part has been increasingly supplemented by **objectives related to Safety, the Environment or Social and Societal impact, for example.**

It is becoming clear that the traditional one-dimensional goal of maximizing returns for shareholders is increasingly at odds with the continuation of a viable future for people and the planet, or the ability of a business to thrive in the long-term (CISL, 2023).

It will be noted that these trends are all consistent and even common with those that we have detailed in the general trends for companies (Sect. 4.3.1).

4.3.5 New Performance Indicators

We have seen that traditional constraints aimed at generating stakeholder satisfaction and performance will always be relevant for the future, and even more so in many cases. This tends to support the KPIs listed in Sect. 4.1.2. We also identified several key trends to consider for the future of organizational performance. It is essential to determine the need to add new KPIs to be future-proof. Here are two summaries of what was presented in this chapter.

- Figure 4.5 is an outline representation that completes Fig. 4.3 with the key constraints to come: the refocusing on the human being and the associated social/societal aspects, environmental consideration, mainly recycling for physical products, digital connection, and data security.
- Figure 4.6 shows a summary of the respective weights of the stress fields for industrial and transactional activities, with future trends. Industrial companies have/will

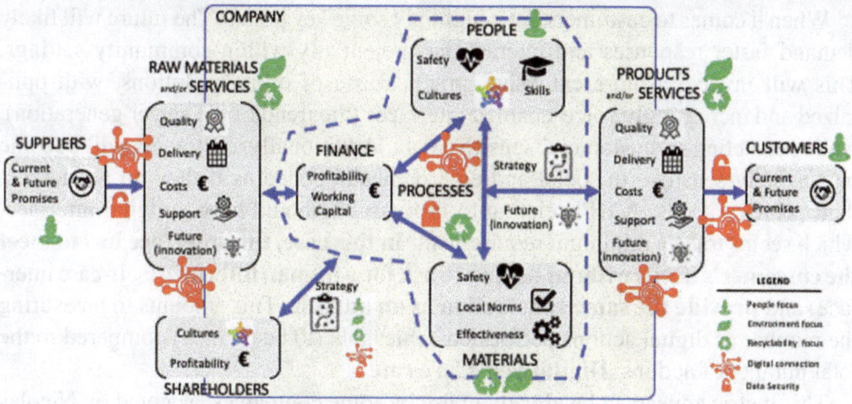

Fig. 4.5 Simplified representation of Contraints' Fields in the future

have **stronger constraints on safety, the management of their industrial assets, and their environmental impacts.**

To list the new relevant KPIs, I suggest that you go back to the categories previously made including satisfaction of customers, employees, companies, and their shareholders. This is to complete the list already provided.

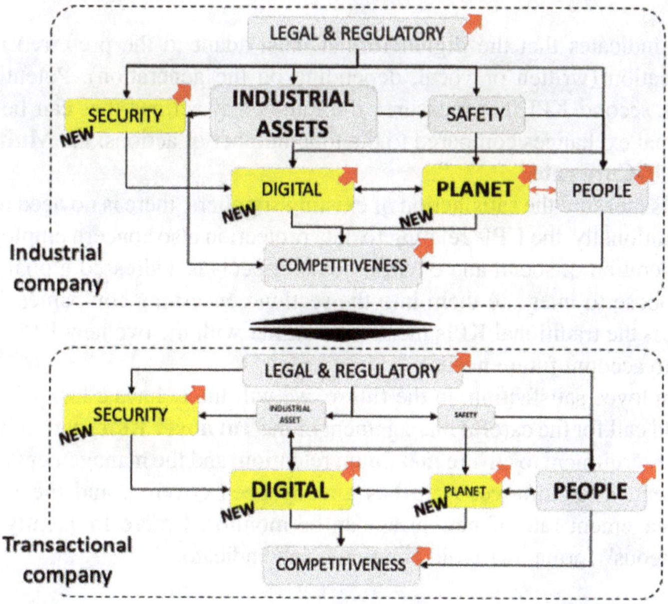

Fig. 4.6 Industrial versus Transactional Business constraints with Future trends

When it comes to customers, let's highlight some key points. The future will likely demand faster responses and interactions, potentially within community settings. This will involve a more extensive, practical use of digital solutions, with optimized and increasingly voice-enabled interfaces (the trend of the newer generation), while protecting its customers' sensitive data. Additionally, customers will become increasingly sensitive to social and environmental concerns if they do not impact their primary needs. A KPI relating to *processes*[29] should have a digital interface, which seems to be a minimum requirement. **In this case, this interface has to meet the customer's needs without having to ask for a human interface** (self-care interface) and **provide the same satisfaction as an advisor**. This amounts to measuring the number of digital actions performed/achievable (to be defined) compared to the total number of actions: **Digital Self-Care rate.**

This metric appears to be already in use by some companies, as noted by Nicolas Siegler:

> *"Our **main metric is Self-Care rate** (what the customers can perform alone, but are fully satisfied). Digital technology is a key driver in improving Self-Care and Internal efficiency to offer the **most efficient IT solutions at the most affordable costs.**"*
>
> From an interview of Nicolas Siegler with Carine Vinardi

This KPI indicates that the digital format must adapt to the preferred modes of communication (written or vocal, depending on the generation). Potentially, this calls for a second KPI that measures the number of actions that can be made in multi-format exchanges compared to the total number of actions: the **Multi-format Digital Self-Care rate.**

As KPIs measure the satisfaction of existing suppliers, there is no need to add any more. Additionally, the KPIs relating to data protection also concern employees, and the consideration of social and environmental aspects is addressed globally. Therefore, I propose to integrate them into the section concerning companies. Table 4.3 summarizes the traditional KPIs mentioned earlier with the two new KPIs proposed to take into account future needs.

For employee satisfaction, in the future, we will likely have a lack of resources. This would call for the careful management of the **Turnover Rate**, particular follow-up during recruitment to ensure maximum retention, and the management of internal development. The Turnover Rate has already been covered, and the monitoring of the engagement rate of new hires can be monitored more frequently using an "instantaneous" format (to avoid creating a new indicator).

---

[29] A more precise definition is found in the Glossary (See No. 2).

Table 4.3 Traditional and New KPIs for customer satisfaction

<table>
<tr><th colspan="3">Customer satisfaction KPIs</th></tr>
<tr><td colspan="2">Customer Lead Time (LT)</td><td>Delivery date/time – date/time of the request</td></tr>
<tr><td colspan="2">Customer satisfaction rate</td><td>Number of (very) satisfied customers/total number of customers × 100</td></tr>
<tr><td colspan="2">Satisfaction/Problem Resolution rate</td><td>Number of (very) satisfied customers with problem resolution/ number of customers with problem × 100</td></tr>
<tr><td colspan="2">Retention rate</td><td>((No. of customers at the end of the period – no. of new customers)/no. of customers at the beginning of the period) × 100</td></tr>
<tr><td colspan="2">Net Promoter Score (NPS)</td><td>Percentage of all promoters (No. of promoters/total number of customers surveyed × 100) – percentage of all detractors (number of detractors/Total number of customers surveyed × 100)</td></tr>
<tr><td rowspan="2">New digital</td><td>New (Leading KPI) (Customer) Digital Self-Care rate</td><td>Number of actions carried out/achievable by the customer without human interface but with equivalent satisfaction/no. of actions achievable × 100</td></tr>
<tr><td>New (Leading KPI) (customer) Multi-Format Digital Self-Care rate</td><td>No. of actions that can be performed without a human interface and in different formats (written, verbal)/no. of actions that can be performed × 100</td></tr>
</table>

Moreover, tracking the average time spent on customizable employee development can be meaningful as a new indicator: **Average Development Time per Employee.**

There is an increasing demand for ethical practices for specific generational expectations (beyond customizable training). This suggests a specific monitoring indicator, which could also be applied at the governance level due to its importance. **Monitoring engagement in real-time using digital solutions will help capture satisfaction levels and enable immediate responses to aspirations.**

Future generations will also expect digital solutions for daily operations within companies. It would not be easy to agree on a single KPI to monitor the deployment of every company *processes.*[30] However, it would be interesting to build up a list of actions developed in a *Bottom Up*[31] approach by the staff, which could be updated as needed. The indicators **Digital Self-Care Rate** and **Multi-format Digital Self-Care Rate** could be relevant, focusing on targeted actions internally (Table 4.4). With future generations, there will be a need to address social development and diversity integration. This can be observed through commitment levels and is increasingly demanded by customers and potential investors. I suggest incorporating at least one dedicated KPI at the company level (Table 4.5) that can make sense, especially for **gender diversity,** because promoting women in the workplace is essential. **Indeed,**

[30] A more precise definition is found in the Glossary (See No. 2).

[31] A more precise definition is found in the Glossary (See No. 6).

Table 4.4 Traditional and new KPIs for employee satisfaction

Employees' satisfaction KPIs		
Safety Frequency rate		Nb of accidents (with or without lost days)/millions of working hours
Safety Severity rate		Nb of lost workdays/total hours worked) × 1000
Unexpected Absenteeism rate		Number of hours of absence during the period/number of theoretical working hours over the period × 100
Turnover rate		Proposed calculation formula = number of departures/ average number of employees over the period × 100
Engagement rate		Percentage according to a reference of questions with two formats: • Full format produced annually • Digital instant format for spot surveys
New digital	New (Leading KPI) (Employees) Digital Self-Care rate	Number of actions carried out/carried out by the employee without a human interface but with equivalent satisfaction/number of actions carried out from a validated list × 100
	New (Leading KPI) (Employees) Multi-format Digital Self-Care rate	Number of actions that can be performed without a human interface and in different formats (written, verbal)/no. of actions that can be performed from an × 100 list
New people development	New (Leading KPI) Average Development Time per Employee	Average number of hours' personalized training/ coaching/mentoring per employee
New people contribution	New (Leading KPI) *Hoshin kanri* [32] Deployment Rate	No. of Leaders engaged in the company's *hoshin kanri* [33]/total no. of leaders × 100

if this diversity is not achieved, neither will other forms of diversity; half of the world's population should not be considered a minority. Given that leadership positions are crucial for promoting diversity, I propose evaluating the number of women in leadership with a **Women Leaders Rate**. Other KPIs could complement the measurement, and some countries, such as France, are asking for more comprehensive sets of indicators.

Lastly, another generational expectation that deserves a specific KPI is a clear contribution to strategic goals and collective results with a short-term focus. This need could be addressed using the *Hoshin kanri*[34] method from *Lean* management: tracking the deployment could be useful with a ***Hoshin kanri***[35] **Deployment Rate.**

Beyond generational aspirations, the fluidity of digital interactions and the psychological safety required for trust and transparency are crucial; here too, conducting

[32] A more precise definition is found in the Glossary (See No. 3).

[33] A more precise definition is found in the Glossary (See No. 3).

[34] A more precise definition is found in the Glossary (See No. 3).

[35] A more precise definition is found in the Glossary (See No. 3).

annual and instant engagement surveys can enable us to respond effectively. Additionally, **data security should** be included as a KPI that we integrate across the entire company, as it also concerns our customers and suppliers.

As far as companies in general are concerned, we have considered social, environmental, and data security indicators that align with future challenges.

At the social level, in the future, we will likely have increased constraints in social matters within companies. This is because generations demand them and also because labor shortages will force this to be taken into account. We will see this in the **Women Leaders Rate** that I mentioned. Another KPI is the **Fair Working Condition Rate,** which measures a company's commitment to paying its employees at or above the adequacy wage determined by each country. These two KPIs ensure the bare minimum only.

It is important to note that these requirements also apply to suppliers. It is vital to know their Corporate Social Responsibility (CSR) approach, which can include environmental aspects, so I propose a KPI to measure it: **Suppliers CSR compliant Rate**. Compliance will depend on the chosen reference approach. In Europe, the environmental aspect often takes precedence, while in the Americas, company CSR expectations are often typically social.

As far as Asian firms are concerned, they seem significantly less inclined to engage in CSR initiatives concerning "organization and management," messages on "CSR interactivity," and those regarding "CSR as a business case" compared with European firms (Vollero et al., 2022).

Concerning the environmental aspects, it is often the ones with new company KPIs for at least two reasons: results are often quicker to demonstrate and easier to communicate (as they are quantifiable) than for social aspects; apart from regulatory requirements, companies want to enhance their image to their customers (at the risk of greenwashing).

Once again, here is a bare minimal proposition.

First of all, as a general KPI, the measurement of Green House Gas (GHG) emissions: **Total GHG emissions**, an indicator many industrial companies have already, with scopes 1, 2, 3:

- **Scope 1** concerns the Direct emissions from the company's operations.
- **Scope 2** concerns Indirect energy emissions (emissions generated by the production of electricity, steam, heat, or cooling consumed).
- **Scope 3** concerns Other indirect emissions (upstream and downstream activities).

The other indicators concern energy, which is a major challenge in the era of digital expansion. At least two KPIs are essential for measuring energy consumption and the proportion of renewable energy consumed: **Total Energy Consumption** and **Renewable energy Consumption rate.**

KPIs concerning water consumption and recycling are just as important in a future where water will become scarce before energy. There are at least two relevant KPIs: **Total Water Consumption** and **Recycled Water Rate.**

For industrial companies that supply physical materials, a recycling indicator is essential. There can be several, such as the rate of recycled raw materials used in

Table 4.5 Traditional and New KPIs for Company Satisfaction

Company satisfaction KPIs		
Product or Service: On Quality Delivery (OQD) rate		Number of compliant goods or services/Total number of goods or services delivered × 100
Product or Service: On Time Delivery (OTD) rate		Number of goods or services delivered in accordance with the agreed date/Number of total goods or services delivered × 100
Equipment Effectiveness (EE) rate		Time to produce products or services at the expected speed/ Total time used × 100
R&D or Innovation rate		R&D or Innovation expenses with internal and external CAPEX and OPEX/Revenues for the period under consideration × 100
Time To Market (TTM)		Date on the market – Decision date
Technology/Customer Maturity rate		TRL/BRL
Economic profit		Net Operating Profit After Tax – (Invested Capital x Weighted Average Cost of Capital)
Working Capital		Current assets – current liabilities
New gender diversity	New (Leading KPI) Women Leaders Rate	Nb of women in leadership roles/ Total Nb of leadership roles × 100
New social	New (Leading KPI) Fair Working Conditions Rate	Nb of employees paid below adequate wages (by country)/ Total employees × 100
New suppliers social responsibility	New (Leading KPI) Suppliers CSR compliant rate	Nb of suppliers CSR compliant/ Nb of suppliers × 100
New environment	New (Leading KPI) Total GreenHouse Gas (GHG) Emissions	Total GHG emissions in CO2 equivalent produced by the company over a specified period for score 1, 2, 3
New environment energy	New (Lagging KPI) Total Energy Consumption	Total energy consumed by the company over a specific period, usually measured in MWh or GJ
	New (Lagging KPI) Renewable energy Consumption rate	Total energy consumed from renewable sources in kWh/Total energy consumed in kWh × 100
New environment water	New (Lagging KPI) Water consumption	Total volume of water consumed by the company over a specified period, usually measured in cubic meters

(continued)

Table 4.5 (continued)

Company satisfaction KPIs		
	New (Leading KPI) Recycled water rate	Total volume of water that was treated and reused within the company's operations over a specified period
New environment product	New (Leading KPI) Recyclable/Recycled Product Rate	Nb of recyclable/recycled product/Nb of products × 100
New environment waste	New (Lagging KPI) Waste produced	Total volume or weight of waste produced by the company over a specified period (kg/unit of product)
	New (Leading KPI) Recycled waste rate	(Total waste recycled/Total waste produced) × 100

production, the recycling rate of finished products, and the rate of finished products that are ultimately recycled. The most transparent and environmentally relevant indicator would be the rate of finished products that are actually recycled, as this signifies a functioning recycling sector… However, given that we are talking about the future, the rate of recyclable end products can also apply: **Recyclable/Recycled Product Rate**. Also, following this logic, the production and recycling of production-related waste could have at least two KPIs: **Total Waste Consumption** and **Recyclable/Recycled Waste Rate.** Table 4.5 summarizes the traditional performance measurement KPIs and the new proposed indicators for the future.

Let us end this governance section with shareholder satisfaction through the governance of the company that it is supposed to have. **The future is not expected to change financial expectations, but in a world where events require quick decisions, there will be increasing pressure on non-financial issues.** Thus, it is essential for governance to have the right expertise and strong teamwork to enable the organization to navigate social, environmental, and digital challenges effectively. At least three KPIs can help to define necessary (but not sufficient) conditions of the proper functioning of a board, by measuring the independence of board members, their expertise, and finally, the board's continuous training: **Independence Rate, Expertise Rate, and Total Training per Board Member.**

To ensure that companies adequately address social and environmental issues, specific standards are/have been established. This is particularly significant for investors (and also for audit and consulting firms that recognize the business opportunities). In Europe, the Corporate Sustainability Reporting Directive (CSRD) has introduced a reference framework for **Environment & Social Governance (ESG)**. I know this might seem complex, but it can be summarized as follows:

- CSR describes a company approach, and depending on the continent, it may or may not integrate the social, societal, and environmental aspects in a balanced way.

- ESG describes a global reference framework with a societal, social, and environmental part and the associated governance. Compliance with this framework is supposed to make it possible to assess whether the data communicated are compliant (in the same way as the financial framework). However, it does not yet apply to all sizes of companies (the overload of work inherent in the production of this report is not neutral).
- Lastly, we have discussed ethics, which is gaining traction with future generations, so it is essential to have at least one indicator that covers training and compliance in the Code of Ethics and Conduct**: the Code of Ethics and Conduct Approval rate.**

Table 4.6 summarizes these various KPIs.

Table 4.6 Traditional and new KPIs for shareholder satisfaction

Shareholder satisfaction KPIs		
Dividend per share		Total amount of dividends that are paid/number of shares outstanding
Capital gain per share		Gain realized by selling the shares if the sale price is higher than the purchase price
New D&I and development	New (Leading KPI) Independence Rate	(Number of board members without affiliations or interests in the company/Total number of board members) × 100
	New (Leading KPI) Expertise (International, business, ESG, digital/AI …) rate	(Number of board members with relevant expertise/Total number of board members) × 100
	New (Leading KPI) Total Training by board member	Total number of training sessions or workshops focused on governance best practices attended by board members per year
New social environment	New (Lagging KPI) Environment & Social Governance (ESG) rating	Letter(s) or results according to referential
New ethic	New (Leading KPI) Code of Ethics and Conduct Approval rate	Nb of employees (including Board members) who approved the Code of Ethics and Conduct/ nb of employees at risks × 100

In 2025, there is still no single ESG benchmark at the global level. Europe was the first to initiate a framework with results being evaluated at the company Board level. However, in May 2024, China drafted its first China Sustainability Disclosure Standards (CSDS), which are based on the "double materiality" concept introduced by the European Corporate Sustainability Reporting Directive (CSRD). These standards, which aim to establish a standardized framework for sustainability reporting, will be implemented between 2026 and 2030. **Therefore, an Environment & Social Governance (ESG) rating indicator could be more relevant than ever.**

As I explained in Sect. 4.1.2, these KPIs are expressed as a value at a given time and variation over a given period.

4.4 Selected Criteria for Assessing Sustainable Performance

Chapters 3 and 4 outlined the key factors to consider when assessing performance, both now and in the future. As mentioned in the introduction, instant performance on a few indicators (especially financial ones) is not easy to achieve but does not necessarily require considering the needs of people and the planet. The "One Shot" approach is possible.

For its part, sustainable performance is characterized by a long-term commitment to improving the transformation of inputs into outputs within a *value creation* [36] approach. **Sustainable performance over time can only be achieved by satisfying all stakeholders, addressing their social, societal, and environmental aspirations, and supporting them through necessary transformations**. The two forms of performance sustainability are interdependent: sustainable performance cannot be achieved over time without performance for the planet and people (essentially all stakeholders), and conversely, achieving positive outcomes for the planet and people requires a long-term perspective, thus necessitating sustained performance over time. **Sustainable performance for the planet also leads to a shift in how we produce our products and services.** We must consider our resources, particularly those we view as abundant or of little value. This mistake has often been made with electricity, which has been cheap for too long. Unfortunately, this happens with water, especially in developed countries where it remains abundant, of high quality and inexpensive. The scarcity of resources drove Toyota to build its sustainable model, which the company has successfully maintained even when resources became plentiful again. **Avoiding or at least limiting water and electricity consumption must guide our actions.** This is the best way, not only to contribute to the health of the planet but also to save money, which we will see.

An evolution is necessary, not just in production but throughout the value chain, from raw materials to the end of use of products or services by customers. Performance is then built during the development phase by considering the complete life

[36] A more precise definition is found in the Glossary (See No. 9).

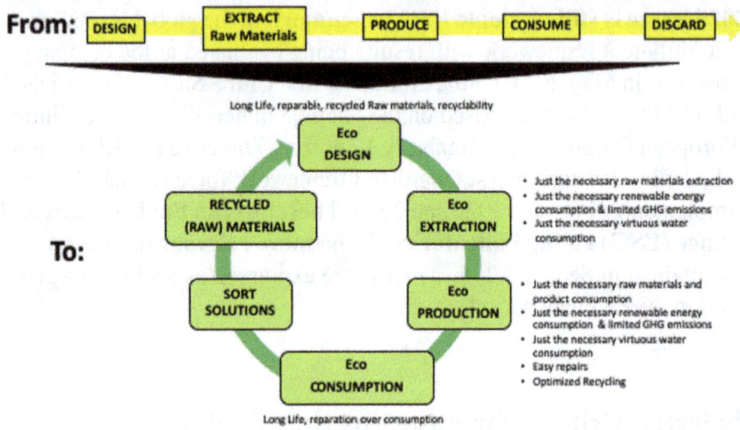

Fig. 4.7 Traditional versus circular models

cycle of a product (Life Cycle Management). When working with existing systems, carrying out "redesigns" is always possible. In addition, material products are affected by two critical phases for the environment: the extraction and recycling of raw materials. It is essential to specify that while developing recyclable products is already challenging, establishing economically viable recycling channels adds complexity. **Despite the numerous obstacles, the transition from the "disposable" model to a circular model should guide companies in their development efforts moving forward** (see Fig. 4.7).

We should also learn from our past mistakes when addressing critical issues. For example, European household appliances were designed to be disassembled, repaired, and even recycled until the late 1990s. It was then that competition from China started to increase, introducing inexpensive products that were often non-repairable and made from materials that were not always healthy. This trend has since spread to many other products, and today, we emphasize the importance of reparability as if it were a new discovery.

To evaluate their ability to deliver sustainable performance, implementing *Lean* and Digital strategies—whether separately or together—must address the challenges mentioned in Sect. 4.3. These should have a positive impact on the listed KPIs (those in Sect. 4.3.4 are a minimum) to satisfy all stakeholders, including customers, suppliers, employees, the company, and shareholders. Everything is interconnected because KPIs measure the satisfaction in meeting these challenges, and they can also be relevant to all stakeholders. A summary for assessing the effectiveness of individual or combined deployments is provided in Table 4.7. It can be completed by Tables 4.3, 4.4, 4.5 and 4.6 for more details with related KPIs.

Table 4.7 Filters (or criteria) for sustainable performance

Assessment criteria		
Positive impact on performance (customers, employees, shareholders, suppliers)	**1**	**Positive impact on the listed KPIs** by considering effectiveness, efficiency and *Benchmarking* [37]
	2	**Positive impact on KPIs confirmed over a period of at least 3 to 5 years** (Empirically, this is the time to change a company's performance culture (Vinardi, 2013)
Environmental issues (customers, employees, shareholders, suppliers)	**3**	Transformation from **linear to circular production**
	4	**Non-generation and/or elimination of waste** related to material, product, energy and water and the carbon footprint more broadly
Societal issues (customers, employees, shareholders, suppliers)	**5**	**Promotion of Diversity and Inclusion, Ethics, Resilience**
	6	**Protection of personal data**
	7	**Effective and efficient cybersecurity**
	8	**Reinforced need for trust** (versus data sharing, changes, etc.)
	9	**Need to act quickly** (action, reaction, decision)
Social and psychosocial issues (customers, employees, shareholders)	**10**	**Elimination of waste linked to i human resources** (illogicality, non-ergonomic tasks)
	11	**Safe working conditions** (limitation of physical risks)
	12	**Working conditions limiting psychological risks** (stress, lack of ethics, meaningless actions, etc.)
	13	**Economically decent working conditions**
	14	**Evolution of the work culture** with a more collective orientation, with more flexible time management, lower hierarchical distance, shorter-term orientation need for feedback and stronger feedback (versus Baby Boomer culture)
Challenges of efficiency and digital sense (customers, employees, shareholders, suppliers)	**15**	**Global data strategy and governance**
	16	**Balance between human and digital**, automation of interfaces when it makes sense
	17	**Customization of digital interfaces, adaptable** according to needs and expectations
	18	**Fluidity and efficiency of interactions** between human and techno/digital

(continued)

[37] A more precise definition is found in the Glossary (See No. 1).

Table 4.7 (continued)

Assessment criteria		
Consistency issues for companies (development in line with the various social, societal and environmental issues	**19**	**Coherent organization** (deed/word), **authentic, assuming its** social and societal responsibilities
	20	**Organization maximizing the retention and development** of its human resources
	21	**Resilient Organization, Leaders, *Process,*[38] Supply Chain**
	22	**Development of remote work and autonomy**
	23	**Development of collaborative work** (including networking) to decide and act quickly
	25	***Simple and efficient S&OP* process**
	26	**Positive impact on the listed KPIs** by considering effectiveness, efficiency and *Benchmarking* [39]
Alignment issues for shareholders and governance	**27**	**Performance and Profitability**
	28	**Expertise for the development of corporate performance in line with** social, societal and environmental issues

References

Adler N. J. (2008). *International dimensions of organizational behaviour* (2nd ed.). From Dr Carol Kovach research (UCLA)

Aggarwal, N., & Bharadwaj, A. (2024). *Why Asia's time is now: What's fueling Asian growth and what does it mean for the rest of the world?* World Economic Forum—Economic Growth. https://www.weforum.org/stories/2024/06/why-asia-s-time-is-now-whats-fueling-asian-growth-and-what-does-it-mean-for-the-rest-of-the-world/

Alshurideh, M., Kurdi, B., Alzoubi, H., Obeidat, B., Hamadneh, S., & Ahmad, A. (2022). The influence of supply chain partners' integrations on organizational performance: The moderating role of trust. *Uncertain Supply Chain Management, 10*(4), 1191–1202.

Barrand, J., & Moïchine, S. (2024). *Histoire de l'agilité dans les organisations.* Harvard Business Review France. https://www.hbrfrance.fr/organisation/histoire-de-lagilite-dans-les-organisations-60415

Beuren, I. M., Dos Santos, V., & Theiss, V. (2022). Organizational resilience, job satisfaction and business performance. *International Journal of Productivity and Performance Management, 71*(6), 2262–2279. https://doi.org/10.1108/IJPPM-03-2021-0158

Bin, A. S. (2016). *The relationship between job satisfaction, job performance and employee engagement: An explorative study. Business Management and Economics, 4*(1), 1–8. https://doi.org/10.15739/IBME.16.001

Binvel, Y., Franzino, M., Guarino, A., Laouchez, J.-M., & Penk, W. (2019). *Future of work : the global talent crunch.* https://www.kornferry.com/insights/this-week-in-leadership/talent-crunch-future-of-work

[38] A more precise definition is found in the Glossary (See No. 2).

[39] A more precise definition is found in the Glossary (See No. 1).

Boers, J. (2020). The Keiretsu advantage: How Japanese automakers Thwarted American competition. *The Yale Undergraduate Research Journal, 1*(1) (Article 33). https://elischolar.library.yale.edu/yurj/vol1/iss1/33

Brassey, J., De Smet, A., Maor, D., & Rabipour, S. (2024). *Developing a resilient, adaptable workforce for an uncertain future.* McKinsey Quarterly. https://www.mckinsey.com/capabilities/people-and-organizational-performance/our-insights/developing-a-resilient-adaptable-workforce-for-an-uncertain-future

Brende, B., & Sternfels, B. (2022). *Resilience for sustainable inclusive growth.* White Paper World Economic Forum in Collaboration with MacKinsey Company. https://www.mckinsey.com/capabilities/risk-and-resilience/our-insights/resilience-for-sustainable-inclusive-growth/

Cascio, J. (2022). *A framework for understanding a turbulent world—BANI–Facing the adge of Chaos.* https://ageofbani.com/2022/04/bani-and-chaos/

Chamorro-Premuzic, T., Hancock, B., Weddle, B., & Rahilly, L. (2024). *Why so many bad bosses still rise to the top.* McKinsey & Company, People & Organizational Performance Practice. https://www.mckinsey.com/capabilities/people-and-organizational-performance/our-insights/why-so-many-bad-bosses-still-rise-to-the-top/

Chiang, A., & Syn, L. (2024). *Navigating the future of board leadership: Insights from the 2024 APAC board leadership forum.* Russell Reynolds. https://www.russellreynolds.com/en/insights/articles/navigating-the-future-of-board-leadership

Coad, A., Holm, J. R., Krafft, J., & Quatraro, F. (2018). Firm age and performance. *Journal of Evolutionary Economics, 28*, 1–11. https://doi.org/10.1007/s00191-017-0532-6

Côté, D. (2024). A reflection on paradoxes and double binds in the workplace in the era of superdiversity. *Humans, 4*(1), 1–21. https://doi.org/10.3390/humans4010001

Darcillon, T., & Rebérioux, A. (2020). Gouvernance d'entreprise: Quarante ans d'évolution. *Revue D'économie Financière, 137*(1), 173–185. https://doi.org/10.3917/ecofi.137.0173

De Monicault, F. (2024). *Génération Alpha, née après 2010: voici ce que l'on peut déjà déceler de leurs ambitions professionnelles.* Figaro, November 28th, 2024. https://emploi.lefigaro.fr/profession-rh/generation-alpha-nee-apres-2010-voici-ce-que-l-on-peut-deja-deceler-de-leurs-ambitions-professionnelles-20241128

De Witt, M. (2024). *8 ways Gen Z will change the workforce.* Stanford Report, Social Science. https://news.stanford.edu/stories/2024/02/8-things-expect-gen-z-coworker

Deepalakshmi, N., Tiwari, D., Baruah, R., Seth, A., & Bisht, R. (2024). Employee engagement and organizational performance: A human resource perspective. *Educational Administration: Theory and Practice, 30*(4), 5941–5948. https://doi.org/10.53555/kuey.v30i4.2323

Drulhe, M. (2011). Usages de l'argent De l'argent spéculatif à l'argent citoyen. *Empan, 82*(2), 41–47. https://doi.org/10.3917/empa.082.0041

Edmans, A., Darcy, P., Chendi, Z., & Lucius, L. (2014). Employee satisfaction, labor market flexibility and stock returns around the world. *Management Science, 70*(7), 4357–4380. https://doi.org/10.3386/w20300

Garrido-Moreno, A., Martín-Rojas, R., & García-Morales, V. J. (2024). The key role of innovation and organizational resilience in improving business performance: A mixed-methods approach. *International Journal of Information Management, 77*. https://doi.org/10.1016/j.ijinfomgt.2024.102777

Gomez, J. Y. (2020). *L'actionnariat familial a-t-il un avenir?* Le Monde. https://pierre-yves-gomez.fr/lactionnariat-familial-a-t-il-un-avenir/

Haykir, Ö., & Çelik, M. S. (2018). The effect of age on firm's performance: evidence from family-owned companies. *Ömer Halisdemir Üniversitesi, 11*(2). https://doi.org/10.25287/ohuiibf.403257

Ionescu, S., Dumitrescu, G., Ioanăș, C., & Delcea, C. (2024). Mapping the landscape of key performance and key risk indicators in business: A comprehensive bibliometric analysis. *Risks, 12*(8), 125. https://doi.org/10.3390/risks12080125

Jakšić, M. M. (2013). *Performance management and employee satisfaction. Montenegrin Journal of Economics, 9*, 85–92.

Kirk, V. (2024). *Neuromarketing—Predicting consumer behavior to drive purchasing decisions.* Harvard Division of Continuing Education. https://professional.dce.harvard.edu/blog/marketing/neuromarketing-predicting-consumer-behavior-to-drive-purchasing-decisions/

Klockner, K. (2017). Developing organisational resilience: Organisational mindfulness and mindful organising. *The Australian Journal of Emergency Management, 32*(4), 47–51. https://doi.org/10.3316/ielapa.222377270002079

Knight, R. M. (2024). *Gen X has had to learn or die': Mid-career workers are facing ageism in the job market.* BBC. https://www.bbc.com/worklife/article/20240320-gen-x-workplace-ageism

La revanche du consommateur. (2025). Courrier International No. 1782, du 9 au 15 Janvier 2025.

Landecy, C. (2024). *Baby-boomers, générations X, Y, Z et Alpha : qu'est-ce qui les distingue?* https://blog.hubspot.fr/marketing/baby-boomers-generations-x-y-z-alpha

Lasida, E. (2013). *Le don fondateur du lien social, le cas de l'économie de marché* (pp. 23–35). Transversalités, 2013/2 No. 126. https://doi.org/10.3917/trans.126.0023

Le Goulven, K., Stabile, M., & Veloso, F. (2024). *Four Global Trends in Business and Society in 2024.* INSEAD. https://knowledge.insead.edu/responsibility/four-global-trends-business-and-society

Levey, J. M. (2019). Mindful leadership for personal and organisational resilience. *Clinical Radiology, 74*(10), 739–745. https://doi.org/10.1016/j.crad.2019.06.026

Loderer, C., & Waelchli, U. (2010). Firm age and performance. https://doi.org/10.2139/ssrn.1342248

Lombardi, S., Pina e Cunha, M., & Giustiniano, L. (2021). Improvising resilience: The unfolding of resilient leadership in COVID-19 times. *International Journal of Hospitality Management, 95*, 102904. https://doi.org/10.1016/j.ijhm.2021.102904

Mauss, M. (1925). *Essai sur le don. Forme et raison de l'échange dans les sociétés archaïques.* Année sociologique. https://gallica.bnf.fr/ark:/12148/bpt6k93922b/f36.item

Mitsubishi Research Institute, Inc. (2022). *Social and economic impacts of carbon neutrality in 2050.* https://www.mri.co.jp/en/news/20221104.html

Orser, B. J., Hogarth-Scott, S., & Riding, A. L. (2000). Performance, firm size, and management problem solving. *Journal of Small Business Management, 38*(4), 42.

Pearlman, R., Kane, K., & Euan-Smith, H. (2024). *2030: The very human future of work.* Korn Ferry Insights—Briefing Magazine. https://www.kornferry.com/content/dam/kornferry/docs/article-migration/Briefings30_Feature_Future_28-37.pdf

Piprani, A. Z., Mohezar, S., & Jaafar, N. I. (2020). Supply chain integration and supply chain performance: The mediating role of supply chain resilience. *International Journal of Supply Chain Management, 9*(3), 58–73.

Raymond, J. L. (2023) *Expérience client d'ici à 2030: quelles attentes consommateurs à prévoir—* Update of experience 2030 (Futurum Research et SAS). https://www.francenum.gouv.fr/guides-et-conseils/developpement-commercial/gestion-de-la-relation-client/experience-client-dici

Renaud, A., & Berland, N. (2007). *Mesure de la performance globale des entreprises* (pp. CD-Rom). Comptabilité et environnement, May 2007, Poitiers, France (halshs-00544875).

Rossi, M. (2016). The impact of age on firm performance: A literature review. *Corporate Ownership and Control, 13*, 217–223. https://doi.org/10.22495/cocv13i2c1p3

Roy, P. (2024). *Engagement au travail: et si on revenait aux choses simples?* Harvard Business Review France. https://www.hbrfrance.fr/management/engagement-au-travail-et-si-on-revenait-aux-choses-simples-60832

Sagawa, J. K., & Nagano, M. S. (2021). Integration, uncertainty and information: How do they affect planning performance? *Revista De Gestão, 28*(1), 23–49. https://doi.org/10.1108/REGE-11-2019-0113

Sammut, J. (2011). *Le modèle mutualiste au défi de l'Europe—Un engagement civique issu de l'histoire*(pp. II et III). https://www.monde-diplomatique.fr/2011/03/SAMMUT/20237

Saragih, J., Tarigan, A., Silalahi, E. F., Wardati, J., & Pratama, I. (2020). Supply chain operational capability and supply chain operational performance: Does the supply chain management and supply chain integration matters. *International Journal of Supply Chain Management, 9*(4), 1222–1229.

Séran, H. (2021). L'impact de la directive européenne Solvabilité II sur la gouvernance des sociétés d'assurance mutuelles françaises. *RECMA, 359*(1), 92110. https://doi.org/10.3917/recma.359.0092

Shukor, A. A. A., Newaz, M. S., Rahman, M. K., & Taha, A. Z. (2021). Supply chain integration and its impact on supply chain agility and organizational flexibility in manufacturing firms. *International Journal of Emerging Markets, 16*(8), 1721–1744.

Singh, S. (2023). *The mega trends that will shape our future world.* Forbes. https://www.forbes.com/sites/sarwantsingh/2023/10/30/the-mega-trends-that-will-shape-our-future-world/

Takeda, Y., Inose, J., Kinehara, Y., Sakai, H., Shimizu, A., Santo, M., Morishige, A., Shindo, K., Shida, R., Ogawa, T., Tanaka, Y., Wataya, K., & Tanguchi, G. (2022). *Future society 2050.* Mitsubishi Research Institute, Inc. https://www.mri.co.jp/en/knowledge/insight/dia6ou000001vs4a-att/ei20200227_Future_Society_2050_en.pdf

Tarigan, Z. J., & Siagian, H. (2021). The effects of strategic planning, purchasing strategy and strategic partnership on operational performance. *Uncertain Supply Chain Management*. https://doi.org/10.5267/J.USCM.2021.2.006

Taşkan, B., Junça-Silva, A., & Caetano, A. (2022). Clarifying the conceptual map of VUCA: A systematic review. *International Journal of Organizational Analysis.*

University of Cambridge Institute for Sustainability Leadership (CISL). (2023). *Future of boards final report: Summary and synthesis (phase 1, part 4).* University of Cambridge Institute for Sustainability Leadership. https://www.cisl.cam.ac.uk/resources/publications/future-of-boards

Vinardi, C. (2013). *Le Lean: Atouts, impacts et limites* (Vuibert ed.).

Vinardi, C. (2016). *Les équipes multiculturelles en entreprise: les comprendre, les gérer, les développer* (Maxima ed.).

Vinardi, C. (2019). Ph.D., Thesis on Industrial Engineering, *Les défis du Lean à l'ère de la mondialisation et de l'industrie 4.0* (Université de Technologie de Compiègne, Français, 2019) NNT :2019COMP2500, tel-03201462

Vinardi, C. (2023). Business Strategy with Hoshin Kanri (SpringerBriefs in Applied Sciences and Technology/Springer Nature, 2023), https://doi.org/10.1007/978-3-031-20963-5

Vollero, A., Yin, J., & Siano, A. (2022). Convergence or divergence? A comparative analysis of CSR communication by leading firms in Asia, Europe, and North America. *Public Relations Review, 48*(1), 102142.

Walsh, S., Johnston, A., Gabriel L. (2024). Building resilient leadership: A study exploring leadership development during uncertainty. Human Resources Management and Services. 6(2): 3474. https://doi.org/10.18282/hrms.v6i2.3474

Walsh, D. (2025). *A new look at the economics of AI.* MIT Management Sloan School, Artificial Intelligence. https://mitsloan.mit.edu/ideas-made-to-matter/a-new-look-economics-ai

Yu, Y., Huo, B., & Zhang, Z. J. (2021). Impact of information technology on supply chain integration and company performance: Evidence from cross-border e-commerce companies in China. *Journal of Enterprise Information Management, 34*(1), 460–489.

Chapter 5
Impact on Sustainable Performance of Individual Key Concepts

Abstract This chapter analyzes Lean and Digital strategies individually to grasp their contributions to the doubly sustainable performance described by the criteria presented at the end of Chap. 4. Building on the definitions and scope ("what") discussed in Chap. 2, we will then continue with the purpose and goals ("why"), the deployment methods with or without automation ("how"), the associated organizations ("who"), the most relevant use cases and application limitations ("when" and "where"). The aim is to assess compliance with the performance criteria.

5.1 The Lean System

5.1.1 *Purpose and Goals*

To specify the purpose and objectives of Lean, specifically the "why" and "what for," it is essential to specify which type of Lean we are considering. This was addressed in Sect. 2.1, where I introduced the Lean System with *Hoshin kanri.*[1] If we refer to another type of Lean, essentially a toolbox, whose teaching is focused on methods more than facilitation, this can mislead Leaders towards other derivative approaches (such as Agile). I experienced this situation with three Top leaders I interviewed who were working in Transactional Activities. If you feel you are in this situation, you should reexamine the definitions I shared and then consider having the "Lean-Agile-Six Sigma" approach within Lean.

It is also important to note that the various components of Lean originated in different regions of the world and at different points in history. Before discussing Lean as a whole system, we should remember that its individual components were **intended to speed up results (performance) and/or performance level in the context of human and/or material problems**. Let's explore a few examples:

[1] A more precise definition is found in the Glossary (See No. 3).

C. Vinardi, *Sustainable Performance in the Digital Age*,
https://doi.org/10.1007/978-3-032-01252-4_5

- The earliest problem-solving tools, such as Quis? Quid? Ubi? Quibus auxilius? Cur? Quomodo? Quando? (WhoWhatWhereWhenHowWhy), were used by the Romans for fast resolution of investigations to satisfy the needs of citizens.
- The first standards found in Europe were for the production of vases in a consistent way. The objective was, therefore, to reduce quality defects, better meet customer expectations, and maximize profits.
- Takt Time, which originated in Germany for aircraft fuselages, was developed to align production with customer demand by minimizing both the physical space required for production and the financial resources involved.

The TPS, which formalized the Lean System, was defined and deployed by Toyota to satisfy customers thanks in particular to the optimal quality products, by ensuring the safety of goods and people at the least cost (with limited resources).

It should be noted that the TPS is complemented by "Toyota Way", which describes the company's values and behaviors and applies to all its employees.

- **Customer performance is the starting point of Lean,** which guides the performance of the company, its employees, suppliers, and shareholders.
- Eliminating **waste and limiting the use of resources to just what is needed is a principle guiding Lean system approaches.**
- While we may reexamine the scope during deployment, it is essential to note that **applying Lean principles throughout the entire company and across the entire value chain is essential for achieving the optimal performance that this approach offers. This comprehensive application is therefore the ultimate goal** (Vinardi, 2019).

Regarding the objectives of Lean, several authors have analyzed the impact of certain Lean principles or tools taken separately, such as TPM or JIT. However, **Lean is a system, and if we separate the components, the expected results will differ.** Some studies show that the correlation between Lean strategy and performance depends on whether it is implemented holistically and with or without *Hoshin kanri*,[2] while others indicate the opposite (Vinardi, 2019). It should be also noted that the impact of Lean deployment on overall operational performance was highlighted in the 1980s in the USA by its implementation in struggling American plants, as they were able to achieve and maintain performance levels equivalent to those of Japanese plants (Vinardi, 2019).

Studies have shown that sustainable results can only be delivered by implementing the complete Lean strategy, i.e., one that does not focus on just a few tools and encompasses the entire value chain of the company. Some studies also highlight the need to include the company of the customer, and the suppliers (Vinardi, 2019).

Insofar as we regard the Lean System as a whole, applicable to the entire company, as described in Sect. 2.1, its objectives are classified in at least six topics:

- **Customers**: some also include Suppliers or Partners, and others a "Growth" market component.

---

[2] A more precise definition is found in the Glossary (See No. 3).

- **Safety and Environment**: includes the objectives of protecting property and people with ergonomics and the protection of the environment.
- **Quality**: includes quality objectives of products and services delivered, up to the end of use, the quality achieved internally and also the quality delivered by suppliers (according to the chosen breakdown method).
- **Delivery**: this includes objectives in terms of customer delivery times, development and production times, and may also include the suppliers' delivery times (according to the chosen breakdown method).
- **Management**: includes employee satisfaction objectives other than Safety and Environment.
- **Cost**: though it is important, placing it last can provide perspective on how objectives are achieved. Chapter 4 highlights that the costs of producing products and services are greatly affected by the failure to address previous objectives.

Lastly, I want to highlight Larry Culp's perspective. I had the privilege of working with him at General Electric, and he contributed to my previous book (Vinardi, 2023). Larry was the President and CEO of Danaher Corporation from 2001 to 2014. He believes the Danaher Business System (DBS), as a Lean System applied to all Danaher employees, gives **holistic and sustainable results** (Anand et al., 2008):

> *"DBS gives all of **our operating executives** the **means** with which to **strive for world-class quality, delivery, and cost benchmarks** and **deliver superior customer satisfaction and profitable growth**"*

This quote comes from a business case made by Harvard Business School in 2008, studying Danaher from 1985 to 2007 (22 years). During this period, Danaher had been **one of the best-performing industrial conglomerates and one of the most profitable** industrials in USA.

The Danaher Business System, like all "Business Systems" described with this term, applies to the entire company. This leads us to deepen the notion of deployment in achieving overall performance.

5.1.2 Deployment

The deployment should clarify the general "how" alongside the "when," "where," and "whose organization."

Let's start with the "when." When is it the best time to start *Lean*? Logically, **as soon as possible** and **ideally, when everything is going well or at least outside a**

crisis period, to structure the *process*[3] in a serenely and orderly manner. This sometimes happens, especially when companies appoint a CEO who is already committed as a "Lean Practitioner." This was the case for Larry Culp at Danaher. Timing nevertheless follows the cycle of change (Fig. 3.1). The constraint can be positive with the factual awareness of a wanted improvement. The "burning platform," describing the feeling of urgency that pushes for radical change, can be achieved by **a strong constraint that you have, or forge for yourself**. This is what athletes or musicians do when they challenge themselves to achieve the best in themselves. It only works with the commitment of most of those at company level. However, this situation is relatively rare and faces several other factors that often lead companies to wait for the moment when they no longer have a choice. Why? Mainly because:

- Despite the spread of Lean over the past thirty years, few leaders in the world are "Lean Practitioners" apart from Chief Operating Officers (COOs) who have an industrial background and a few CEOs (such as Larry Culp, former COO).
- CEOs often have a financial profile, so Lean focuses on costs as a starting point. Unfortunately, this is rarely the most sustainable way to deploy Lean, as we will see later.
- When the company has a crisis, assuming it does not come from its customers but from its finances, this often strengthens cost as the starting point of Lean.

When companies delay implementing *Lean* until a strong need or crisis arises, it can still be effective. We have seen how constraints, in whatever form, drive change and innovation. Toyota is the best example. However, constraints should not be "limited to financial factors" and must always begin with the customer.

After committing to deploying Lean, the following question is often "where?"

For industrial companies, deployment often starts with Operations and production (historical, visible flows, etc.) before spreading **to other departments, or suppliers or even to companies acquired through M&A**. This is also why, within industrial companies, some start with a (Performance) Production System before switching to a (Performance) Business System when the deployment is done throughout the company, from executives to operators, for all activities or departments. That's what Danaher did, for example.

The company started with the Danaher Production System (DPS). Later the Danaher Business System became established to reflect more global and business-oriented deployment, and it has largely helped the company through its many acquisitions. Between 2001 and 2006, the firm made over 50 acquisitions and acquired companies were comprehensively integrated into the core business through DBS deployment. **Successful implementation of DBS across its acquisitions had resulted in rapid growth** (Anand et al., 2008).

In another example, Toyota began focusing on production yet continues to talk about the Toyota Production System (TPS). This is despite TPS being a longstanding concern for the entire organization (with *Hoshin kanri*,[4] as we have seen). The

---

[3] A more precise definition is found in the Glossary (See No. 2).

[4] A more precise definition is found in the Glossary (See No. 3).

company could have rebranded TPS as the Toyota Business System (TBS), but it has chosen not to do so. Based on my observations working alongside the management, there are possibly several reasons for this decision:

- The Toyota Way uses parts of TPS to describe the values and behaviors that apply to all employees in the company.
- Focusing on the content and avoiding changing terminology simplifies the message and communication.
- Even though the name has remained the same, the content of TPS has evolved since its inception and is the foundation of a standard.
- The importance of production for the automotive sector. Apart from meeting the customer's needs, the key performance driver is the ability to optimize activities and production outputs. Innovating and selling are important of course, but when the target is to manufacture tens of thousands or even millions of vehicles, production performance is essential.

To finish on the case of Toyota, it should be noted that the company also supports the deployment of its suppliers with a dedicated support team with the Toyota Supplier Support Center (TSSC) in the different areas of its suppliers.

Now that the context is clarified, it is necessary to start with a reduced scope model or pilot step (test and learn before scale-up) before considering global deployment. In Chap. 7, we will see what steps are needed to tackle scaling up and sustainability using learning. Apart from the *process* [5] that guides the scale-up and global deployment, taking into account culture is a key element that impacts performance as seen in Sect. 4.2. This is, therefore, to be assimilated when deploying Lean. According to **Masaaki Imai,** one of the world's most well-known and recognized Japanese Lean experts and consultants (Vinardi, 2023):

> "***Training Lean only can get caught up in cultural differences***, *for instance, in certain communities, it is seen as rude to expose the flaws and problems in processes as this is embarrassing to the people implicated. Visual management's transparency can consequently be counterproductive without an awareness of the broader cultural influence. However, employees can overcome their cultural objections when they gain a better understanding of the concepts and its connection with the tools and the holistic improvement system.*"
>
> *From an interview of Masaaki Imai with Carine Vinardi (2023)*

Effective deployment involves taking into account various aspects of culture, including national, company, sector-specific, and the organization's size.

---

[5] A more precise definition is found in the Glossary (See No. 2).

Table 5.1 Example of average positions for a few countries versus critical cultural dimensions to ease Lean deployment (*Source* Adapted from Vinardi, 2016)

	CULTURAL DIMENSIONS	JAPAN	CHINA	USA	RUSSIA	FRANCE	GERMANY
1	Flexible (low) standard application	--	++	-	++	+	-
2	Strict (high) standard application	++	--	+	--	-	+
3	Learning by doing first	++	+	++	--	--	++
4	Focus on a collective approach	++	++	--	++	+	-
5	Long term orientation	++	++	--	++	++	++
6	Low hierarchical distance	--	--	+	--	-	++
7	Direct negative feedback	--	--	-	++	++	++
8	Low expression of emotions	--	--	+	++	++	-
9	Flexible or synchronous time management	--	++	--	++	+	--
10	Low uncertainty avoidance	--	++	++	--	--	-

LEGEND	++	fully aligned	--	fully not aligned
	+	somewhat aligned	-	somewhat not aligned

To begin with the national aspect, we will utilize cultural dimensions to compare and illustrate the relative positions of different cultures. **Ten dimensions are identified to facilitate the deployment of Lean**, with an ideal extent for each (Vinardi, 2019), as summarized in Table 5.1. **Deployment becomes more complicated and requires more effort** when a country's position on specific dimensions is less aligned or not aligned with the expected position to facilitate the deployment of Lean (and vice versa). The examples in Table 5.1 show that different cultures must make varying efforts.

Regarding corporate culture and business sectors, organizations can be either industrial or transactional. **Even before the industrial era, Lean and its components were already more easily deployed where there are physical flows**. This is simply because **we can correct or improve what we can see**. However, Value Stream Mapping (VSM), as illustrated in Fig. 2.10, can help visualize *processes*[6] that may not be immediately apparent, providing solutions to facilitate Lean deployment in transactional activities. Each of the two types of activity has advantages and

[6] A more precise definition is found in the Glossary (See No. 2).

Table 5.2 Summary of the main advantages and disadvantages of industrial and transactional activities in the deployment of Lean

INDUSTRIAL ADVANTAGES	INDUSTRIAL DISADVANTAGES
• Physical flow that visualizes the impact of actions and guides the organization. • Lean credibility is already established • *Benchmarking*[1] is done.	• Large stress fields (Fig.30). • Evolution of flows and physical equipment, in addition to the human dimension, is to be considered. • Prioritization of issues is often complex.
TRANSACTIONAL ADVANTAGES	**TRANSATIONAL DISADVANTAGES**
• Deployment depends mainly on people. • Faster implementation when the prerequisites are in place. • Few or no preconceived ideas about Lean if it is properly started.	• Difficulties in perceiving problems and waste, and a strong temptation to digitize before understanding • Reluctance to formalize and comply with standards. • Less of a performance culture. • Adapting Lean principles to the business.

Table 5.3 Summary of the main advantages and disadvantages of SME and big firms in the deployment of Lean (*Source* Vinardi et al., 2019).

SME ADVANTAGES	SME DISADVANTAGES
• Closeness to customers. • Fast decision-making and high responsiveness - few hierarchical levels. • Direct communication. • Flexibility. • Number of Leaders to hire.	• Little diversity of Lean expertise. • Difficulty accessing resources. • Difficulty comparing oneself to the outside. • Lack of management expertise. • Lack of "system" and "performance" culture.
LARGE COMPANY ADVANTAGES	**LARGE COMPANY DISADVANTAGES**
• Access to expertise (*Lean* in particular). • Access to resources (materials and training). • Testing of deployment possible before scaling up. • "System" and "performance" culture.	• Decision-making is often slower because of more actors and several hierarchical levels. • More complicated Leaders' Engagement (number, diversity, etc.) • Access to external resources (too easily).

disadvantages in the deployment of Lean. I previously mentioned these aspects when discussing how the nature of activities impacts performance. The main advantages and disadvantages are summarized in Table 5.2.

Another important factor is company size, which influences corporate culture and can also have benefits and challenges during Lean deployment. In 2019, I co-authored a study that examined the support provided for Lean deployment in over 400 companies within the European aeronautics industry. Based on this research, the main advantages and disadvantages related to company size are summarized in Table 5.3 (Vinardi et al., 2019).

While Lean does not utilize technically complex concepts, its deployment is a subtle art similar to the arrangement of musical notes or synchronized movements in sports. Chap. 7 will explore how to use roadmaps and transition from steps to an action plan to support the implementation *process*.[7]

[7] A more precise definition is found in the Glossary (See No. 2).

The deployment of Lean calls for its own strategy to be assimilated into the company's global strategy. As with any strategy, it has a failure rate. **Independently of the rate itself, it seems that the number of strategic initiatives that fail is higher than expected,** considering the energy given to define them particularly (Vinardi, 2023).

? What is the success rate of deploying the Lean System? Are there any statistics available?

It is not easy to obtain studies with reliable figures regarding success in deploying strategy or Lean (with, but often without, *Hoshin kanri*).[8] Many studies are carried out by basing their data collection from companies on a system of interviews. It is possible to compare the performance with the KPI values and definitions of the different companies. However, it is much more complicated for managers to self-evaluate objectively to understand the factual causes of their failures or successes when they themselves are often in training; this is the difficulty of being judge and jury.

Similarly, for the results to be meaningful, managers and researchers should be able to assess whether the Lean approach that has been proposed to them is complete and, therefore, compare the "How" (Vinardi, 2019). This is what was done in 2019 with the study I co-authored. However, if we want to give some figures, since 2017, the percentage of Lean deployment has varied between 60 and 70% according to studies by the Fraunhofer Institute, Mac Kinsey, and Lean Enterprise. In 2023, the **latest updates showed 70–75% deployment on average** worldwide, with **many countries below 50%:**

- While the deployment of Lean is widespread, **only a minority of companies reached the anticipated outputs and sustain it.**
- Different studies confirm that **15–30% of worldwide companies have successfully and sustainably deploy Lean, I.e., seven out of ten fail** (Vinardi, 2013).

While these estimates may debatable, it seems certain that there is still a significant **potential for traditional performance to be obtained with a well-deployed Lean approach, as described in** Sect. 2.1.

Finally, the effect of the increase in stress related to the application of Lean itself is not scientifically confirmed. **Negative or positive impacts seem to be a result of managerial choices during deployment. These can be exacerbated during deployment**, but the responsibility cannot be attributed to the principles of Lean themselves (Vinardi, 2019). This does not prevent Top leaders from being vigilant during deployment.

[8] A more precise definition is found in the Glossary (See No. 3).

We have briefly examined the general aspects of deployment. I will now shift the focus to **automation, which relates to both Lean methodologies and digital *processes*.**[9] In terms of Lean, there are two special cases that affect specific indicators of sustainable performance measurement, as described in Chap. 4.

5.1.3 Automation in Lean Systems

As we saw in Table 3.1, the concept of automation has existed for centuries. **Humans learned how to automate long before robotization.** Let's clarify or remind some definitions that we have previously discussed, as some of them will be useful for the rest of this chapter:

- **Automate**: Transferring the human handling of a task to a device or equipment:
  - Low Tech like Karakuri (without energy).
  - High Tech, such as the robot for a physical or transactional *process*[10] (RPA) or the cobot (which we introduced in Chap. 2)
- **Robotize**: Automate with a robot (physical or RPA).
- **Bot**: A computer program connected to one or more servers.
- **Robot**: Physical equipment incorporating bots.
- **Chatbot:** A bot programmed to interact with visitors or customers.
- **RPA** (Robotic *Process*[11] Automation): it is a virtual robot (with software) to perform a transactional task
- **Cobot** (Collaborative Robot): a robot with simplified programming that allows you to work with humans without physical separation.

With this clarification in mind, the following general considerations regarding the purpose and deployment of automation apply to the high-tech automation we will see in Sect. 5.2.

There are various issues that are frequently raised but they are often material (Technical, financial, and regulatory: technical feasibility, cost to automate, cost of operators versus automation (benefit), applicable regulation, etc.). When it comes to human resources, the issue is often economic (scarcity of human resources) or to express concerns about implementation (resistance of the operators concerned).

These points are important. However, they should be addressed when automation is chosen with a consideration for the customer, the employees, and the company.

---

[9] A more precise definition is found in the Glossary (See No. 2).

[10] A more precise definition is found in the Glossary (See No. 2).

[11] A more precise definition is found in the Glossary (See No. 2).

Therefore, **you have to start with the need, as is traditionally done for Lean. Automation must make sense**, because in effect it can remove a task from a person.

? In your experience, when would you accept seeing equipment perform tasks for you?

There are at least three situations:

- When the task is dangerous and represents a substantial risk to physical integrity.
- When the task is physically strenuous (repetitive or ergonomically unadaptable) and/or psychologically (not very rewarding, generating stress, etc.)
- When the task presents a high risk of error for the product or service as nobody likes to do their job badly.

Moreover, **the Czech word *robota*, which means "drudgery, hard work', is the origin of the word "robot." This should always be kept in mind.**

Tasks can be divided into two categories: Value-Added (VA) tasks and Non-Value-Added (Non-VA) tasks, the latter of which includes waste as described in Chap. 2.

- **Waste that cannot be eliminated must be automated first**. Automating in this manner makes it easier to explain to those involved, and they tend to appreciate the support provided in the *process*[12]. **When something is given, it is easier to ask as in the donor cycle, which also generates social ties,** as seen above.
- When we consider **waste, we must also consider its frequency.** Automating *processes* [13] that are **High Volume and High Speed** is sensible; however, it is much less rational to automate one-off or exceptional tasks, especially if significant investments are required.
- To effectively automate a task, you need to **understand its nature, role, and usefulness within the ecosystem**—considering Suppliers, Inputs, Outputs, and Customers. This understanding forms the basis for identifying waste. Therefore, a **preliminary study is essential**, and utilizing a VSM—a key tool in Lean methodologies—can provide clarity, particularly when analyzing *process*.[14] This approach using VSM becomes **increasingly valuable when assessing transactional *processes*[15] that may not be immediately visible**.
- In the same way, it is easier **to automate what already exists** than to automate a task not yet known. If internal or external *Benchmarking*[16] is made, you can plan for automation and implement it if it is a particular waste that cannot be eliminated.

[12] A more precise definition is found in the Glossary (See No. 2).

[13] A more precise definition is found in the Glossary (See No. 2).

[14] A more precise definition is found in the Glossary (See No. 2).

[15] A more precise definition is found in the Glossary (See No. 2).

[16] A more precise definition is found in the Glossary (See No. 1).

Once "permanent" waste is automated, it will make more sense for the staff concerned to take care of value-added tasks. However, it should always be motivated **first by the elimination of risk relating to physical and psychological safety, and second, by the quality of the product or service.**

By automating with the same logic we use for ourselves, we can meet employees' expectations and give purpose to automation. Several studies have stated that **the highest care should be taken to select suitable tasks** (Doguc, 2020; Ribeiro et al., 2021).

However, automation is often driven primarily by costs, and this is seen increasingly. This is followed by speed, compliance, and error reduction (Doguc, 2020). From experience, it is **almost always possible to combine this need for automation with the logic guiding its implementation, which has been shared previously.**

Not all *processes*[17] can be automated. For those that can, deployment must be prioritized. The **high level of automation may result in the problem of over automation**. It is necessary to automatize only to a certain extent to prevent over automation and avoid frustrating results and lack of easy adaptation.

Highly automated production imposes a problem of high initial and operating costs but also **extremally skilled labor** whereas **the just necessary leads to relatively unskilled labor and more flexibility to lately adapt** (Schreech, 1996). To address the issue of labor shortages through automation, it's important to recognize that, aside from the basic Karakuri methods we'll discuss later, implementing **technical automation requires skilled labor that is also in high demand**. Therefore, we cannot simply **eliminate one resource problem while inadvertently creating another**. Note that this concern **applies equally to automation strategies that utilize digital solutions.**

Regardless of whether we employ low-tech or high-tech approaches, automation is a key component of the Lean System. Let's start with the case of Karakuri which is a significant example of Low Tech and Cost Automation.

5.1.3.1 Low Tech and Cost Automations with Karakuri

Building on the history of Table 3.1, I will now elaborate on the concept of Karakuri and its implementation. In Japanese, Karakuri means "moving something by an elaborate mechanism consisting of strings and coil springs like clockwork, or the thing that is moved that way" (Schreech, 1996). It is also often summarized by "mechanism" or "system" (Yokota, 2009).

Karakuri is **considered as the predecessors of modern automation and robotics in Japan** (Linner & Book, 2023). It is mentioned in a literature dating back approximately 1200 years and first formalized record found in around 720 in The Chronicles of Japan (Yokota, 2009).

[17] A more precise definition is found in the Glossary (See No. 2).

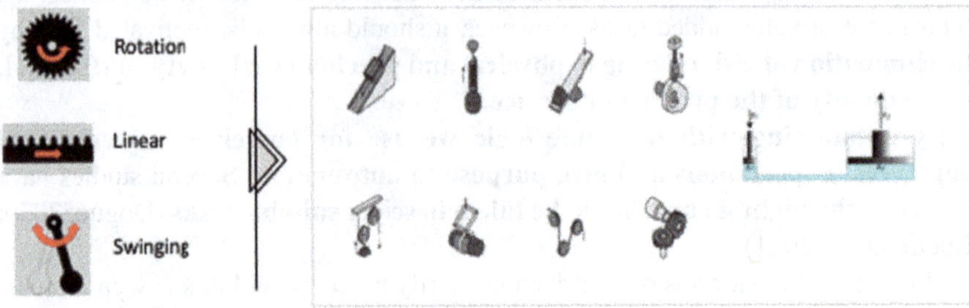

Fig. 5.1 Karakuri basic movements and parts

Chinese invention and technology greatly influenced Japan and those mechanisms, amongst other forms of technology which came from China via the Korean peninsula, helped begin the Japanese Karakuri tradition (Schreech, 1996).

Sakichi Toyota, the founder of the Toyota chain of companies, was a Karakuri Master. **Karakuri became part of Lean System but often neglected in favor of traditional automation** (using energy and complex systems) (Schreech, 1996).

Karakuri aims to utilize natural physical phenomena, such as gravity force, wind, etc. and **elemental mechanisms or moves** (Hatayama et al., 2014). A summary is proposed on Fig. 5.1.

It is also characterized by the absence of a power source or a control device such as a motor or a sensor. The mechanisms created are **simple, hand-crafted, cost-effective** and allow to **solve many problems** in the workplace (Schreech, 1996). Two examples are illustrated in Figs. 5.2 and 5.3.

- In Fig. 5.2, two illustrations of a mechanism highlight a before-and-after scenario. This mechanism successfully addresses the postural problem in the back of the arms by incorporating a pedal system and utilizing the operator's body weight, all for under €2000.
- Fig. 5.3 show two illustrations of a table that uses gravity, weights, and various rotation systems. The photo is located on the left, while the schematic diagram is presented on the right (source: Hatayama et al., 2014)
- With Table 5.4 (source: Hatayama et al., 2014), the bill structure of Improvement by Karakuri Mechanisms used for Example of Fig. 35.1 et 2. It illustrates how the composition of Karakuri can be broken down.

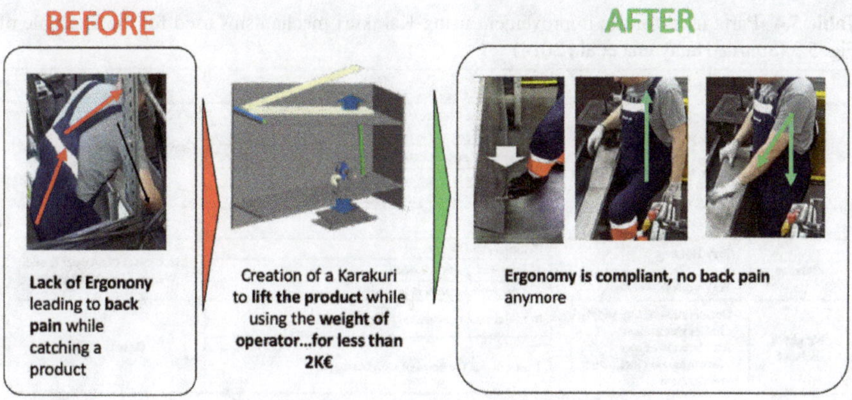

Fig. 5.2 Karakuri industrial example No. 1

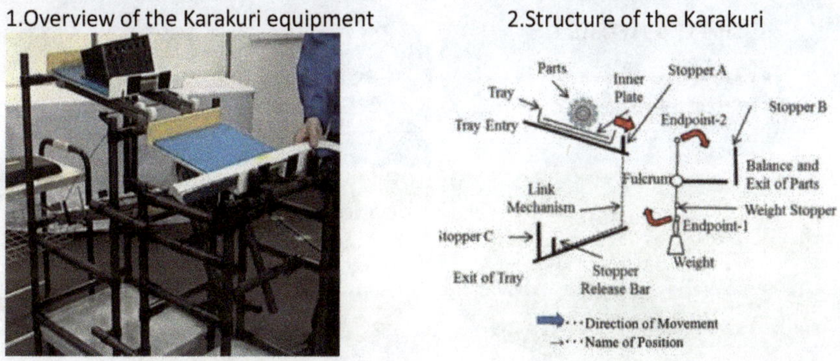

Fig. 5.3 Overview of a working table using Karakuri (left) and table of mechanisms (*Source* Hatayama et al., 2014)

Awareness and familiarity with Karakuri is significantly **higher in Asia than in Europe**, where **technical automation** is greater. This difference is mainly due to Japan's influence and the need to limit investments in low-cost, competitive countries in Asia. In Europe, there are numerous manufacturers of automated machines and equipment lines, and the labor cost sometimes justifies higher investments.

Several benefits of Karakuri are highlighted such as (Yuji & Sandström, 2018):

- low cost (Karakuri are also called in English "LCA" for "Low Cost Automation"),
- low skills required
- new opportunities for improvement:
  - exerting workers' creativity and innovation,
  - attracting and engaging young workers,
  - positively impacting engagement through co-design.

Table 5.4 Parts used for the improvement using Karakuri mechanisms used for the Example of Fig. 5.3 (*Source* Hatayama et al., 2014)

Elements	Functions	States (Code No.: Description of Status)	Way of State Transition/Mechanism
Tray A	-Work Holding -Work Transportation	0: Initial state (Before parts entry)	Gravity Force and Implemented Mechanisms
		1: Tray with parts is at the entry point	
		2: Tray with parts is at the exit point	
		3: Empty tray is at the exit point	
		4: Empty tray is at the tray return point	
....	....	....	....
Balance	-Tray Holding -Tray Transportation	0: Initial state (Vacancy)	Mechanism of Stopper B and Weight Slide Mechanism
		1: Tray with parts is loaded	
		2: Tray without parts is loaded	
Weight of Balance	-Deceleration of tray supply by buffer mechanism -Acceleration of tray forwarding by inclination mechanism	0: Initial state (Located at the endpoint-1)	Gravity Force
		1: Located at the point of weight stopper	

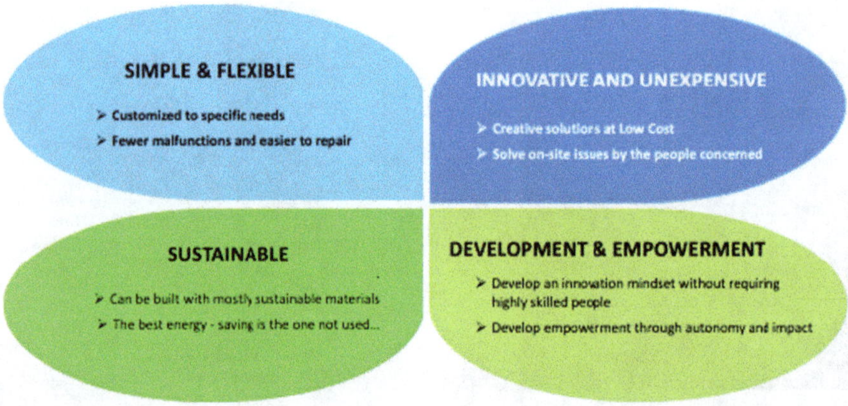

Fig. 5.4 Summary of the Karakuri key benefits

A summary is proposed in Fig. 5.4. **The Karakuri concept meets the need for sustainable performance in both design and longevity.** It avoids energy consumption, is made from simple materials, and may be recyclable and durable. **Karakuri not only has significant potential as a standalone concept** (similar to any component of Lean) but can **also be enhanced when combined with digital solutions such as Cobots, AGVs, and SGVs**. We will explore this further.

5.1.3.2 Traditional High Tech Automation

Although robots are the most well-known form of high-tech automation, they have various other forms. In France for example, the BOA train can unload 80 to 120-m rails directly from the steel mill, weld them electrically to the track, replace old rails, and then reload the replaced rails. It can achieve speeds of up to 200 m per hour when replacing long welded rails and performs two electrical welds every 25 min. This type of automation is essential for meeting the growing traffic demands and the corresponding need for track maintenance. Such tasks could not be performed manually as efficiently, making this innovation more acceptable to employees.

Robots first emerged in the automotive industry during the 1960s, coinciding with the formalization of Lean principles through Toyota's Toyota Production System (TPS). High-tech automation, a Lean component, includes both physical robots and Robotic *process*[18] Automation (RPA), excluding the latest digital advancements.

Robots are well-suited for **fast repetitive tasks** (with a high hourly or daily frequency), whether in terms of **product output** (for physical robots) or **data processing** (for RPAs). They can operate 24/7 if necessary, making them efficient for large volumes. Additionally, automation provides **consistent task completion times**, enabling better planning. However, this reliability depends on the maintenance of the robots, as they require regular upkeep and can be prone to breakdowns.

Physical robots typically need to be **fixed to the ground, which limits their flexibility**. They also **often require protective barriers for safety**, resulting in a larger footprint than manual workstations. Consequently, robots are not **ideal for production environments that necessitate frequent human interaction**—such as recharging or making adjustments—since operations must be systematically paused or slowed down to secure the equipment before resuming full speed. On the other hand, robots are fast (1000 mm/s) and are **potentially quicker and more consistent** than humans at the same **short task, provided they do not require any particular dexterity.**

Physical **robots are intended for a specific function,** such as painting or packaging, and **execute a limited series of precise tasks** that are complicated to change but at no additional cost. In contrast, humans physically go to workstations where they can be trained as required.

Humans require training. **Ergonomists estimate that it takes about one week of training for individuals to reach their full capacity to perform tasks with optimal timing and quality for tasks with a cycle time of one minute that involve multiple activities**. For the same one-minute cycle time, physical robots also require about a week's programming by a qualified technician, involving maintenance, safety, and quality technicians. The **additional cost of installing a robot compared to human training can be significant and is typically justified only for sizable production volumes**.

Finally**, robots cannot automate all tasks**, especially those that require special dexterity or unique skills that they cannot replicate.

In the 1990s, the idea of automating tasks that shifted from physical activities to transactional processes was partially introduced with **Robotic *process*[19] Automation** (RPA), also known as "software robots". The first RPAs were created in the 2000s before their widespread deployment gained momentum in the 2010s, with the emergence of the RPA Back Office market (Doguc, 2020).

---

[18] A more precise definition is found in the Glossary (See No. 2).

[19] A more precise definition is found in the Glossary (See No. 2).

Table 5.5 Main use, prerequisites and specificities for RPA deployment (Tounkara, 2020; Doguc, 2020)

Typical uses	Prerequisites and implementation specificities
• Data transfer across multiple applications • Opening and closing applications • Form filling • Opening content from a document database • Automatic billing • Automation of administrative processes (leave, payroll, etc.) • Real-time process monitoring • Creation of customer or employee databases • Data extraction and reporting • Data comparison and validation	• Requires prior analysis and selection of the tasks concerned • Easily integrates into the company's IT ecosystem • Can be **learned with a few weeks of training**; hence no extensive programming experience is needed • **"Outside-in" approach:** it **controls the computer on the user interface level,** without **disturbing the underlying computer systems** • Low Impact on IT infrastructure, however, connection and integration is one of the key task • Low investment for a large company • Software provides tools for users to build workflows **in a visual way (entirely independent of coding knowledge)**

The expected objectives with RPA are the same as for physical robots. It is frequently claimed that RPAs take away the low-value tasks of employees, allowing them to focus on more meaningful work. While this objective can sometimes be genuine and commendable, there are instances where the main aim is to optimize workforce numbers rather than enhance employee roles, which can inevitably lead to resistance against implementation. Despite being less known and utilized than their physical robot counterparts, RPA has proven to be a **simple and mature technique** (Tounkara, 2020). Effective deployment requires adherence to certain principles, with key rules that govern its operation (Doguc, 2020):

- 5 decisions max per bot,
- 5 application calls using APIs (Application Programming Interface),
- 500 clicks maximum.

Table 5.5 gives an overview of the most typical uses, prerequisites, and barriers to implementation. Figure 5.5 illustrates the transition between a *manual process*[20] and a fully automated *process*[21] in RPA. Three operators manually input data into new systems before the RPA automates these tasks. Table 5.6 summarizes the main advantages of robots and RPAs compared to human workers.

[20] A more precise definition is found in the Glossary (See No. 2).

[21] A more precise definition is found in the Glossary (See No. 2).

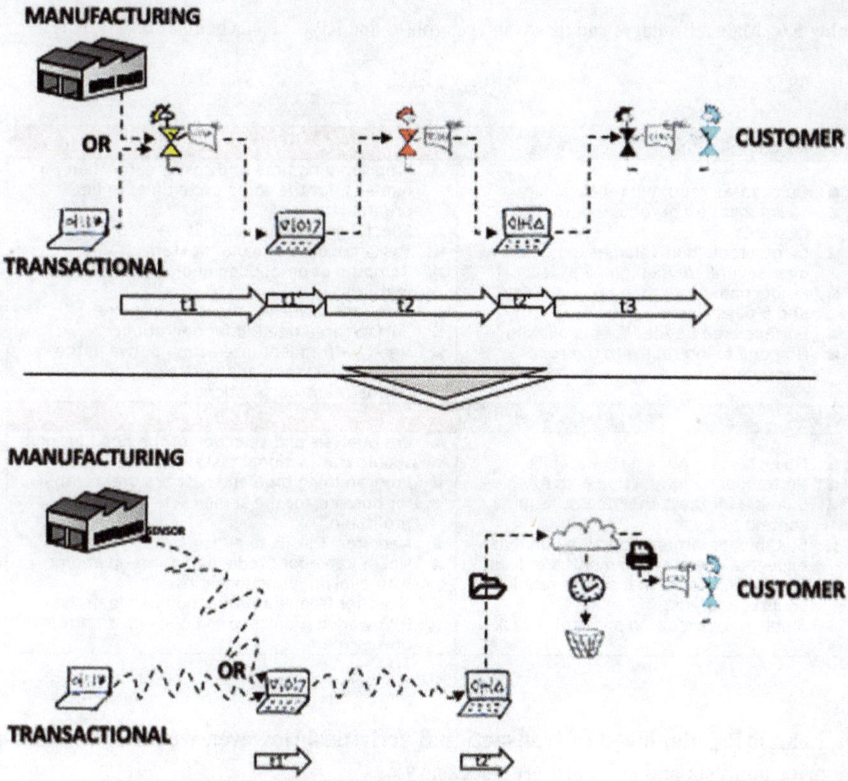

Fig. 5.5 Illustration of manual to full automation with RPA

It's important to note that some specialists differentiate between RPAs, which concern a single *process*,[22] and BPAs (Business Process Automation), which involves a more complex set of different *processes*.[23] To simplify, we will consider both simple and multiple approaches when talking about RPAs.

Global RPA market was estimated $22 bn in 2024 and expected to be 178 bn in 2033. However, possibility of automation does not imply that automation will indeed take place (Doguc, 2020). In 2016, some consultants said that RPAs would be the next revolution in corporate functions (Tounkara, 2020). Despite being used in banks and insurance companies, RPAs are far from reaching this expectation. This is for at least three reasons from my point of view:

[22] A more precise definition is found in the Glossary (See No. 2).

[23] A more precise definition is found in the Glossary (See No. 2).

Table 5.6 Main advantages and disadvantages robots and RPAs versus humans

PHYSICAL ROBOT ADVANTAGES	PHYSICAL ROBOT DISADVANTAGES
Quick tasks requiring repeatability. Tasks that do not require particular dexterity. Large production volumes expected over several many months at least. Production: more than 16 hours/day and 5 days/week. Surface area needed for installation. No need to interact with the robot frequently.	Programming time and cost greater than for humans, for the same cycle time, to be profitable. Special programming skills needed. Tasks need a particular dexterity. Ramp-up of production and/or with evolving features. Frequency: weekly, monthly, or more. Surface area needed for operation Need for frequent interaction between the robot workstation and upstream and downstream workstations.
RPA ADVANTAGES	**RPA DISADVANTAGES**
Quick tasks requiring repeatability. No frequently interaction with RPA. Suitable for tasks that do not require context analysis. Suitable for large production volumes expected over several months at least. Suitable for tasks that do not require context analysis. Suitable for recurring and stable tasks.	Pre-analysis and selection of the need for non-visible transactional tasks. Programming time and cost are greater than for humans for the same cycle time, to be profitable. Need for IT skills to anticipate. Not suitable for production ramp-up and/or with evolving functionalities. Need for frequent interaction between the RPA workstations and the ones upstream and downstream.

- **Lean is less deployed in transactional activities.** However, we have seen that a prior analysis and standard are necessary.
- Analyzing, selecting, and prioritizing tasks to automate is **more difficult because you physically can't see them.**
- The **hidden objective of job reduction by implementing RPAs** are often understood by the employees concerned, and do not become the driving force.
- **High technical constraints are often put forward** with the failure to Consider Software Licensing Restrictions. In many cases, critical systems like Enterprise Resource Planning (ERP) platforms, which are required for RPA capabilities, operate on a Software as a Service (SaaS) basis with per-user licensing arrangements (Precedence Research, 2024).

We will see these points along with digital solutions.

5.1.4 Organization to Design, Deploy and Support

Lean is primarily a **managerial approach, led by leaders, especially top leaders** in relation to *Hoshin kanri*.[24] There are a few general organizational considerations:

- **The management and facilitation** for the Lean System to work sustainably is carried out by team leaders, i.e., Management. They can be aided by dedicated people who have Lean expertise, **whom we will call Lean Leaders, for simplicity**.
- **The organizational principles** for the Lean System to work sustainably are to be relayed by the Hierarchical Line Leaders on the impulse given by Top leaders (with Lean Leaders if necessary).
- **Methods and tools, including automation:** I have briefly mentioned a few of them, including automation. These play a crucial role in **quickly and sustainably eliminating waste** based on the needs of both customers and employees. Depending on their maturity, leaders may initially require specific support to implement these methods before they learn to apply them autonomously. It is essential for Lean Leaders to develop their expertise in these methods and tools to ensure effective implementation. **Global mastery** is needed for significant and lasting deployment. This enables the selection of the most appropriate tools for specific contexts and situations while establishing the conditions for success and sustainability. Otherwise, there is a risk of relying solely on familiar techniques, similar to a musician who only knows a few notes or a painter who has mastered only a couple of colors. This may not be sufficient for the situation.
- **As with the managerial experience of Lean, experience with methods, tools, or automation is acquired by being trained and practicing**, either by being a Leader (supported in the deployment) or by being a dedicated support function but supervised by a more senior person.
- How **leaders are coached must be suited to the need** (stronger at the beginning when leaders are less trained in Lean and less so when they only need to be challenged) **and also suited to the local culture.**
- **The support organization (Lean Leaders)** can be **internal to the company, outsourced, or hybrid** (limited internal resources using external consultants). Depending on the size of the company, internal resources can be:

[24] A more precise definition is found in the Glossary (See No. 3).

  - Dedicated to Lean only (this is the case for larger companies)
  - Combined with other functions to be supported, such as Safety, Environment, Quality, or Supply Chain. In this case, ambitions must be in line with the organization.

- **The support organization** to be implemented for the support of Leaders in the deployment of Lean will depend on the organization's maturity**, the Lean skills of the leaders, and its ambitions** (for more details: Vinardi, 2013). However, let's take the case of *Hoshin kanri*,[25] which concerns the Top leaders, illustrated with key points. This can be extrapolated for the entire Lean approach (Vinardi, 2023):

  - To be considered real support, the expert considered need to be able to **allocate at least 30%, but preferably 50% of their time** in this role.
  - When the company already has trained expertise, this support can be done in-house. Otherwise, support is preferably external.
  - The expert assistant can adapt the help with various key steps. Let us take another look at the change cycle (Fig. 3.1) implemented in *Hoshin kanri* :[26]

    Step no 1: the experts help the top executive **become aware** of any gaps in practice as regards strategy with the deployment of *Hoshin kanri*.[27] They can also help **to ask the right questions about motivation.**
    Step no 2: this is where the experts have a **consulting role** to help the top executive get into action to involve their teams. It is also where the expert can **give "action-training"** if necessary.
    Steps no 3 and no 4: the support may be lighter here because top executives have sufficient training and practical knowledge. They will **need to be encouraged and challenged**. Steps 3 and 4, therefore, call on the expert to have a more **coaching role.**

Even if coaching is fashionable, the role of the coach is not necessarily the one to be privileged in all situations. Asking questions to raise awareness rather than just giving answers as a consultant is something to be learned. Moreover, it is essential to adjust the type of support to the needs of the person who is accompanied. **In situations of great stress, people often need more guidance and direct advice than being asked questions** (which is what coaching is about). In a "stable" context, coaching is the most instructive role. From experience, this is often **a very delicate point because the higher the level of the Leader, the less likely they are to transparently share with their peers their lack of competence on subjects** (Vinardi, 2023).

- Finally, **the support and managerial organization should aim for uniform Lean skills across the company**. Indeed, challenges arise in larger organizations when the Lean expertise of Leaders, including Lean Leaders, varies according to

---

[25] A more precise definition is found in the Glossary (See No. 3).

[26] A more precise definition is found in the Glossary (See No. 3).

[27] A more precise definition is found in the Glossary (See No. 3).

country or department. In this case, **everything undertaken can collapse when a Leader changes position**. To establish a common foundation, Leaders can be seconded in some companies for a few months or years to entities dedicated to (internal or external) Lean support. This also accelerates the development of a common Lean expertise. This is also the case for Toyota and Danaher, which I will discuss further at the end of this chapter.

Now that these guidelines have been given, the organization can start "from the top" of the pyramid (Top leaders) or "from the bottom," i.e., those who produce value in companies, the Front Lines. I suggest starting from the need of Front Lines or "inverted pyramid."

It should be noted that these teams are not the decision-makers of the Lean System and the organization. However, all of these components must meet their needs.

It should also be noted that, "just like chameleons", the downstream hierarchy will do everything to satisfy its Top leaders, with more or less exemplarity and authenticity.

Table 5.7 summarizes the key components of Lean, the associated support to be adapted with the general considerations discussed above, and the associated Front Line needs.

Let's focus on the Top leaders at the other end of the Inverted Pyramid. We have seen that change management progresses through various stages, which must also be considered when implementing the Lean methodology. **For Top Leaders, abruptly halting or reversing course can be more detrimental than never starting the initiative in the first place**.

Additionally, the **urgency often seen in Lean implementation** can stem from the impatience of Top leaders. While it is essential to be ambitious and to challenge teams to step outside their comfort zones, it is crucial to avoid the opposite extreme of demanding everything to be completed overnight (Vinardi, 2013).

? What organization at the head of companies is best suited for Lean deployment?

Twenty-five years ago, many industrial companies had a representative on their Group Executive Committee (ExCom) responsible for the Lean System, often titled "**Chief Performance/Competitiveness Officer**" or "**Company Business Excellence Officer**."

Having such a role in the Executive Committee is a strong indicator of a company's ambitions, even if the rest of the organization may not always align with those ambitions.

Like any functional position at the C-level, this role involves strategy, challenges, advice, support for execution, coordination, alignment, and developing skills and learning.

The operational deployment of Lean should remain the responsibility of operational leaders. Top leaders need to understand, validate, and support the Lean System, which is the foundation for deployment. The Lean System must also adapt over time

Table 5.7 From Lean key component to front line teams' needs

Lean key components		Support	Front lines needs
Management and facilitation	*Hoshin kanri* [28]	Top leaders with Lean leaders	Understand contribution to the strategic objectives and make proposals.
	PDCA cycle management	Hierachical Line Leaders with Lean Leaders	Understand, formalize, correct, improve and learn.
	Standard visual management		Participate in the definition, maintenance and development of its *processes* [29] and standards.
	Problem solving and escalation management		Correct and use standards.
	CI and BI management		Improve standards.
	Project and action work out management		Participate in projects and action work outs that are meaningful.
	social and cultural change management		Understand and participate in the changes that affect them.
Organization	Minimum hierarchical levels	Hierachical Line Leaders	Understand the management line.
	Learning teams		Draw inspiration from what has been done elsewhere.
	Transparent and rapid flow of information		Have the confidence to dare to share data and problems with superiors.
	Autonomous teams and employees		Control the expectations of positions and their scope.

(continued)

[28] A more precise definition is found in the Glossary (See No. 2).

[29] A more precise definition is found in the Glossary (See No. 2).

Table 5.7 (continued)

Lean key components		Support	Front lines needs
	Hands on top leaders		Meet with top leaders, talk about problems (solved or unsolved) and their ideas.
Methods and tools		Lean leaders with hierarchical line leaders (at the beginning)	Understand the environment and the priorities chosen using the tools and methods.
			Understand and participate in the implementation of tools and methods that are meaningful to them.

and will be overseen by the C-level function responsible for supporting Lean teams. The Lean System can be assimilated in the Safety and Quality System for simplicity and consistency.

Lean at the highest level of organization is becoming increasingly uncommon, as numerous examples show. This trend is partly due to some companies believing that after a few years, they have "finished their Lean deployment", which may lead to the misconception that their Lean skills are uniform across the organization. However, the successful deployment figures I gave in Sect. 5.1.3 suggest otherwise. **Many companies struggle to implement Lean sustainably and effectively despite its potential performance benefits.**

In some cases, when there are still Lean functions, Lean Leaders have been "demoted" from C-level positions to Vice Presidents or have even taken on roles one or two levels below the Group Executive Committee. This reflects the unfortunate trend of Lean being overshadowed by other functions, particularly Sustainability and Digital (I will be back on it), which can **undermine overall coherence.**

Company Lean Leaders are often associated with Operations or, in some cases, Finance.

- However, when Lean is tied **to Finance**, there is the risk of prioritizing cost reduction over customer satisfaction, which is not advisable.
- Reporting **to Operations** may be more effective, especially when there is a **Chief Operating Officer (COO) on the Executive Committee with proven experience in Lean**. This approach makes sense, provided that Lean deployment extends beyond just Operations, as **only a company-wide implementation can yield maximum sustainable performance. The COO can collaborate with the CEO** and other members of the Executive Committee to align on the Lean strategy and assist with the deployment of *Hoshin kanri* ,[30] assuming they have the experience to do so.

[30] A more precise definition is found in the Glossary (See No. 3).

Organizations with a Lean function at the C Level still exist, but this is becoming increasingly rare. Let's look at two such companies. Toyota (a Japanese company) and Danaher (an American one inspired by the Japanese model while internalizing its system with its own culture) have proven their longstanding sustainable performance: over fifty years for Toyota and more than thirty years for Danaher. Both still have an organization represented at the company's highest level in support of operational leaders.

- After more than fifty years, Toyota still has an entity that meets its internal needs called OMCD (Operations Management Consulting Division with TPS consultants). OMCD's primary responsibility is to help senior managers move their organizations toward the ideal. They are "learner-leader-teachers." OMCD is at all levels of the organization, from the Top Leaders to the Front Lines.

  The second entity mentioned is the TSSC (Toyota Supplier Support Center (TSSC). It plays the same role outside the company with suppliers.
- Danaher, for its part, has a DBSP team (Danaher Business System Organisation). DBSO's role was to train Leaders, both in acquired companies and existing Danaher businesses. DBSO Leaders are executives, physically located in the businesses, and had a senior operating executive role before joining (Anand et al., 2008).

5.1.5 Best Use and Limitations

As I described in Chap. 2, Lean is **primarily a managerial approach**. The tools and methods enable organizations to operate more efficiently and faster, which is especially important given current and future trends.

This approach applies to all types of companies, not just those in manufacturing. Additionally, it can include the principles of Agile methodology into its facilitation actions (see Sect. 2.2). Table 5.7 summarizes the main factors contributing to failure and the conditions for Lean's success as a comprehensive system. In contrast, Table 5.8 focuses specifically on *Hoshin kanri*,[31] which uniquely influences overall strategy (Table 5.9).

To summarize, Lean produces its best performance when:

- It is **integrated into a complete system** (distinguishing clearly between **the tools and the system**) and **is coherent** in its deployment, aligning promises with actions.
- **It engages everyone** and is **visibly supported by its Leaders** including the CEO.
- **It shows commitment through actions** encouraging employees to make proposals, co-design solutions, and participate in the implementation.
- And finally, it fosters the **development of a learning organization.**

[31] A more precise definition is found in the Glossary (See No. 3).

Table 5.8 Summary of failure factors and the conditions for success in implementing Lean (Vinardi, 2013, 2019)

Failure factors	Conditions for success
• **Failure to take into account customer needs.** • **In deployment, not taking into account every cultural dimension** (national, company, activity, business) of managers and local teams. • **Lacking preparation.** • **Focusing on tools first** and foremost. • **Focusing on costs** as a primary objective. • **Combining Lean implementation with a cost and resource reduction approach.** • **Lacking means of monitoring** implementation to ensure sustainability and **learning cycles.** • **Lack of efficiency in problem solving** that is not sufficiently developed. • Recruitment of a **Lean team whose role is to implement improvements**, which "drip feeds" the organization with these resources. • Deploying Lean **only in Operations or Production and in the field.** • **Extending internal performance actions** (Lean, Six Sigma, Agile, Commercial Excellence, etc.) **without any overall coherence.**	• Sharing a **common *vision***.[32] • Including the Lean System and the deployment of Lean as a **contributor to the overall strategy.** • **Creating conditions so that (Top) Managers are ready to train and are always connected with the field** • **Progressive commitment** of all Leaders. • **Knowledge and training development.** • **Reframing the role of Leaders** so that they can develop their teams. • **Having dedicated Lean teams** to support i.e., **train, advise, mentor, coach** (Top) Managers to develop their Lean skills to create a new culture. They will also have to challenge actions and results. • **Implement a systemic approach** in overall coherence (meaningful to each). • Establishing a **learning organization.** • Onboarding suppliers **as soon as possible, once** *processes*[33] **and organizations are clarified.**

For leaders, management mistakes or errors can be damaging, whether they are operating within a Lean framework or not, even if some authors have attributed these issues to Lean. **Micro-management and a lack of leader involvement with their teams and in the field are two extremes that can jeopardize the sustainable implementation of the Lean approach.**

It should also be noted in the cultural field that those who support Leaders in the deployment must adapt to the different business cultures of each of these people. Lean is not deployed in Production or Operations in the same way as in R&D or Sales. This must be done, for example, **by working with a credible and recognized leader of the profession.**

Business mastery (mastering the fundamentals of sales when you are a salesperson, for example) **is one of the basics for Lean deployment.** As Richard Vinardi reminds us, we tend to forget this by taking it for granted but it should be emphasized.

---

[32] A more precise definition is found in the Glossary (See No. 4).

[33] A more precise definition is found in the Glossary (See No. 2).

Table 5.9 Summary of failure factors and the conditions for success in implementing *Hoshin kanri* [34] specifically (Vinardi, 2019, 2023)

Facteurs D'Echecs	Conditions de Reussite
• **Delegating *Hoshin kanri* [35] by** the Directors and Top Leaders to support functions. • **Lacking transparency** and *catch-ball* (alignment exchange sessions) that is not really carried out. • **Confusion between daily and strategic facilitation.** • **Lacking data to understand current** and on-the-ground maturity and/or a lack of feedback on what is really happening on the ground. **Lacking reflection and learning**: little sharing of successes and failures of teams.	• **Simple and clear mission and *vision*** [36] that direct and focus the disruptive actions undertaken by the teams on a daily basis. • Knowledge **of competitors' performance levels.** • **Collective objectives that engage,** and coherence in objectives. • Objectives with **non-financial indicators.** • **Fact-based decisions and robust data.** • **Maturity of Leaders at all levels**, including at the highest level of the company, who accepts to be challenged. • A **maturity of Leaders and employees** to manage their **performance on a daily basis,** to solve problems at their level and to understand how the organization works so that it is effective. • Rigor **and determination** to implement the defined breakthrough actions.

> *"Before developing Lean skills, future users must meet **the expectations of their job.**"*
>
> From an interview of Richard Vinardi by Carine Vinardi

An essential aspect to consider during scale-up is the company's commitment to a Lean approach within pilot entities or sites. Companies must invest the necessary energy and resources into these deployments to minimize resistance and foster enthusiasm across the organization. **However, achieving success on a small scale—whether it be in a department, call center, factory,** etc.**—is challenging, and transposing successes to the entire organization is equally difficult**.

It is essential to establish how to use pilot entities, how leaders will be supported, who will provide that support, and how skills will be developed, all the while considering the company's unique constraints. Gradually, this approach should extend to encompass the entire organization, as well as any acquired companies (through M&A: Merger & Acquisitions) and suppliers. An end-to-end strategy is necessary to ensure that all parties speak a common language of performance.

[34] A more precise definition is found in the Glossary (See No. 3).

[35] A more precise definition is found in the Glossary (See No. 3).

[36] A more precise definition is found in the Glossary (See No. 4).

While roadmaps **can facilitate the scale-up *process*,**[37] **it's important to remember that significant changes cannot be accomplished quickly (see** Chap. 7). Generalizing insights across the entire company or beyond can take years, especially for larger organizations. This timeline may be further extended by management changes that can disrupt the *process*.[38]

As such, one of the limitations to the successful and sustainable deployment of Lean, as with any approach that requires the involvement of Leaders, is the staff turnover of teams. The development of common expertise takes time. If Leaders and/or support functions leave the company once they are trained or change positions every two years before they start to perform (in the third year), then expertise is difficult to acquire, the impact is more difficult to demonstrate, and sustainability more difficult to ensure.

As far as best use cases for automation are concerned, remember that **Karakuri is the first step to consider in automation activities with a physical flow.** It meets the double challenge of sustainable performance by avoiding the consumption of energy and resources, avoiding waste, being flexible, and engaging employees. For high-tech automation requiring minimal investment and fixed costs, **common sense should prevail,** including **taking advantage of lessons learned**.

In 1979, Toyota launched the Lexus LS400 with the most advanced automation in the company at its Tahara, Japan, plant, including robots in assembly doing jobs normally done by people. Sales were below expectations and the plant was under-utilized. Toyota's reflection was that the high capital costs were fixed and could not be adjusted to match demand (Ehrenfeld, 2018). Toyota continues to automate its factories, however, when ramping up production, human involvement is often preferred until volumes justify full automation. **Before investing in automation, it's wise to fully utilize existing resources to achieve initial performance levels.** When automation allows you to go further, choose solutions that first meet customer satisfaction, such as product or service quality and employee working conditions. This approach **adds purpose to automation efforts**. Lean practices can quickly reach limitations if they are viewed solely as cost-cutting measures. Eventually, organizations may abandon this type of Lean if it fails to reduce costs.

5.1.6 Compliance with Selected Criteria

Following the highlights of the Lean deployment and after clarifying how to initiate this implementation successfully, let's look at the selection criteria defined in Sect. 4.4 (Table 5.10).

[37] A more precise definition is found in the Glossary (See No. 2).

[38] A more precise definition is found in the Glossary (See No. 2).

Table 5.10 State of the art assessment of Lean deployment using sustainable criteria

ASSESSMENT CRITERIA			STATUS
POSITIVE IMPACT ON PERFORMANCE (Customers, employees, shareholders, suppliers)	1	**Positive impact on the KPIs** by considering effectiveness, efficiency and *Benchmarking*[1]	++
	2	**Positive impact on KPIs confirmed over time over a period of at least 3 to 5 years** (Empirically, this is the time to change a company's performance culture (Vinardi 2013).	++
ENVIRONMENTAL ISSUES (Customers, employees, shareholders, suppliers)	3	Transformation from **linear to circular production**	++
	4	**Non-generation and/or elimination of waste** related to material, product, energy and water and the carbon footprint, more broadly.	++
SOCIETAL ISSUES (Customers, employees, shareholders, suppliers)	5	**Promotion of Diversity & Inclusion, Ethics, Resilience**	+
	6	**Protection of personal data**	No impact
	7	**Effective and efficient cybersecurity**	No impact
	8	**Need for greater trust** (versus data sharing, changes, etc.)	++
	9	**Need to act quickly** (action, reaction, decision)	++
SOCIAL & PSYCHOSOCIAL ISSUES (Customers, employees, shareholders)	10	**Elimination of waste related to human resources** (illogic, non-ergonomic tasks)	++
	11	**Safe working conditions** (limitation of physical risks)	++
	12	**Working conditions limiting psychological risks** (stress, lack of ethics, meaningless actions, etc.)	++
	13	**Economically decent working conditions**	+
	14	**Evolution of the workplace culture** being more collective, with more flexible time management, lower hierarchical distance, need for feedback and stronger meaning (versus Baby Boomers culture)	+
CHALLENGES OF EFFICIENCY AND DIGITAL MEANING (Customers, employees, shareholders, suppliers)	15	**Global data strategy and governance**	+
	16	**Balance between human and digital**, automation of interfaces when it is meaningful	++
	17	**Customization of digital interfaces, adaptable** according to needs and expectations	+
	18	**Fluidity and efficiency of interactions** between human and tech/digital	++
COHERENCE ISSUES FOR COMPANIES (Development in line with the various social, societal and environmental issues	19	**Coherent organization** (actions/words), **genuine, true** social and societal responsibilities	++
	20	**Organization maximizing the retention and development** of its human resources	++
	21	**Organization, Leaders, *process*[2], Supply Chain**	++
	22	**Development of remote work and autonomy**	+
	23	**Development of collaborative work** (including networking) to decide and act quickly	++
	25	***Simple and efficient S&OP process*[2]**	++
	26	Leaders: **role models, psychological safety basics, with a greater focus on soft skills**: change management, resilience, ability to create links, to give meaning	++
ALIGNMENT ISSUES FOR SHAREHOLDERS & GOVERNANCE	27	**Performance and Profitability**	++
	28	**Expertise for the development of corporate performance in line with** social, societal and environmental issues	+

? What are the impacts of deploying (right) Lean based on these success conditions?

Here are a few remarks regarding the impacts indicated in yellow and marked with a single plus (+):

- Criterion 5: The impact is indirect, facilitated by multidisciplinary teams that enable the deployment of Lean principles throughout the organization.
- Criterion 13: The impact is also indirect, as it optimizes production costs by minimizing waste.
- Criterion 14: By expanding Table 5.1 with examples of expected cultural dimensions for Lean, only "short term orientation" is not aligned.
- Criterion 15: The impact is indirect, supported by *Hoshin kanri*[39] to implement an effective strategy.
- Criteria 16 and 17: These criteria are facilitated by implementing VSM and the basic elements of Lean beforehand.
- Criterion 22: *Lean* fosters the development of autonomy by establishing Autonomous Teams. This is outlined in Chap. 2.
- Criterion 28: The impact remains indirect, considering collective and social issues, and an alignment with *Hoshin kanri* .[40].

5.2 Digital Technologies

We have seen that implementing digital technology requires **a subtle balance between human and digital interactions**. In the business context, digital elements can pertain to products and/or services and how they interact with customers, who are also active participants on social networks and can influence others through their opinions. Additionally, digital *processes*[41] involve both the physical flow of information within the organization and the flow of information between customers and suppliers.

There are **four key differences between Lean principles and Digital strategies** that require careful management for successful and sustainable implementation:

- Lean require a **human investment** in skills development but **do not necessitate prior material investments**. For instance, investments in automation are seen as advancements rather than prerequisites.

  In contrast, a digital approach should be part of the initial strategy and requires **substantial human and material investments** upfront, which we will outline later. These **investments often call for partnerships**.

---

[39] A more precise definition is found in the Glossary (See No. 3).

[40] A more precise definition is found in the Glossary (See No. 3).

[41] A more precise definition is found in the Glossary (See No. 2).

- **Both Lean and Digital approaches begin with the needs of markets and customers** while relying on consistent data. However, Lean can begin its deployment using various forms of internal data—whether verbal, physical, analog, or digital. In contrast, **Digital approaches are solely driven by digital data**. Therefore, there is a specific section in Sect. 5.2.2.1 to discussing the role of Data in our deployment strategy.
- Lean is a system with three main foundations: "Managerial and Facilitation", "Methods and tools", and "Behavior". When starting with a Pilot or Model, it can be implemented by facilitation, through methods, or using tools. However, its main purpose is to test **the system's application as a whole**.

  On the other hand, Digital systems encompass **technological solutions that need to be tested, though not all solutions will fit** within the same Pilot or Model. We will discuss this further in this chapter.
- **Unlike Lean, except for a few passionate advocates, Digital technology is integrated into our daily lives**: we are all, to varying degrees, exposed to and interact with it. This trend is expected to accelerate with younger generations, as noted in Chap. 4. This can **help the adoption of Digital systems in professional settings, but it can also lead to heightened expectations and misunderstandings**. Individuals who use smartphones are not just users; they are customers, suppliers, employees, and shareholders. This latter point is made by Richard Vinardi from the L'Occitane En Provence Group.

> *"Tech/Media communication on digital creates **expectations or resistance, prejudices that require encouraging and de-escalating corporate debates**."*
>
> From an interview of Richard Vinardi by Carine Vinardi

5.2.1 Purpose and Goals

In Sect. 2.3, I proposed a definition "Digital" (the "what") based on my experience and academic research. This definition is a prerequisite before we delve into the "Why" and "What for," which are our current focus.

The maturity of companies in industrial and transactional sectors regarding digital implementation varies significantly. Their motivations for adopting digital technologies and their objectives also differ.

While **digital technology deployment is well underway in transactional organizations**, it is **just beginning or still finding its way among manufacturers**.

This point is emphasized by the insights of three partner companies associated with these industrial groups, as noted by Olivier Drean from Dassault System, Eric Ciampi from Oliver Wyman, and Marie-Ange Giuseppin from Inetum.

*"**Industrial companies are still at the beginning of their journey** as their **business model is not based on digital.** They are **progressively learning to make their business revolution.**"*

From an interview of Olivier Drean with Carine Vinardi

*"For the last 5 years, **Transactional Activities have been deploying Digital, whereas Industrial Activities have been discussing Digital deployment**. Transactional Activities are fully impacted, **not yet industrial ones**. It is still difficult to mention a digital reference for industrial companies."*

From an interview of Eric Ciampi with Carine Vinardi

*"**In the industry, we can also notice a poor adoption of digital tools**. Very often, once the digital tool is deployed and the project team is demobilized, users are changing little by little their way to use the tool, resulting in a deterioration of the data quality and thus less trust in the analysis performed by the system."*

From an interview of Marie-Ange Giuseppin with Carine Vinardi

This can be explained in several ways. As discussed in Chap. 2, **transactional companies offer services, and their Information Systems (IS) are central to their performance**. Therefore, they were directly affected by the arrival of digital technology, which transformed their IS. Additionally, as explored in Chap. 4, the constraints faced by industrial companies differ from those of transactional companies. This distinction influences their priorities.

Furthermore, industrial companies typically provide products that are not inherently technological or closely tied to services. As a result, they are not directly impacted by digital technology. This leads to a competitive landscape that remains largely unaffected by digital advancements. This point is emphasized by Fabrice Barthelemy from Tarkett:

*"**Digital has not (yet) disrupted our business**: our products and road to markets are essentially the same. **Today, digital is not the main driver of performance, to be honest.** I am **convinced that it could be the case tomorrow.***
*For the moment, I am surprised by how few **industrial companies have managed to leverage Digital.** The **only ones who seem to benefit clearly are the hardware and computing power providers.***

*In **fairness, we are not competing against very advanced competitors** in this respect, and our executives believe they have more pressing issues, and have short-term priorities…"*

From an interview of Fabrice Barthelemy with Carine Vinardi

In my interviews with **various manufacturers, I found that digital technology is not yet a primary driver** of performance. However, they all expressed expectations and ambitions for this technology to become a key factor in the future. Currently, their goals regarding Digital initiatives mainly focus on enhancing **internal efficiency**, even if there are significant ambitions related to business development in their future projects and objectives. This was also a common insight from the Senior Consultants I spoke with, including Eric Ciampi from Oliver Wyman, Olivier Drean from Dassault System, Philippe Roger from Orange Consulting, Marie-Ange Giuseppin from Inetum, and José Carlos Pires from the Kaizen Institute.

*"For the last 5 years, **Transactional Activities have been deploying Digital, whereas Industrial Activities have been discussing Digital deployment. Transactional Activities are fully impacted, not yet industrial ones**. It is still difficult to mention a digital reference for industrial companies."*

From an interview of Eric Ciampi with Carine Vinardi

*"**Industrial companies are still at the beginning of their journey** as their **business model is not based on digital** . They are progressively learning to make their business revolution.*
*For **Industrial companies, Digital is not yet the main driver** of performance, but it will definitely be a significant one in the future. Most of our industrial customers are requesting automation solutions **to reduce costs, proceed faster, and with better quality.** Therefore, the **priority is still internal efficiency.** Industrial companies need to consider **digitizing their products and services as well as their activities.***
*However, it is accelerating in its **transactional services with environmental regulations.** Companies need more and more to **collect and capture data to provide the required reports**."*

From an interview of Olivier Drean with Carine Vinardi

"The generic benefits of digital solutions implementation often depend on the sector in which you operate:

- ***In the mass market,*** *the objectives are* ***primary: improving your customer experience*** *and* ***secondary: increasing your productivity, which together contribute to increasing your margin****.*
- ***In manufacturing,*** *the objectives are to improve* ***your operations*** *(productivity of direct and indirect workers, which also means a better relevance of the decisions they make on the shopfloor, safety, capital expenditures optimization, time to deliver,* etc.*)* ***and your employee experience,*** *which contributes to your company's attractiveness."*

From an interview of Philippe Roger with Carine Vinardi

"The Top 3 expected benefits of digital implementation for our industrial customers are:

- **N°1 Internal efficiency** by reducing wasted time through automation and immediate access to the information needed—automating analysis and facilitating more accurate decision making
- **N°2 Go To Market acceleration** by reducing cycle times and working in a systemic and parallel approach rather than sequentially. This is mainly for the New Product.
- **N°3 Secure Customer On Time Delivery**: In the context of a supply chain crisis, getting a robust end-to-end planning chain and decision-making tools to anticipate and reduce missing parts is critical."

From an interview of Marie-Ange Giuseppin with Carine Vinardi

"The digital solutions are expected to provide ***faster, safer, and error-proofing processes*** *.*[42] ***Well implemented solutions*** *can bring* ***more accurate, productive, safer and robust processes.***[43] *We have more and more requests from our industry."*

From an interview of José Carlos Pires with Carine Vinardi

Olivier Drean from Dassault System points out that manufacturers increasingly consider digital technology to help them draft reports, particularly for requirements like the Corporate Sustainability Reporting Directive CSRD) in Europe. However, while the deployment of digital technology does have an environmental impact—which we will discuss later—it's important to consider whether relying on this same digital technology to gather the data needed for reports that evaluate environmental impacts is acceptable.

[42] A more precise definition is found in the Glossary (See No. 2).

[43] A more precise definition is found in the Glossary (See No. 2).

Philippe Roger from Orange Consulting draws a parallel between various situations based on company activity. These companies are directly impacted, especially those that provide digital services. Digital technology, much like the Information System in the past, serves as the **driving force behind their business and growth in transactional activities. This shift also indirectly affects their internal performance**, as Nicolas Siegler from Maif and Bruno Echardour from La Poste as for transactional companies like those of Eric Ciampi from Oliver Wyman and Olivier Drean from Dassault System).

"Digital has already impacted the Information System and connected organization *with a* ***focus on the customer experience, easy to be scaled****, the* ***inclusion of Designers to (re)build the process***[44] ***with the best ergonomics*** *(removing pain points), and for the* ***best user-friendly experience****. Before we had a team in the company to design and optimize the process*[45] *for the customer, but it was not as efficient.*
In the past, customers interacted with our teams mainly in our offices. It transitioned to phone calls. Even if phone calls still represent 80% of the flows, ***digital interfaces are increasing and could be the next transition."***

From an interview of Nicolas Siegler with Carine Vinardi

"Information Systems are strategic for retail businesses; therefore, ***digital systems are the primary driver of performance for us****, both for the front and back office. In 2023, IT spent € 2.7 Bn, which is truly significant. The* ***expected benefits of digital implementation are both Business Growth and Internal Efficiency. Business Growth is the red thread,*** *even if it is* ***expected to be done without increasing costs.*** *It implies working on the efficiency to make business and being efficient with our internal processes.*[46]

[44] A more precise definition is found in the Glossary (See No. 2).

[45] A more precise definition is found in the Glossary (See No. 2).

[46] A more precise definition is found in the Glossary (See No. 2).

The business model has completely changed*. Historically, business was based on mail, representing 80% of the revenues versus less than 15% today (mainly for business). Attendance at the Post Office is continuously decreasing. At the same time, there is a continuous increase in e-commerce and new opportunities for specific services such as electronic signatures and trusted cloud services. The induced internationalization is also new. It represented €4 bn in the recent past versus €16 bn today. It is the central part of e-commerce.* ***The customer path and our employees' ways of working have also completely changed****. For example, the postman is now equipped with a digital tool to make his round and could not make it without it.* ***Digital impacts should continue to grow.*** *Note that it is not* ***the new technologies but the complexity that is the most important change."***

From an interview of Bruno Echardour with Carine Vinardi

"For us, ***Digital is already a driver of performance****. Now we benefit from it for business growth, efficiency to make business, and driving/supporting internal efficiency.*

From an interview of Eric Ciampi with Carine Vinardi

We started to develop PLM and purchased MES. Now, we are providing digital software and connected solutions, which ***was definitely a game changer for our business****. Digital implementation is* ***already generating our business growth****, helping us make our business efficient* ***and optimize our internal efficiency****."*

From an interview of Olivier Drean with Carine Vinardi

While internal efficiency has become the primary focus for industrial companies, transactional companies prioritize multiple objectives centered around customers and business needs. Philippe Roger from Orange Consulting highlights this. Similarly, Nicolas Siegler from the MAIF Group, Bertile Burel from Wonderbox, Marie-Ange Giuseppin from Inetum, and Olivier Drean from Dassault System, also emphasize this.

"Digital is one key driver in achieving our goals, *which are to satisfy our customers through the best customer experience, satisfy our employees through the best employee experience, and satisfy our company with the fastest lead time (with the right automation) and the best cost efficiency. In addition,* ***our Digital Transformation aims to address three challenges****: empowerment, Just Necessity, and Transparency."*

From an interview of Nicolas Siegler with Carine Vinardi

"Digital will not make the marketing plan. However, ***it can help to save time*** *with product standards and coding, for example."*

From an interview of Bertile Burel with Carine Vinardi

"In our company, digitalization helps us get a ***consolidated view of business*** *opportunities so that we can federate the best experts, regardless of their geography or business unit, to answer our customers' needs. We can also* ***adjust our business model*** *regularly to the market's fluctuations. It also helps to* ***monitor our forecasts for delivery processes***[47] ***more precisely****."*

From an interview of Marie-Ange Giuseppin with Carine Vinardi

"Inside our company, we are also using our digital platform with Digital Twins (the one we commercialize) to:

- ***Share explicit knowledge*** *(documents, good practices…),*
- ***Share tacit knowledge*** *through a network facilitation supported by AI,*
- ***Model our Operations.***"

From an interview of Olivier Drean with Carine Vinardi.

Olivier Drean from Dassault System emphasizes the importance of **digital technology in sharing and leveraging company knowledge**. This topic is discussed further in Chap. 7. It highlights a crucial sustainable performance criterion: nurturing **a learning organization**. Additionally, it addresses the growing demand for skilled human resources, as educated talent is becoming increasingly scarce. This necessitates efforts to **capture and develop internal expertise**.

David Briggs from Naarea has also indirectly touched on this topic. Notably, Naarea is a "Deep Tech" start-up focused on producing small modular nuclear reactors by 2030. From the outset, the company has invested in digital technology, which is somewhat unusual. This approach was inspired by manufacturers that had previously experimented with Lean and Digital methodologies.

"We initially invested in Digital because of the speed of deployment, ***faster time to market****, and* ***increased efficiency, with fewer interconnection errors****. It is an* ***enterprise strategy****; we* ***are already saving time****.*
Based on our experience, we have estimated that with the top leadership team, ***we will also maximize our efficiency later****. We* ***expect benefits from a Seamless experience, Full traceability, and Interoperability****."*

From an interview of David Briggs with Carine Vinardi

[47] A more precise definition is found in the Glossary (See No. 2).

Like Lean principles, Digital should not be viewed as an end but as a means. Even when a digital application or platform is at the heart of the service provided to customers, it should be seen as a means to respond to market needs through a new) business model.

With that in mind, I wanted to share a final insight from Eric Ciampi at Oliver Wyman, as we have discussed internal efficiency extensively, whether as a primary or secondary objective.

> *"As for Lean, it is easier for Digital to communicate and engage people* ***when internal efficiency is not a focus but a means.****"*
>
> From an interview of Eric Ciampi with Carine Vinardi

The success of digital technology relies on its integration into the company's overall strategy, particularly in "how" it is deployed.

Before discussing deployment, it's important to reflect on AI's purpose and impact. As I mentioned in Chap. 2, Luc Julia, whom I interviewed, prefers to refer to it as Augmented Intelligence, and I share this view. **AI should serve people for sustainable development by enhancing human potential and capabilities.**

AI is based on mathematics and uses **computing power to perform tasks faster than humans**. Unlike people, IT does not require sleep or food. However, AI **lacks intrinsic creativity, empathy, dexterity, and the ability to think critically and understand the context** (see Sect. 5.2.6).

5.2.2 Deployment

Let's discuss deployment in the context of Lean methodology. This approach helps clarify the "how" in general along with the "when", "where" and "whose organization".

While the time to commit the deployment of Digital tools can be significant, managers are often strongly tempted to make this commitment without a justified reason. This can stem from a "Fear Of Missing Out" FOMO) or a "Fear Of Being Obsolete" FOBO).

A study shows that **62% of CEOs are excited about AI's potential** for new revenue streams. **At the same time, the same CEOs are afraid to become obsolete**, vs competition moving faster (Batwa et al., 2024).

Since then, the media has relentlessly communicated about AI, leading to a daily barrage of information. However, it's important to note that this information is not always presented in a consistent manner. Curiosity and a willingness to learn are essential. To engage in a successful and sustainable deployment, it is crucial to resist the pressure and approach this *process*[48] thoughtfully.

Even without a formal decision to commit, it's possible to Prepare (the P of the PDCA). This involves **understanding maturity** by asking a few essential questions to identify the human and material prerequisites:

- How do I get started?

  Based on my interviews with various people, it's clear that Top Management plays a crucial role in ensuring that digital deployment aligns with key elements of the company strategy (Vinardi, 2023). This alignment involves:
- The ***vision***:[49] What we aspire to be in the future.
- ***Strategic Objectives***:[50] Indicators for measuring future performance that will help sustain the company's future.
- Strategy: The actions that will be taken to achieve the *vision* [51] and meet the *Strategic Objectives* .[52]

As mentioned, it is important to approach this *process*[53] pragmatically rather than making it overly complex. When Top leaders engage in this exercise, they not only gain a better understanding of the deployment but also organize themselves to support it effectively. These elements will also serve as a foundation for training initiatives.

David L. Rogers, an internationally recognized expert in Digital Transformation, **emphasizes the significance of having a clear strategy:**

> *"Digital Transformation is not about Technology,it is a about **strategy and new ways of thinking.**"*

- Do we have the necessary IT infrastructure to get started?

  This is a crucial prerequisite that can often be underestimated. Implementing a solid **infrastructure**, which includes tools that align with the company's digital strategy, is essential. This infrastructure is the foundation for connecting components such as IoT, application software, communication tools (like videoconferencing), API management (Application Programming Interfaces), ERP systems, and more. It's important to note that these solutions can be tailored for small and medium-sized enterprises (SMEs).

---

[48] A more precise definition is found in the Glossary (See No. 2).
[49] A more precise definition is found in the Glossary (See No. 4).
[50] A more precise definition is found in the Glossary (See No. 5).
[51] A more precise definition is found in the Glossary (See No. 4).
[52] A more precise definition is found in the Glossary (See No. 5).
[53] A more precise definition is found in the Glossary (See No. 2).

– **Storage and Computing Capacity**: Investing in one or more Data Centers can be a strategic decision, particularly if you handle very sensitive data. A private Cloud solution can complement this approach. However, this option is generally more expensive and less flexible when adapting to fluctuating data volumes and computing needs. Alternatively, you can utilize one or more Cloud service providers to manage the Data Centers.

  A cloud solution is essential for handling large amounts of data generated by applications and enabling AI. The choice of cloud service type (e.g., private, public, or hybrid) will depend on the company's strategy and the data sensitivity. Sensitive data can also be managed using local servers, and any solution must ensure scalability and security.
– **Cybersecurity**: This involves implementing firewalls, antivirus software, and VPNs to secure communications, as well as establishing access control measures. Employee training is also critical in maintaining strong cybersecurity.
– **Network**: For optimal performance, a fast, reliable, and secure broadband connection, such as fiber or 4G/5G, is essential.

Note that the importance of sovereign or trusted clouds, which operate under a country's legal and regulatory frameworks, was emphasized by Nicolas Siegler from Maif and Bruno Echardour from La Poste. The extraterritorial claims of certain countries raise concerns for organizations that handle sensitive data.

"To mitigate security risk, it is key for us as we manage sensitive data for our customers. ***We are waiting for a national Cloud*** *and have partnered with two companies."*

From an interview of Nicola Siegler with Carine Vinardi

"In AI, Networks and the Cloud are the top digital elements to implement. The Network is key (we have around 17,000 Post Offices). We had to partner with Orange to make the right transition. The ***Cloud is both Sovereign and secure,*** *provided by big players."*

From an interview of Bruno Echardour with Carine Vinardi

• Do we have trained and motivated resources ready to get involved?

This is always *the* right question to ask. Does the organization have "Digital Natives" eager or predisposed to support the deployment? The **management of internal skills** must allow for a quick answer to this question.

- How can we organize and support the deployment?

The **sponsors and Top leaders are responsible for organizing and coordinating** the *process* [54]. We also need to consider the local relays, who will provide hands-on support in larger companies. The **operational deployment is carried out by the relevant leaders**, who **must have adequate support**.

- What partnerships should we initiate?

This topic complements the previous one, as local support often requires a **specific ecosystem**. This includes **infrastructure**, **IoT**, **technologies/software**, and implementation assistance from **consultants**. Additionally, companies can select **startups** based on their needs. This is emphasized by Olivier Drean from Dassault Systemes and Eric Ciampi from Oliver Wyman:

> ***"Our customers often think they can deploy Digital by themselves.*** *Unless they have experimented people to lead it and/or they have internalized the skills to make it,* ***external partner is necessary."***
>
> From an interview of Olivier Drean with Carine Vinardi
>
> *"As a consulting firm,* ***we have also created an ecosystem of external partners (start-ups)*** *to stay state-of-the-art and provide the best customized support."*
>
> From an interview of Eric Ciampi with Carine Vinardi

- Have the needs of customers and teams already been expressed? Are these meaningful?

This is an important consideration because, like any strategy, a digital approach should be driven not only by *Top Down*[55] actions but also by *Bottom Up* [56] proposals. This aligns with the principle of *Hoshin kanri* .[57] Similar to Lean methodology, ideas must be gathered from the field. If these ideas align with the strategy and objectives regarding Safety, Quality, Delivery, Cost & Cash, and Management, they can enter the pipeline of potential actions and be listed in a priority matrix see Fig. 2.5) based on what makes sense for the company.

---

[54] A more precise definition is found in the Glossary (See No. 2).

[55] A more precise definition is found in the Glossary (See No. 7).

[56] A more precise definition is found in the Glossary (See No. 6).

[57] A more precise definition is found in the Glossary (See No. 3).

Once the previous questions have been answered, the starting point is already established, and the next "where" question begins to be addressed through strategy.

As with Lean practices, it is generally wise to consider beginning from within the organization. However, the Digital system relies on data that is not immediately visible. The ***processes***[58] **supporting this data must be well understood and formalized to be effective**, as discussed in earlier chapters. This makes it challenging to start from the production phase. Additionally, digital systems must utilize the same shared data to avoid redundancy of identical data, negatively impacting performance and the environment (as noted in Sect. 5.2.2.1). Therefore, **maintaining consistency in data management** is crucial for organizing efforts and guiding decisions about start-up locations:

- For all companies, deployment should be a priority and aligned with the overall strategy. This includes:
  - Solutions that benefit not only customers but also employees and suppliers.
  - Solutions that empower sales and marketing departments to foster customer engagement throughout the development and production *processes*[59] and facilitate the sharing of reliable and seamless data between the company and its customers.
  - Solutions that streamline Sales and Operations Planning (S&OP) as it intersects with customers, Commerce, Supply Chain, Production, Purchasing, and suppliers.
- For industrial companies, factories and logistics platforms must align with the strategy and their level of maturity. The classic industrial objectives—Safety/Environment, Quality, Delivery, Management, Cost and Cash—along with *Strategic Objectives* [60] will guide priorities. We previously discussed this in the context of Industry 4.0 and 5.0, and we will revisit examples of implementing these solutions in Sect. 5.2.3.

With the scope clarified, similar to Lean methodology, it is essential to begin with a reduced model or pilot project (testing and learning before scaling up) and to evaluate the technologies before considering a global deployment. Chapter 7 will revisit the steps necessary to address scale-up and sustainability through learning and knowledge sharing.

When discussing deployment, we are inevitably referring to organizations or companies ("whose organization"), and it is essential, as with Lean, to understand the advantages and disadvantages of deployment in SMEs compared to large corporations, as well as in service companies versus industrial firms. We will revisit the points that have already been discussed to facilitate this understanding (Tables 5.11 and 5.12).

---

[58] A more precise definition is found in the Glossary (See No. 2).

[59] A more precise definition is found in the Glossary (See No. 2).

[60] A more precise definition is found in the Glossary (See No. 5).

Table 5.11 Main advantages and disadvantages of industrial and transactional activities in the deployment of digital

INDUSTRIAL ADVANTAGES	INDUSTRIAL DISADVANTAGES
• Physical flow that visualizes the impact of actions and guides the organization. • Deployment of Lean that can support facilitation and digital deployment. • Additional performance is accessible, whereas traditional performance levers have reached their limits.	• Large stress fields (Fig.30). • Prioritization of issues is often complex. • Potentially significant investments for scale-up and IT infrastructure. • Evolution of flows and physical equipment in addition to the human aspect, to be considered. • Still too little *Benchmarking*[1] and competitive pressure.
TRANSACTIONAL ADVANTAGES	**TRANSATIONAL DISADVANTAGES**
• Strong and direct customer culture and relationships. • Strong constraints of the market and competitors that force you to be creative. • Deployment depends primarily on the human factors (less investment). • Faster implementation when the prerequisites are in place (no industrial assets).	• Lower Lean maturity and lower formalism of *processes*[2] so there is a greater risk of errors. • Difficulties in perceiving problems and waste. Strong temptation to digitize before understanding. • Less culture of performance and often no Lean basics, even if Agile is a brick.

Table 5.12 Main advantages/disadvantages sof SME/big firms in the deployment of digital

SME ADVANTAGES	SME DISADVANTAGES
• Close to their customers. • Fast decision-making, high responsiveness – few hierarchical levels. • Flexibility. • The number of Leaders to engage is smaller. • Lower upfront costs.	• Limited resources and difficulty in accessing specialized resources. • Limited financial resources and sometimes long return on investment. • Accessible Cybersecurity Solutions Still Limited. • Lack of "system" and "performance" culture.
BIG FIRM ADVANTAGES	**BIG FIRM DISADVANTAGES**
• Expertise is accessible. • Access to resources (materials and training). • Testing of deployment possible before scaling up. • "System" and "performance" culture. • Prerequisites often already in place (including cybersecurity).	• Decision-making is often slower because there are more players and several hierarchical levels. • Organizational cumbersomeness. • Complicated scale-up (Engagement of more Leaders) and more expensive. • Access to external resources (too fast and easy).

It is important to note that implementing digital solutions requires a financial investment from the start of the deployment *process* .[61] This initial cost is often cited as a significant roadblock, especially by smaller companies, as Christian Cornille from Mecachrome highlighted.

[61] A more precise definition is found in the Glossary (See No. 2).

> *"One of the main roadblocks for SMEs is linked* ***to financial means*** *to develop digital solutions despite the cash needs."*
>
> From an interview of Christian Cornille with Carine Vinardi

We will conclude this general approach with two issues to address during deployment: Data Management and ERP integration.

5.2.2.1 Data Management

The key point about digital transformation is that by definition, without data, it does not exist. Companies must be capable of **collecting and utilizing data to make informed decisions that create value and generate assets**. It is essential to understand data throughout its entire lifecycle, from collection to disposal.

Figure 5.6 illustrates the main phases that need to be integrated.

As previously described, several principles must govern this cycle to ensure sustainable performance.

- Rule no. 1: Ensure reliable and straightforward **Data Collection**; this is the starting point. While collecting information is not a new concept, leveraging the potential of digital solutions requires that data be accessible and available in a digital format. This can involve automation through equipment or input from individuals who need to be trained to make it meaningful. This matter is crucial for all companies, as highlighted by Christian Cornille from Mecachrome.

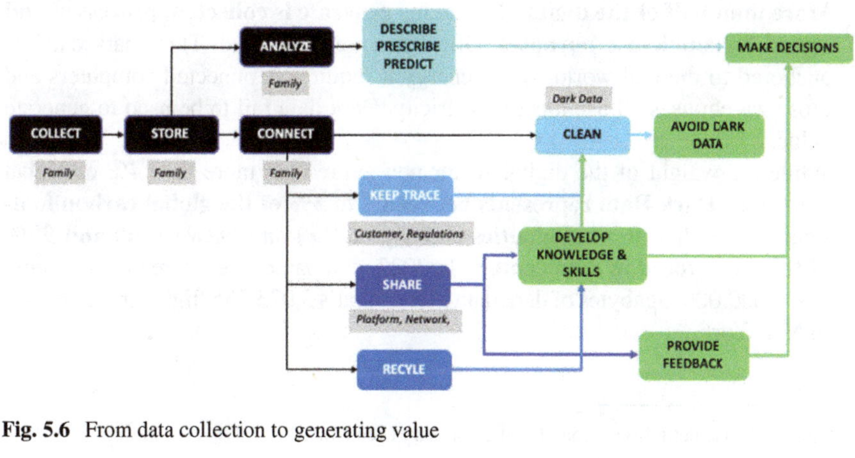

Fig. 5.6 From data collection to generating value

> ***"Access to data is now a must** to improve our performance within the Industry. In the Aerospace Industry, the quality of our blue collars is key to our performance. Nevertheless, digital is essential, especially access to data to understand issues at the shop level."*
>
> From an interview of Christian Cornille with Carine VINARDI

- Rule no. 2: Ensure **Data Continuity**. To achieve this, data circulation and storage must be organized to avoid controlled or uncontrolled duplication.
- Rule no. 3: Organize a **Data Recycle or Reuse**. To do this effectively, it is essential to identify the knowledge that needs to be shared and can generate value.

Empirical data indicates that information workers **in western Europe are losing 50% of their time every week searching for, governing and preparing data (30%) and duplicating work (20%).** A critical consequence of this is that employees and decision-makers within organizations will repeatedly search and store the same digital data over time, with data being duplicated. **Identifying ways to facilitate knowledge reuse in organizations can reduce the repetition of data searches. Pioneering organizations will adopt sustainable digital strategies that will focus on the effective reuse of existing knowledge resources** and the conscious development of the knowledge pool for future knowledge (re)use.

- Rule n°4: Avoid creating **Dark Data.** When it cannot be avoided, eliminate it quickly and automate the *process*[62] if possible.

  Let's take a moment to discuss dark data. This term refers to all stored but unused data, including received emails that remain undeleted. Dark data "sleeps" and does not generate value. Its impact on the environment is significant, and it can also incur costs for companies (Jackson & Hodgkinson, 2022a, 2022b; Jackson et al., 2023c):
- **More than half of the digital data firms generate is collected, processed and stored for single-use purposes.** Often, it is never re-used. This "dark data" is anchored to the real world by the energy it requires. Connected computers and those warehouses all use lots of electricity. Dark data fail to be used to generate value.
- While the weight of the digital sector now represents more than 4% of global emissions, **Dark Data represents between 1 to 2% of the global carbon footprint,** more than **the aeronautics activity (2.1%)** and **between 60 and 90% of the data stored in data centers**. In 2022, companies were already producing 1,300,000,000 gigabytes of dark data a day – that's 3,023,255 flights from London to New York.

[62] A more precise definition is found in the Glossary (See No. 2).

- Storage of Dark data cost **millions to companies every year around the world out** of the costs linked to the regulation risks (lack of compliance) and cybersecurity risks.

  Addressing Dark Data is one of the actionable solutions highlighted by experts to reduce digital impact on environment (Feng & Ali, 2024; Jackson et al., 2022a, 2022b). Technological development and social change are accelerating digital advancement and changing the ways organizations obtain and use information. A critical consequence of this is that employees and decision-makers within organizations are **repeatedly searching and storing the same digital data over time**, with data being duplicated and generating huge drains on power and energy.

  Effective knowledge reuse reduces dark data and positions organizations in a stronger position to maintain **sustainable knowledge** ***processes***[63] **and, thus, decrease their digital carbon footprint.** (Jackson et al.,)

- Rule no. 5: **Ensure a quick and efficient feedback loop** with customers, employees, and suppliers. This feedback from the collection can be personalized suggestions or responses to problems.
- Rule no. 6: Implementing **Data Governance** is essential. This key point, emphasized by Nicolas Siegler from Maif in Chap. 4, serves multiple purposes:
  - It helps organize **Data Continuity**, which is necessary to avoid data duplication and the creation of **Dark Data** and supports the implementation of an *omnichannel* [64]
  - To limit or eliminate dark data, it is essential to implement and share effective policies. One recommended approach is to use Edge Computing, which involves processing data locally before sending it to the Cloud or Data Centers. A significant challenge in analyzing data from numerous machines is the sheer volume of raw data, which can be challenging to assess efficiently. **Edge computing can** help **reduce communication overhead, ensure applications remain functional in remote areas,** and **filter data, thus minimizing the information sent to centralized servers**. Additionally, it can enhance **security**, reduce latency, and protect **data privacy** (França Teles et al., 2022).
  - Before collecting data, **organize it into families for more efficient processing** and to determine the necessary sensitivity and protection levels.
  - This involves **deciding on applicable regulations for each data family**, determining which data should be retained for compliance purposes, and establishing suitable storage methods and data cleaning protocols.
  - Lastly, **supporting and structuring decision-making based on data analysis** and facilitating skills development through recycling is crucial.

[63] A more precise definition is found in the Glossary (See No. 2).

[64] A more precise definition is found in the Glossary (See No. 10).

5.2.2.2 ERP Integration

ERPs were developed in the 1990s, building on the foundations laid by MRP (Material Resource Planning), which originated in the 1960s. These systems have been implemented across various types of companies, not just in industrial sectors. As mentioned in Chap. 2, ERPs are not strictly digital tools, as they existed before the digital age, nor are they considered Lean tools; it's possible to manage and improve a company's *processes*[65] without an ERP.

ERPs are designed to centralize the integration and management of business *processes*[66] across all departments within a company. They provide a comprehensive view of key aspects such as the order book, inventory levels, raw materials, finished products, and cost prices. However, they are not sufficient on their own to manage operational activities, such as R&D, sales, or production. In production, ERPs are typically connected to MES (Manufacturing Execution Systems), which emerged in the 2000s to address the limitations of the initial ERP systems.

? How does the integration of ERP software into an IT system work?

Implementing an ERP system involves a crucial integration phase with specific protocols that must be followed. One key step in this phase is **data collection and migration**, where all information needs to be transferred from one system to another. This *process*[67] carries several risks if not executed properly, as it involves transferring critical information such as supplier details, customer data, invoices, orders, accounting records, and structural data.

An effective IT architecture ensures seamless data exchange between applications, reducing complexity and facilitating smooth integration of the various tools involved. Consequently, ERPs, designed to centralize essential data, must be effectively interfaced with modern digital and scalable technologies to address organizations' evolving needs. Modern ERP web pages should be Responsive to change their layout and content depending on the device used to access them (Feng & Ali, 2024).

Here are some methods to manage the integration of ERPs with digital ecosystems:

- **Connecting ERPs with Other Applications** via **APIs** (Application Programming Interfaces):

APIs allow real-time data sharing, making them ideal for **simple, synchronous interactions**. They are particularly effective in environments that require external connectivity in a lightweight and flexible manner, often organized around cloud-native architecture.

- **Connecting Multiple ERPs together and with other software ESBs** (Enterprise Service Bus) **or Middleware**:

[65] A more precise definition is found in the Glossary (See No. 2).

[66] A more precise definition is found in the Glossary (See No. 2).

[67] A more precise definition is found in the Glossary (See No. 2).

ESBs act as a "communication bus" that connects various ERPs and serves as a centralized intermediary for handling messages between applications. This method is suitable for scenarios where **many complex and legacy systems** need to communicate **asynchronously**.

It's worth noting that ESBs and APIs can be combined. An ESB also benefits companies with different ERPs across various sites that prefer not to switch to a common ERP solution. However, this approach has advantages and disadvantages, including the management of multiple systems and considerations regarding their long-term sustainability. It should be noted that a question arises for companies that have grown through acquisition.

? Should they switch to a standard ERP or connect the different ERPs with ESBs and APIs?

- Switching to a standard ERP is almost like setting up a new ERP which is a fairly complex project, but it has the advantage of managing a single ERP from a maintenance and cybersecurity point of view.
- Connecting the different ERPs avoids making this switch and provides modern and flexible interfaces for users. Still, it has significant disadvantages: firstly, in maintaining and securing different systems, and secondly, Data Discontinuity needs to be understood.

As Eric Ciampi from Oliver Wyman notes, there is no universal choice; companies must ask the right questions to determine the best path forward.

> *"The question of a centralized ERP as prerequisite vs use of more agile solution is also a key question."*
>
> From an interview of Eric Ciampi with Carine Vinardi

- **Integration with IoT Sensors**

ERPs must be capable of integrating data from IoT sensors to enhance production *processes*[68] es, enable predictive maintenance, and improve logistics.

- **Connecting to the Cloud**

In the past, ERP systems were typically implemented on-premise, utilizing local servers. While this approach is still available, new options have emerged with the advent of digital technology. There are two main choices if you want to connect your ERP system to the Cloud for seamless integration with other cloud services. The first is a Cloud ERP accessible via the Internet. It allows for natural interfacing with

[68] A more precise definition is found in the Glossary (See No. 2).

various cloud services, as it is hosted on the provider's cloud platform and managed by their IT team. The second is a hybrid ERP, which combines both on-premise and cloud elements, where some components of the ERP are hosted on internal servers while others are managed through cloud services. This hybrid model enables companies to maintain tight control over sensitive data while benefiting from the agility and flexibility offered by cloud-based features.

Integrating Cloud technology is part of the ERP advancements needed to implement digital solutions. Other developments are also to be expected.

- **Enhanced Cybersecurity**

As the number of interfaces between customers and suppliers increases, ERPs must enhance the protection of sensitive data.

- **Customer Experience aligned with expectations of the time:**

Users today expect intuitive, personalized, and customizable interfaces—features that are "user-friendly" and similar to what they experience in their daily lives. This demand will be even more pronounced among younger generations.

The role of Enterprise Resource Planning (ERP) systems with digital transformation strategies have become an important aspect of modern businesses to stay competitive in the fast-paced digital landscape (Feng & Ali, 2024).

- **Old versions of ERP systems** may not have the functionality required to meet the evolving needs of organizations.
- **Users may find the ERP system user interface outdated** or difficult to use, leading to decreased productivity and user adoption.
- **Integrating outdated ERP with other systems and technologies can be** difficult and time-consuming, leading to increased costs and decreased efficiency.
- ERP systems from various IT infrastructure hardware and software could have been assembled over the years. Due to this, **the systems could have several unknown vulnerabilities** that can certainly be taken advantage of by hackers.
- **Old versions of ERP system Support and maintenance can cost an organization thousands** of dollars every year, and that cost may overtake the benefits provided by the system.

ERP will, therefore, have a key role in the deployment of digital data through data centralization. However, companies with a financial holding or even very decentralized companies tend to favor local deployments to give full responsibility to the Leaders of the entities concerned. **This choice**, even if it is still possible, **does not reduce the complexity of the IT infrastructure and the fluidity of data flow,** as Olivier Drean from Dassault System points out:

> *"To provide the same data to all people and ensure connected data continuity,* ***it is key to centralize and standardize the consolidation****. However,*

it is not done in many companies that prefer local customization. In addition, a common ***Digital platform enables digital continuity with a single data collection point.****"*

From an interview of Olivier Drean with Carine Vinardi

Some ERPs also integrate AI features for predictive analysis. However, this is not mandatory because other solutions exist.

5.2.3 New High Tech Automation and Solutions

When discussing the deployment of high-tech solutions, it's essential to consider not only artificial intelligence (AI) and software access but also virtual and physical solutions.

Throughout my interviews, **AI consistently emerged as *the* pivotal solution impacting businesses across various sectors. Many companies highlighted the significance of generative AI (GenAI),** with notable insights from Nicolas Siegler from Maif, Eric Ciampi from Oliver Wyman, and Philippe Roger from Orange Consulting.

"AI should continue to impact, ***especially GenAI****. Natural Language Model could help us to better interface. New generations talk more than they write, so making voice instead of typed entries could be a game changer."*

From an interview of Nicolas Siegler with Carine Vinardi

"All our consultants have been ***trained on GenAI*** *and we also have* ***developed our internal Oliver Wyman Gen AI solution****".*

From an interview of Eric Ciampi with Carine Vinardi.

"Generative AI will open new opportunities and enable*, among other possibilities,* ***the use of these solutions by less trained workers*** *and by direct and indirect* ***workers who have no available hands, and so use voice interaction."***

From an interview of Philippe Roger with Carine Vinardi

Companies are still trying to figure out how to take advantage of GenAI. Computer coding seems to be the only serious avenue at this stage since 40% of coders use it (The Economist, 2024). This is a question raised by Fabrice Barthelemy from Tarkett:

> *"For the moment, I am surprised by how few companies have managed to leverage digital and AI. The* ***only ones who benefit clearly are the hardware and computing power providers****."*
>
> From an interview of Fabrice Barthelemy with Carine Vinardi

In Chap. 2, I briefly described the key definitions and their principles. I also gave some application cases for different types of AI: ML, DL, and GenAI. We'll come back to their best uses and limitations in Sect. 5.2.5. It should be noted that the study of current possibilities could be the sole subject of a book in itself, so the goal here is not to give exhaustivity but the major families of applications and their prioritization. To prioritize usage, note that there are three main types:

- **Use cases you can customize are "mindful" AI**, where AI and its algorithms are developed in collaboration with partners who assist in training the model and integrating it with your software and IT infrastructure. This approach is not limited to very large companies, as demonstrated by Fabrice Barthelemy from Tarkett**.**

> *"We experimented with* ***two AI applications in R&D, developed with external partners,*** *one with Machine Learning for formulation support, one with Deep Learning (GenAI) for new design generation (with the R&D Department)."*
>
> From an interview of Fabrice Barthelemy with Carine Vinardi

- **Standard Use Cases: These involve "off-the-shelf" AI solutions** that are readily available through software or platforms offered by suppliers, such as ERPs. AI algorithms can be integrated with various tools, including those that manage equipment and enhance cybersecurity. Often, users interact with these AI solutions unknowingly; for example, ChatGPT is a well-known illustration of this type.
- **Hybrid Use Cases: These allow for a customizable version** of standard platforms tailored to specific needs. This includes "Low Code/No Code" solutions, which are designed to empower a broader audience to create software using intuitive graphical tools instead of programming languages. A notable example is Banco do Brasil, where employees utilized Low Code/No Code platforms to develop applications that assist customers in finding insurance products (The Economist, 2024). The primary benefit of "Low Code/No Code" solutions is that they enable **business teams to create solutions that address their needs without relying on developers**.

Deploying any use case requires **prior diagnosis and a clear definition of needs**. This step can be done in collaboration with a solutions provider, however, it's important to remember that the provider may primarily aim to sell you a solution that benefits their interests.

As Bertile Burel mentioned in Chap. 4, AI does not solve every problem and can sometimes create new issues. Therefore, remaining **curious and informed (and thus trained) is advisable while exercising caution**. Understanding the principles of AI will enable you to have constructive discussions with your provider and help you avoid potential misunderstandings.

In the case of a fully customizable application, the service provider will develop the algorithm in collaboration with you. However, based on experience, **it's often more practical to start with machine learning (ML), which has effectively popularized AI and already addresses many challenges**.

Here are a few key steps to consider when deploying AI with a completely personalized use case:

- Step 1: Definition of the need and use case.
- Step 2: Data collection and preparation: This involves cleaning, standardizing, transforming, and formatting the data to ensure consistency with the training model.
- Step 3: Define the algorithm.
- Step 4: Implementation and validation of the algorithm through the training phases.
- Step 5: Complete integration into existing systems or applications by assembling software and technological components to create a coherent and functional system.
- Step 6: Ensure after-sales service by monitoring the model's performance and optimizing it post-deployment (including updates, improvements, and troubleshooting).

The definition of the algorithm that meets your needs is rarely developed in-house, except in very large companies that have dedicated teams for this purpose. However, **partnering with traditional suppliers or universities** can be a viable option. These providers can help you make the best choice by balancing the model's accuracy, complexity, and the required human and financial resources. You can hire the same vendor for algorithm development and integration into your IT infrastructure or choose separate vendors for each task, and manage the project with its interfaces.

Both the development phase of the model and its integration are crucial. A model can be "perfectly" designed but may fail if it is not integrated correctly, with a smooth and user-friendly interface, and if after-sales support is not adequately managed. I have encountered a situation where a DL model, despite being well developed, failed to be deployed on a large scale for these reasons.

Conversely, if the right model is not selected, the data is not cleaned, or the model is not properly trained, it will not be reliable. All the effort you invest in system integration won't be effective if these foundational aspects are not addressed.

Access to software is central to developing digital solutions and is typically offered through three main approaches: open source, closed source and hybrid solution.

- **Open source Model:** In this model, the architecture, code, and model are available for anyone to view, modify, and distribute. Open source allows for the widespread sharing of technology. Start-ups and private companies can offer models that can run locally. Examples include Le Chat by Mistral AI and Deepseek.

– Pros: Generally, more affordable and promotes innovation by enabling creativity.
– Cons: It is often less secure, and data may be hosted on GAFAM Cloud services. Additionally, documentation quality can be inconsistent, and generating revenue can be challenging unless it involves developing related support services or hosted solutions.

- **Closed source Model**: In contrast, the closed source model keeps the architecture, code, and model inaccessible and under the strict control of the owning organization. An example of this is ChatGPT from OpenAI.

– Pros: Closed source models offer configured filters and enhanced cybersecurity, which are key marketing elements.
– Cons: Development and maintenance costs are high, and the options for customization are limited. Moreover, this model can create a dependence on the partners involved in the development.
– **A hybrid model** is emerging, with more organizations adopting an approach that **combines both models** with open source and proprietary applications.
– **Another alternative is to develop a custom model yourself** or outsource the development to a competent third party.

Large companies such as Google, Apple, and Facebook have voluntarily restricted access to their generative AI (GenAI) technologies, which they could otherwise control.

Why is this the case? If ChatGPT or DALL-E generate racist, discriminatory, or offensive content, they will be excused due to their status as emerging companies still in the learning phase. However, **this "right to make mistakes" would not apply to the GAFAM**, as their reputations would be severely damaged, not to mention the potential legal repercussions (Rayna & Erwan Le Pennec, 2024).

Additionally, training and developing AI algorithms is incredibly costly. **Allowing open access to their AI systems would mean giving up their competitive edge**. This situation is paradoxical because these same companies initially thrived by promoting the use of technologies like search engines, web platforms, e-commerce, and app SDKs, which were tightly controlled by other established players at the time (Rayna & Erwan Le Pennec, 2024).

Furthermore, as I am writing this book, a Chinese open source GenAI model, Deepseek, has emerged to compete with ChatGPT, a closed source model for text generation. This competition is noteworthy, especially considering Le Chat by Mistral AI already is an open source project. We will see this topic in Sect. 5.2.5.

In the interviews I conducted, I gathered insights from a diverse range of individuals across various companies, including experts and consultants. They shared experiences related to both industrial and transactional activities.

AI is a common thread in many solutions, but the most frequently mentioned tools for specific activities include:

- For transactional activities: RPA and APIs.
- For industrial activities: The options are more extensive, encompassing Industry 4.0 solutions discussed in Chap. 2. Key technologies in this category include the Internet of Things (IoT), digital platforms, Cloud-based Enterprise Resource Planning (ERP) systems, exoskeletons, various forms of robots, Digital Twins, and Virtual Reality (VR) and Augmented Reality (AR).

Additionally, I will address 3D Printing, also known as Additive Manufacturing. While the interviewees did not say so, it has the potential to be a disruptive solution in the future.

Most of the solutions we will explore in this chapter are particularly relevant as they often integrate with AI.

For instance, when discussing APIs with the chapter dedicated to the role of ERPs, we highlighted their significant role in connecting software systems, thus facilitating digital transformation.

Among others, Google, Facebook, and Amazon have developed their digital ecosystems around robust APIs that enable millions of developers to create applications and services that integrate with their platforms. These **APIs** ("AI APIs") often facilitate the integration of **AI** models into apps, covering areas such as speech recognition, image processing, and predictive analytics. Many of these models can also be open source. While it's essential to understand the use cases for APIs, we should also consider several precautions:

- Cybersecurity Risks: Connecting APIs to external systems increases the company's exposure to cybersecurity threats.
- Complex Interfaces: The proliferation of APIs can lead to **increasingly complex interfaces and compatibility issues**, which can hinder the desired fluidity of integration. Updating, maintaining, resolving, and improving the performance of APIs are key elements (the API should be considered an essential "standard" in the Lean sense and treated as such).
- Integration Challenges: Incorporating modern APIs into **existing, particularly legacy systems, can be complex or even impossible**. In such cases, using an Enterprise Service Bus (ESB) in conjunction with APIs can be beneficial.

Overall, a **holistic approach** is necessary. **API integration should consider the whole system**, be as **fluid and straightforward** as possible, and facilitate **maintenance, troubleshooting, and performance enhancements**.

We discussed the case of RPA in Sect. 5.1.3.2. Though RPAs have existed since before the digital age, their integration with AI, especially ML, allows for more advanced levels of automation (Doguc2020; Ribeiro et al., 2021).

Some experts call this combination IPA (Intelligent *Process*[69] automation), highlighting an association between RPA and AI, particularly ML. The primary advantage of merging RPA software with AI is the ability to provide enhanced capabilities beyond simple automation. This includes predictive analysis and the generation of decisions based on the inputs and outputs of automated tasks.

However, it is crucial to understand and share the underlying AI model when AI is involved in an automation *process*[70] that is not visible to operational staff and was previously performed by humans. **This transparency is necessary to build trust in the relevance of the AI-generated proposals**.

For more traditional industrial solutions, let's **begin with IoT**, which plays a key role in digital transformation. As Philippe Roger from Orange Consulting points out, installing sensors and actuators is essential in this context:

> *"For Industrials, **IoT enables the capture of the 'Operating Technology' data** (often real-time) coming from automated machines and to **combine it with 'Information Technology' data** (ERP, MES, CMMS) to bring the right information and propose relevant decisions to workers, through the most appropriate device."*
>
> From an interview of Philippe Roger with Carine Vinardi

It is important to note that the potential for low-cost installation of sensors and actuators should not encourage you to put them everywhere. The production of these devices has environmental implications, as does the associated AI (see Sect. 5.2.5). We must avoid the rebound effect, which occurs when lower costs lead to increased consumption without always being justified. In this context, I'd like to share a relevant counter-example without giving away the author:

- In 2024, a message circulated on LinkedIn from a start-up CEO, which was also shared by the World Economic Forum. This CEO touted a revolutionary platform aimed at reducing manufacturers' carbon footprint, which caught my attention. The accompanying video showed that the solution involved installing sensors everywhere on products and machines, connecting them to an AI-based application for value stream analysis. The goal was to enhance productivity and reduce electricity consumption, boasting environmental benefits.

  However, a closer look reveals a more nuanced picture. The platform comes at a cost, requires significant resources for implementation, and takes weeks to

[69] A more precise definition is found in the Glossary (See No. 2).

[70] A more precise definition is found in the Glossary (See No. 2).

generate actionable proposals. Furthermore, the analysis and proposal generation depend on a Closed-Source AI model, meaning the underlying algorithm is not available for scrutiny.

- What alternatives do manufacturers have? I previously mentioned the Value Stream Mapping (VSM) approach used in Lean methodologies, which can be initiated with just pencil and (recycled) paper.

  With few operational resources and a trained facilitator, it is possible to co-develop a diagnosis and define a target along with actionable proposals for a small factory or a department in as little as three or four days. This collaborative *process*[71] fosters team engagement in proposal development.

  If further action is needed, as we'll discuss in Chap. 6, it is feasible to supplement this approach by installing sensors (and associated actuators) at strategically chosen locations to capture real-time data.

From the perspective of achieving sustainable performance, the clear **preference should be to start with the classic VSM approach**.

In addition to immaterial solutions, Digital can also offer very physical solutions, such as **new-generation exoskeletons.** Since the 1970s, exoskeletons have been used by the army to assist soldiers. In the early 2010s, companies began using these devices. Technology advancements have significantly reduced their weight and made them more affordable, with many models now available for less than €1000. Exoskeletons can be for **specific parts of the body**, such as the upper or lower body, the shoulders, or back, providing support where it's needed most.

Basic exoskeletons use mechanical or electromechanical systems to assist movement and do not always integrate AI. However, **AI has the potential to enhance performance by personalizing support to individual users and adapting in real-time**. These systems are **designed not to automate tasks but to assist humans in performing tasks**, especially when these are physically difficult. Before investing in an exoskeleton, follow these steps:

- Step 1: Diagnose the job and assess its difficulty, considering factors such as frequency, type of work, and impact on the body.
- Step 2: Identify potential solutions in a prioritized manner:
  a. Eliminate any tasks deemed wasteful.
  b. Completely automate (100%) the task if it is Value-Added. This is a case that all employees can understand, especially when employee safety is a concern.
  c. If neither cases a) or b) are possible, consider partial automation (such as using a cobot) and/or evaluate the feasibility of an exoskeleton.
- Step 3: Analyze the need for an exoskeleton and test it with the intended user.
- Step 4: Develop and validate the solution in collaboration with the user.

[71] A more precise definition is found in the Glossary (See No. 2).

- Step 5: Deploy the exoskeleton once the tests are successful on identical workstations.

Physiotherapists can also utilize exoskeletons to assist patients in rehabilitation, providing targeted and temporary support.

Let's shift focus from relieving tedious tasks to discussing "hardware and static" automation. Many solutions now enable the automation of physical tasks, with or without the assistance of AI, particularly in the case of RPAs. It's important to note that basic robots do not utilize AI. Still, AI is increasingly being integrated to enhance their autonomy, adaptability, and the complexity of the tasks they can perform. For example, traditional robots can evolve into "**Smart Robots**" through AI integration, allowing them to adapt to various conditions and carry out a wider range of complex tasks, including trajectory planning and visual recognition.

Another category of robots integral to digital solutions, particularly in Industry 4.0, is **Cobots** (collaborative robots already mentioned). Developed after the early 2000s and marketed since 2010, these robots are equipped with advanced technologies, including cutting-edge control systems, the latest mechatronic solutions, and different types of sensors. While Cobots, too, do not incorporate AI, it is logical to enhance their efficiency similarly to Smart Robots.

The main advantages of deploying Cobots include **their flexibility** (since they do not need to be fixed to the ground), their **compact size** (which eliminates the need for safety barriers), their **ease of programming** (which is intuitive and does not require specific skills), their **ability to collaborate safely with humans and other Cobots,** and their **lower cost**.

In Table 5.13, you will find a summary of the characteristics that affect their respective usage and deployment, along with a comparison of their use cases. Figure 5.7 illustrates the levels of automation provided by Robots and Cobots, highlighting the potential for **collaboration and partial automation using only Cobots**.

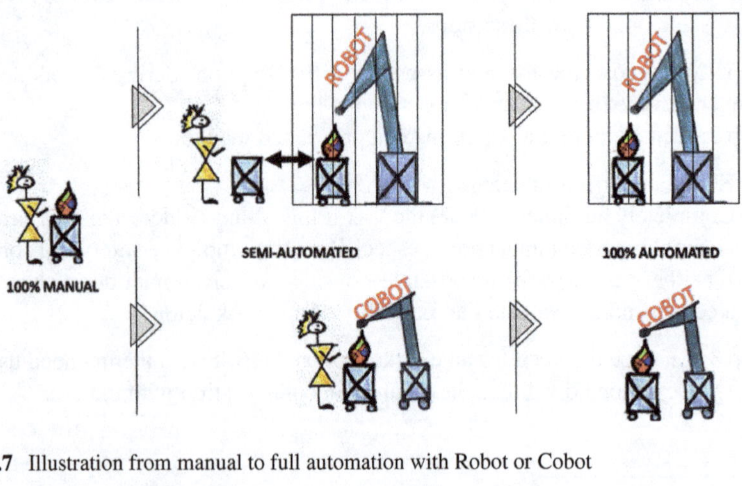

Fig. 5.7 Illustration from manual to full automation with Robot or Cobot

Table 5.13 Key features of robots and cobots impacting their use and deployment

Robot	Cobot
Handle of loads up to 2000 kg possible	Handle of loads up to max. 30 kg possible
High speeds 1000 mm/s	Average speed 250 mm/s and power speed near operators
Requires space for robot amplitude and guards	Takes up little space because it is compact
Requires physical protection of people	Does not require physical protection (built-in)
Fixed installation	Mobile installation
Weight > 50 kg	Weight < 30 kg
No built-in load cells	Detection of external forces
Technical programming (expertise)	Simple programming (no prior expertise)
Used for a particular task	Multi-function installation are possible
Cost > 50 K, and up to 10 × more	Cost between 10 and 50 K€

Other forms of robots have evolved into new **digital solutions that are dynamic**—that is, they can move. One such example is the Self-Guided Vehicle (SGV), which can navigate its environment independently by adapting to changes with the help of AI. SGVs are designed to collect or deliver parts, components, finished products, or even meal trays in restaurants. You might be familiar with similar technology at home, such as autonomous vacuum cleaners that adapt to their surroundings (Table 5.13).

SGVs were developed based on AGVs, which originated in the 1960s and were used in Japanese factories starting in the 1970s. AGVs are still around today. They are rolling robots that require guidance to move, traditionally using rails and, more recently, magnetic paint. They follow a predetermined path and are best suited for repetitive, stable movements in environments with few unexpected obstacles. This limitation led to the emergence of the SGV, which offers greater flexibility by **navigating through changing and uncertain environments**. It's important to note that both SGVs and AGVs can be fitted with Cobots to enhance their capability, including **object manipulation**.

Moving on from rolling robots to **flying robots, we have drones, another form of dynamic technology**. Drones have been used by the military since the 1990s; these earlier types are not associated with Digital Technology. They became widely deployed in civilian applications starting in the 2010s, and thanks to advancements in technology, they can be considered in their current form as a digital solution. Although the simplest versions of drones use predefined algorithms for stabilization and returning to base, the more advanced models integrate AI, enabling autonomous navigation similar to SGVs. Sophisticated drones can do image recognition, which helps detect people, identify fires, or conduct inventory checks. Drones can operate independently or collaborate in formation known as swarms.

Additionally, drones are valuable for creating Digital Twins by scanning environments, work areas, and machinery.

My interviewees mentioned **Digital Twins** several times. The concept was highlighted by the NASA Apollo program in the 1970s, where the idea of constructing identical space vehicles on Earth to simulate and predict the behavior of those in space was first explored. Digital Twins rely on having a centralized database, AI (therefore Cloud-based), and also often use VR (Virtual Reality) and AR (Augmented Reality). The need may pertain to **products and components, interactions between components, equipment or production lines,** ***processes***[72] **or even macro*process***[73] (workshop, factory, city). DT has been applied in several areas of sectors such as energy, manufacturing, transporting, health, aviation, building, and agricultural sectors (Moshood et al., 2024), (Zhang et al., 2024).

Digital Twins is not about replicating everything; instead, it's about their deployment in a **comprehensive perspective**. We should start with **a clear need**, particularly **when traditional solutions reach their limits**. The potential uses for Digital Twins can vary widely and include anticipating or correcting ergonomic issues, conducting value stream analysis (including digital VSM or eVSM), analyzing competitors' products, simulating design, production, or supply chain *processes*,[74] and training operational staff in new work environments, especially when integrated with Virtual Reality (VR).

Some researchers are exploring using Digital Twins to assess companies' information systems and enhance defenses against cyberattacks.

The development *process*[75] can take several months, especially for larger projects. Depending on the complexity, costs range from $50,000 to $400,000. For smaller factories, investment typically falls between $100,000 and $150,000. Implementing a **centralized data platform is crucial to ensuring efficient deployment and maintaining data continuity**.

The real-to-virtual mapping implies that various types of data are perceived and transmitted to the digital space, the virtual-to-real mapping implies that decision-making and optimization instructions are sent from the digital space back to the physical space (Zhang et al., 2024). The main implementation steps are:

- Step 1: Analyze the scope to be modeled and clarify the expectations.
- Step 2: Define the 3D model mathematically for the relevant perimeter.
- Step 3: Collect data using a product, components, or existing physical equipment (including sensor installation).
- Step 4: Configure the system using a learning algorithm (referring to previous steps in AI) and run simulations.
- Steps 5 and 6: Connect with existing systems, such as ERP and MES, similar to AI applications.

Richard Vinardi from L'Occitane En Provence, who uses this technology, and Olivier Dream from Dassault Systems, who provides it, share their experience:

[72] A more precise definition is found in the Glossary (See No. 2).

[73] A more precise definition is found in the Glossary (See No. 2).

[74] A more precise definition is found in the Glossary (See No. 2).

[75] A more precise definition is found in the Glossary (See No. 2).

> *"For the Engineering & Innovation department, the key steps concern the ability to develop solutions from design to ramp-up phases through* ***a digital suite integrating digital twins, VR and AR tools****. All of this facilitates tools to simulate flow, ergonomics and energy.*
> *Similarly, trade and consolidation flows are digitized in a suite developed in-house to move towards zero paper, zero non-standard interfaces."*
>
> From an interview of Richard Vinardi with Carine Vinardi

> *"We use a* ***strong scientific simulation model*** *to provide a Digital platform with Digital Twins solutions which:*
>
> - *Enable* ***digital continuity with a single point of data collection*******
> - *Provide a* ***virtual replica of the real world*** *to make all the necessary simulations*
> - *Provide* ***solutions to optimize equipment efficiency*** *for Smart Process*[76] *Control and inspection, Predictive Maintenance, and virtual VSM using a what-if scenario approach.*
>
> ***Connected with AR and VR****, it is also a way for Front Line people* ***to project themselves into the future and reduce their stress facing the unknown****."*
>
> From an interview of Olivier Drean with Carine Vinardi

Digital Twins, when combined with Augmented Reality (AR) and Virtual Reality (VR) technologies, **can significantly enhance health and safety in the workplace**. They improve ergonomics, foster a commitment to change, and reduce uncertainty-related stress. Additionally, these technologies benefit **the environment by providing accessible content without the need for printed materials and optimizing consumption**.

Now, let's examine current solutions that utilize only **AR or VR** for their deployment. The transition to digital resources such as software, IoT, and advanced processors has made the advancement of these technologies possible. However, many foundational technologies date back several decades (see Sect. 3.1). For instance, the first VR headsets designed for gaming, which gained traction in the mid-1990s, played a crucial role in accelerating the rise of VR and the initial AR applications during that time (Vinardi, 2019).

- VR requires an immersive headset similar to those used for the Metaverse. This technology is suited for **standardized training** programs that are recurrent and engage a large number of participants, especially since development costs can

---

[76] A more precise definition is found in the Glossary (See No. 2).

be high. It is also suited to **information** and **decision-making** change-related projects, such as those involving new equipment or products.

When combined with AI, training courses can be tailored to the complexity and content that best suits each learner's level and curiosity. This level of personalization fosters a stimulating learning experience for individuals. Moreover, these technologies can facilitate immersive meetings during **new product development, allowing teams to share designs and features** more effectively.

- AR can be activated using video projectors (suitable for large products or equipment), headsets, glasses, or even contact lenses. AR is particularly useful for developing new products and **supporting executing the standard of lengthy tasks** that are difficult to memorize, especially those **requiring specific quality or safety considerations**.

  When combined with AI, AR enables adaptive training through simulations involving complex tasks with multiple options—such as pre-surgical simulations for surgeons. Additionally, AR can facilitate immersive meetings, similar to VR.

The implementation *process*[77] for AR is generally straightforward: it begins with defining the need, followed by selecting the appropriate tools, creating the content, and then testing the solution before validation. However, some challenges need to be addressed:

- Common Challenges: cybersecurity, efficient network performance
- VR-Specific Challenges: development costs must be managed to ensure profitability, mental and physical health concerns such as addiction and the discomfort caused by the effects on eyesight and brain.
- AR-Specific Challenges: the media used must account for storage capacity.

It is important to note that the Metaverse discussed in Chap. 2 is fundamentally built on VR and AI technologies and operates in a networked environment. As such, I will not expand on applications similar to those mentioned previously.

Let's close this brief summary with the production solution mentioned in Sect. 3. 1: **3D Printing, also called Additive Manufacturing**. Although the technology has its roots in concepts dating back to the 1860s, it gained traction in the 2010s. The 3D printing technology involves **creating products, decorations, tools, or objects to help production**. This *process*[78] uses various materials, which are added layer by layer or shaped by removing material. A digital medium is required to use this technology, along with one or more machines (such as printers or specialized robots) connected to energy sources and appropriate materials, either in liquid or solid form. In 2011, 3D printing entered the food sector, particularly in sugar and chocolate. Then, in 2014, the Chinese company Win Su introduced the idea of constructing

[77] A more precise definition is found in the Glossary (See No. 2).

[78] A more precise definition is found in the Glossary (See No. 2).

low-cost houses using 3D printing, generating significant interest within the construction industry and paving the way for creating complex structures. Following this promising start, **3D printing was heralded as *the* transformative technology for the industrial world**.

Indeed, its ability to produce locally means that factories and transportation may no longer be necessary, potentially leading to faster production times and significant environmental benefits, provided that suitable raw materials are available. However, **machine manufacturers and raw material suppliers may also need to rethink their business models**. The key steps to produce products or objects, beyond confirming the initial need, are as follows:

- Step 1. Design. Create a 3D model using CAD software for new product development or scanning an existing product.
- Step 2. Material Selection. Choose the appropriate material and/or printer and/or robot.
- Step 3. Adjustment and Post-Processing: Manufacture the parts and do post-processing, as the raw parts may require sanding and smoothing, depending on the chosen technology.

Most universities and factories have small machines, and many private individuals now purchase these devices for just a few hundred euros.

I haven't mentioned cybersecurity, which emerged with the advent of the Internet before the rise of Digital technologies. However, cybersecurity must be hightened for the smooth deployment of Digital initiatives. This can be achieved through various methods:

- Training and Information: To better defend against threats, this can be done through traditional methods, role-playing scenarios, fake phishing exercises, and "citizen hacking" training. **More advanced techniques involve neuroscience and social engineering**, which are increasingly used to understand customer behavior. Integrating these with AI and machine learning can help identify and counter risky behaviors.
- Cryptography: Though not a new concept, cryptography is crucial for protecting the confidentiality and integrity of data through encryption. It is important to note that Quantum Computing has the potential to revolutionize this field.

When deciding to make or buy, the entire digital landscape must be considered. Organizations should evaluate what to develop internally—allowing control over content and ownership—versus what to acquire (services, software, robots, etc.) or even lease. A **clear and adaptable strategy** is imperative, as Bruno Echardour from La Poste emphasizes.

Table 5.14 Main use cases for selected solutions (from AI and Software)

Use cases	Selected high tech solutions	
Automation	Software and AI	RPAs, Smart Robots, Cobots, Drones, SGVs
Simulation		Digital Twins, AR/VR Digital Platforms, Drones
Connection		APIs, ERP Cloud, Digital Platforms
Design		Digital Twins, AR/VR, 3D Printing
Production		AR, 3D Printing, Exosqueletons
Training, exploring		Digital Twins, AR/VR, 3D Printing
Protection (of people)		Exosqueletons, Digital Twins, AR/VR, 3D Printing

> *“**Digital Make or Buy is also key**: for example, we have our own Cybersecurity center and our own Data and AI center.”*
>
> From an interview of Bruno Echardour with Carine Vinardi

Table 5.14 lists the main application cases for the solutions discussed in this chapter regarding Digital New High Tech.

5.2.4 Organization to Design, Deploy and Support

Organizational structure and strategy are closely interconnected. Together, they enhance performance through effective collaboration. **A key requirement for this is coherence between the strategy and the organizational structure, whether it is Lean or Digital**. This coherence becomes even more critical when considering the company within the context of its overall strategy (Vinardi, 2023). **Both Lean and Digital strategies should be integrated into this overarching strategy, which can be optimized for performance through *Hoshin kanri* .**[79]

It is crucial for top-level managers to support Digital and Lean initiatives. My interviews with executives for this book focused on digital strategies; they all expressed similar views on this matter.

? Beyond the observation, what principles should guide the organization’s definition to support digital deployment?

In this chapter, I will follow a similar approach as I did for Lean, discussing general considerations, the importance of Front Lines, and the organizational structure at the level of Top leaders. Here are some general considerations:

[79] A more precise definition is found in the Glossary (See No. 3).

- The **deployment of digital initiatives involves more stakeholders** than Lean, which is primarily a management approach that does not require significant material investment upfront.
- In contrast, Digital implementation is a bit more complex:
  - It **necessitates specific investments from the beginning** for the **solutions chosen, and potentially at the IT level**, especially if the existing architecture and resources in areas like network security, cybersecurity, and cloud are lacking.
  - Furthermore, the Digital **approach must focus on creating value** to ensure that investments are profitable over time. While this focus is also relevant for Lean, it is even more critical for Digital initiatives, as digitization can **directly affect products or services and thus immediately impact customers**. Consequently, Sales and Marketing **resources** must be actively involved in this ***process.***[80]
- **Management and facilitation actions** must be **meaningful**. There is nothing revolutionary about Digital implementation; it aligns with what **we have discussed regarding Lean, but it is specifically focused on the *processes*[81] impacted by digital transformation**. If Lean practices are already in place, no additional commitments are necessary. However, if they are not established, we must implement the fundamentals of performance in terms of management and team dynamics. This is where the Agile approach has sometimes emerged, as noted in Chap. 2. Similarly to Lean, the management style must be executed by team leaders, referred to as the managerial line. These leaders can be supported by dedicated resources with **Lean expertise, whom we will call Digital Lean Leaders** for simplicity. It should not be necessary to create a new position if Lean is already in practice.
- The **organizational principles** discussed in the deployment context must also be understood by the Hierarchical Line Leaders, guided by the inspiration of Top leaders and, if necessary, assisted by Lean Digital Leaders.
- Regarding solutions—and, by extension, the equivalent tools in Lean—there are **two main blocks to consider: Data and the solutions that utilize this data**. **Mastering both areas is challenging for a single individual** due to their technical complexities (in contrast to Lean). For simplicity, we will consider both blocks **as if they were internalized within a single organization**, although this scenario is rare and typically found only in very large groups.
  - Regarding the Data aspect:
    - Data Governance: As mentioned it in Sect. 5.2.2.1. It must be led by a **Data Governance Leader**; we will look at this later.
    - Data Storage, Cleaning, and Analysis: This can be handled by a Data Engineer and/or by a **Data Analyst** for the Analysis part alone.

---

[80] A more precise definition is found in the Glossary (See No. 3).

[81] A more precise definition is found in the Glossary (See No. 2).

  - Data Recycling: This part encompasses using data for learning and *value creation*.[82] It is the responsibility of the Hierarchical Line, which coordinates with the **Digital Lean Leader.**
- For the Virtual Solutions part, with Software (with RPAs), APIs and AI:
  - Development: this includes developing and maintaining software and AI models apprehended by a **Development Manager** who can focus solely on AI as needed.
  - Integration: This includes integrating and maintaining software, AI models, and AI into the existing IT architecture, which is carried out by an **Integration Manager.**
  - User-Friendly: Development and integration must allow for fluid, intuitive, and easy-to-use interfaces. A **Digital Designer** is responsible for this task.
- Other physical solutions (such as IoT, smart robots, cobots, etc.) can be bought or rented, but they must be mastered and offered in a relevant way by an **Industry 4.0 or Digital Product Manager**.
- For the IT part, what I just mentioned could also merge with existing resources:
  - Risks and Mitigation, including Cybersecurity: This function exists in most companies (excluding SMEs) with an IT department, and it is accessed via the Internet.
  - **Digital Help Desk**: This is the Hotline for problems related to digital deployment. When solutions are implemented, they must be aligned with users' pace of work.
  - Material & Software Infrastructure: IT infrastructure, networks, and the cloud. This generally exists in companies as long as there is an IT department.

As mentioned, I have separated the main areas, but some functions can be grouped according to individual skills available and already in place.

Some companies also appoint "**(IT/Digital) Business Architects**" to ensure the alignment of all stakeholders from strategy to execution with a Business approach. This is what Marie-Ange Giuseppin from Inetum points out:

> *"This is the role of emerging functions called Digital Acceleration Offices, where* ***Enterprise Architects*** *are arbitrating and defining the right project portfolio to align the digital roadmap to the Company Strategic Objective, Industrial and Digital Maturity, and Budget Constraints."*
>
> From an interview of Marie-Ange Giuseppin with Carine Vinardi

[82] A more precise definition is found in the Glossary (See No. 9).

It is important to note that this function already exists and is referenced in several job offers on dedicated sites. However, **when *Hoshin kanri* [83] is implemented, Leaders handle the alignment and coordination of the overall strategy, which should not require additional roles**. The IT Architect is a role that was in place long before the introduction of digital initiatives.

The proliferation of interfaces seldom guarantees improved performance. The deployment of digital technologies involves various stakeholders. It will necessitate the identification of key contacts to streamline interfaces, **ideally through a single point of contact**, as is commonly practiced for customer interactions.

Similar to the Lean methodology, before addressing organizational structure at the senior leadership level, we should first consider the necessity of Front Line Teams based on the aspects outlined in Chap. 2, which detail essential aspects of deployment and possible support mechanisms. A summary is provided in Table 5.15.

Several top leaders I interviewed for this book mentioned the need to **integrate a cultural element into their organizations**: specifically, the inclusion of "**Young Digital Natives,**" which refers to the upcoming generations, such as Gen Y, Gen Z, and Alpha. Key figures like Bertile Burel from Wonderbox, Fabrice Barthélémy from Tarkett, and Delphine Asseraf from Harvest specifically emphasized the importance of these individuals and their strong needs in the workplace.

"To develop our Digital Skills, we ***recruited external for a young Digital Engineer*** **who previously made his internship with us**."

From an interview of Bertile Burel with Carine Vinardi

"Young Digital Team members *will be key to drive Digital Transformation.*

From an interview of Fabrice Barthelemy with Carine Vinardi

Upskilling is key (…) in addition to ***recruiting Digital Natives*** *of course."*

From an interview of Delphine Asseraf with Carine Vinardi

As we have discussed, diversity is a source of value and performance for an organization, **provided it is accompanied by inclusion**. Recruiting **Young Digital Natives** makes sense if at least two conditions are met (Vinardi, 2016):

- They **contribute to the organization**, which in turn fosters their integration. This integration is facilitated by social connections, as when Young Digital Natives assist older employees with digital skills, the older generation can, in return, offer business knowledge and experience.
- They **represent a significant source of diversity**, meaning that for a group to have a meaningful influence, at least 25 to 30% of its resources should reflect this diversity.

---

[83] A more precise definition is found in the Glossary (See No. 3).

Table 5.15 From digital key component to front line teams' needs

Digital key components		Support	Front line contributions
Management and facilitation	Digital strategy	Top leaders with Digital Lean Leaders	Understand its contribution to and *Strategic Objectives* [84] make proposals.
	Data recycling	Hierachical Line Leaders with Digital Lean Leaders	Anticipate, formalize, correct, improve and learn.
	Process [85] selection and Digital standard Management		Participate in the definition of the *processes* [86] to be prioritized and in the facilitation of digital standards (of the solutions used).
	Problem solving and escalation management		Refer to and change digital standards.
	CI and BI management		Improve digital standards.
	project and action work out management (Agile Mode)		Participate in projects and projects that are meaningful within their scope.
	Technical, social a cultural change management		Understand and participate in the changes that affect them.
Organization	Promoting centralization and sharing of data	Top leaders with Digital Lean Leaders	Understand local needs. When they are not aligned with collective needs, and give them up.

(continued)

[84] A more precise definition is found in the Glossary (See No. 5).

[85] A more precise definition is found in the Glossary (See No. 2).

[86] A more precise definition is found in the Glossary (See No. 2).

Table 5.15 (continued)

Digital key components		Support	Front line contributions
	Promoting the transparent and rapid flow of information	Hierarchical Line Leaders	Have the confidence to dare to share data and problems.
	Autonomous teams and employees		Manage the expectations of the position and the scope with digital solutions.
Digital Tools	Data governance	Data Governance Leader	Understand the workings of data management.
	Data storage and cleaning	Data Engineer	Abide by the rules of storage and cleaning.
	Data Analysis	Data Analyst	Provide the data and understand how it are used.
	Software and AI model Development and maintenance	Development Manager	Understand the main principles of development.
	Software, AI Models and APIs Integration and maintenance	Integration Manager	Understand the main principles of integration.
	User-friendly development and maintenance	Digital Designer	Contribute feedback as a user.
	Physical digital solutions development and maintenance	Industry 4.0 or Digital Product Manager	Understand and participate in the implementation of appropriate tools and methods.
	Risks and mitigation including cybersecurity	IT Risk Manager	Observe the rules of storage and cleaning.
	Digital help desk	Help Desk Manager	Request and receive help with issues about digital solutions.
	Material and software infrastructure management	Infrastructure Manager	Request and receive help with network, server, and storage issues.

While this idea is interesting, one study has deemed it insufficient: a study on employee upskilling finds that relying on junior workers to educate senior colleagues about emerging technology is **no longer a sufficient way to share knowledge, particularly when it comes to the use of generative artificial intelligence** (Somers, 2024).

Diversity and Inclusion are social and societal issues that are part of the performance criteria defined in Sect. 4.4. Managers must take into account including Young Digital Natives, as with other types of diversity.

Skills development is a significant challenge for Top leaders, as it involves more than just integrating Young Digital Natives into their organizations. It's important not to overlook the contributions of those who have helped build the company.

For smaller organizations and SMEs, incorporating a few Young Digital Natives can enhance the development of existing resources.

The approach must be more structured in larger companies, involving internal support and/or partnerships. This enables the co-development of solutions and action-training programs and the utilization of digital training tools (such as apps) to go Digital. Additionally, **discussions around the evolution of certain roles** will be necessary to meet future needs.

To illustrate this challenge, I would like to share insights from two Top leaders in large companies who have experience with this issue: Luc Julia, a worldwide recognized expert in AI at Renault, and Bruno Echardour from La Poste.

*"When we develop the **digital skills for small teams**, co-development is the best way to engage people and make them understand and apply.*
*When we develop the **digital skills for bigger teams/lots of people**, we provide the content to the company's training department, which selects the best tool and support to give the training correctly."*

From an interview of Luc Julia with Carine Vinardi

*"We have an Internal University to develop skills on Cybersecurity, Cloud, Architecture, data & AI. **Internal skills' development is our preference. Some of our Postmen became Web Developers.** We recruit outside when needed on selected jobs."*

From an interview of Bruno Echardour with Carine Vinardi

It is indeed possible to transition from being a postman to becoming a web developer; you have to dare to take risks. The French public-sector postal company La Poste made this transition under challenging circumstances. So, why shouldn't you?

This brings me to a cultural aspect that Top leaders must embrace. Certain cultures—whether national, corporate, or professional—are more inclined to take risks. This willingness often stems from a learning model that prioritizes empirical evidence over theoretical knowledge, a lack of concern about losing face, and a favorable balance between potential benefits and risks. I've outlined some of these aspects in Table 5.1 and addressed this topic in various publications (Vinardi, 2016, 2019).

I also had the opportunity to interview Luc Julia, who emphasizes that digital transformation increasingly requires daring and courage in risk-taking. In addition to his role as Renault's CTO and start-up executive, he is also a worldwide AI expert with dual French and American cultural backgrounds. These two cultures have different stances toward risk-taking.

> ***"I am still struggling to define tips to help rather Low Risk Taking culture as French to become High Risk Taking culture as USA.***
> *Empirical first, when you learn, means that you envisage the mistake as part of the game, as the true life is not perfect. It is one of my explanations that French people have difficulties embracing mistakes/failures as a learning process,*[87] *starting from the fact that Theory does not help. However, the* ***dare to express and share mistakes can be developed,*** *especially by the Leaders of the French teams."*
>
> From an interview of Luc Julia with Carine Vinardi

Top leaders can foster an environment conducive to risk-taking by not punishing those who take them and make mistakes. Instead of sidelining or distancing themselves from these individuals, they should **value sharing experiences, whether positive or negative.**

Additionally, **both centralization and data governance** require attention to a crucial cultural aspect: teams' willingness to accept a centralized organization is closely related to the concept of "Power Distance," which reflects how cultures view hierarchical relationships. Cultures accustomed to formal hierarchical structures, whether due to their national context or professional norms, tend to have a High Power Distance and will find it easier to implement data centralization. Special efforts will be necessary, especially if this is the first step in a digital transformation, as Bruno Echardour at La Poste demonstrated.

[87] A more precise definition is found in the Glossary (See No. 2).

> *"**The first step was linked to governance**. Our IT teams were decentralized, and we needed to organize cross-activities and collaboration **through more centralization and a common lead**. It was the **first step of the roadmap**. Then we could work on more tactical aspects such as interoperability, shared service, and the omnichannel [88] implementation.*
> *A third step was to build a robust Cybersecurity: working on the culture, implementing "must-rules" to mitigate major risks."*
>
> From an interview of Bruno Echardour with Carine Vinardi

Centralization and, therefore, the provision of data by everyone for everyone requires a **significant amount of trust that cannot be imposed** and has been discussed multiple times. Trust will be fostered particularly by:

- A **clear understanding of the purpose of data collection and processing**.
- **Using data to assist those who provide it**, rather than risking its misuse against them, even unintentionally.
- A **healthy spirit of competition that promotes emulation can be meaningful** as long as it does not lead to decisions that could negatively impact the teams involved.
- An **appropriate acknowledgment of the effects of data** dispossession and the **influence of digital technology** on the individuals concerned.

Beyond the technical aspects, **Top leaders must recognize the sources of resistance and work to eliminate them, creating an environment conducive to digital deployment**. As for others, they must also **educate themselves and seek out the necessary information to achieve this**.

A lack of deep understanding of the emerging technology's capabilities drives to focus on **change to human routines rather than system design,** and focus on interventions at the project level rather than system deployer or ecosystem level. **It's up to senior leaders to better understand how Gen AI works and how to use it** (Somers, 2024).

If many leaders are informed, few are adequately trained. Based on experience, I would like to share several direct or indirect reasons for this: some leaders feel that they already know enough and lack curiosity; others claim they don't have the time for training despite its strategic importance (indicating a need to reassess their priorities); and many believe that it's someone else's responsibility to handle it (leading to inadequate delegation).

However, training can be provided by experienced colleagues or employees who have mastered the subject. This could take the form of existing academic courses or Learning Expeditions. A Top leader who is well-trained and capable of training their teams, as does Luc Julia at Renault with an innovative approach to training can be significant in reaching this need, as suggested by Nicolas Siegler at Maif:

---

[88] A more precise definition is found in the Glossary (See No. 10).

"I also am a ***role model*** *and* ***directly train*** *my (small) teams when needed."*

From an interview of Luc Julia with Carine Vinardi

"Ten years ago, we made a ***learning trip*** *to San Francisco (California) to see other companies to* ***understand and see how we could benchmark.***
Each member of the Executive Team was the sponsor of the Start-Up we decided to invest in, which helped to demystify it. They were also *Board Members in their respective boards."*

From an interview of Nicolas Siegler with Carine Vinardi

? Which organization at the head of companies is best to support digital deployment?

Similar to Lean, there isn't just one organization but multiple possible organizations. However, the variety of stakeholders has also led to broader choices:

- Digital advancements have sometimes resulted in the emergence of new dedicated roles: CDO for "Chief Data Officer" or "Chief Digital Officer" but also "Chief (Digital) Transformation Officer" or more recently with GenAI (Bajwa, 2024) "Chief AI Officer".

  CDOs' responsibilities tend to vary, but commonly include the development, refinement, and execution of an overarching digital strategy for the company and leading the required change management efforts to prepare the business for the digital era. Interestingly, CDO positions are frequently created with a direct reporting relationship to the CEO which is something that CIOs have traditionally failed to obtain. The coexistence of CDO and CIO also demands a demarcation of responsibilities, many of which would have traditionally become part or continued to be part of the CIO role (Haffke et al., 2016)

  Whereas several businesses have introduced a new position in the top management team—the CDO, other organizations have diverged from this phenomenon due to concerns and doubts about whether adding one more layer of centralized responsibility at the upper management level is the most appropriate strategy for businesses when coping with DT. Despite its visibility and increasing trend in practice, the phenomenon of CDO appointments is still new, a fact that is verified by the results of a recent global study which showed that 84 % of appointed CDOs are the first to hold this position in their companies (Christofi, 2024).

  Eric Ciampi at Olivier Wyman also states that some CDOs have been named, in particular in large companies:

> *"Our customers are* ***mainly big companies with a. CDO*** *(D for digital or data), they report mainly to a CIO and a few to a COO."*
>
> From an interview of Eric Ciampi with Carine Vinardi

- The Head of Digital Transformation is often at the company's top level, such as the Group Executive Committee, to display their ambitions (as it was already the case for Lean initiatives) in transactional companies. When they are not at this highest level, they typically report to a C-Level executive, such as the Chief Information Officer (CIO) if they are part of the Group Executive Committee, the Chief Operations Officer (COO), or, less commonly, the General Counsel or the CFO.
- Unlike Lean, the role can be either functional—focusing on strategy, consultation, execution support, coordination, alignment, and skill development—or operational, especially when managed by the CIO's IT teams.
- Sometimes, the transformation is directly led by the Chief Technology Officer (CTO) or the CIO. There are also hybrid arrangements, where the CIO develops and implements solutions while a COO or CTO defines the requirements and supports the deployment.

The interviews I conducted reflect the various organizational structures I've described, except for any mention of a "Chief AI Officer." **There is no one-size-fits-all organization; it's essential to understand the dynamics at play**. When the coordination of Digital initiatives is entrusted to the CIO, they tend to approach it from an IT perspective. If entrusted to the COO, the focus is more on local operational implementation, while if assigned to the CTO, the emphasis may be on pure technology. Ultimately, achieving a balance and taking a global approach is crucial, as Marie-Ange Giuseppin from Inetum highlights:

> ***"Several actors are involved in the digital strategy with different objectives and constraints resulting in a power game from which a fragile balance emerges:*** *innovation director or CTO push for new technologies, not always mature and stable enough, CIO push for rationalizing and securing the Information System thus slowing down innovation and COO push for best of breed solution that could solve their operational problems as quickly as possible, so against harmonization and rationalization* ***What is important is to find the most appropriate balance between those objectives & constraints."***
>
> From an interview of Marie-Ange Giuseppin with Carine Vinardi.

However, several key elements need to be considered:

- A C-level executive is necessary to **oversee the company's digital strategy at the highest level.**
- This executive must be **credible within the organization and skilled in listening to the needs and proposals of the Front Lines.**
- In addition to a **solid Business Orientation**, this individual should possess excellent soft skills and be an **effective educator and communicator**. Daniel Hager from Hager emphasizes this point:

> *"Our CIO is* ***modern and business-oriented****, actively participating in digital deployment and data optimization. The administrative aspects of IT have been delegated to the CFO, while the transformation and Added Value component is considered the core of IT. He is also an excellent* ***educator who provides training****. Furthermore, the question of organizational structure and its evolution will undoubtedly arise. We transformed our organization in 2016, establishing a functional structure, and it* ***will likely need to evolve to keep pace with digital transformation****."*
>
> From an interview of Daniel Hager with Carine Vinardi

To begin with, we should assess the existing organization and each individual's skills and interests. This assessment should be compared to the ideal target that aligns with the company's strategy. **The global coordinator's role will be to integrate the diverse** perspectives within the company, starting from the customer's viewpoint, rather than solely focusing on the typical priorities of their role.

The ideal target may represent a trend rather than an absolute model. Furthermore, the organization's evolution **can occur in phases**, adapting to the **company's foreseeable opportunities,** such as employee departures and successions.

To conclude the organizational section, let's discuss the success rate of digital transformations. As with Lean, it is challenging to find scientific studies that consistently compare success rates using the same indicators across different deployment methods. Additionally, digital transformations are still relatively new, which makes it difficult to assess overall impact. Despite this, before 2020, prominent American firms like Gartner, BCG, and McKinsey reported that **success rates for digital transformations did not exceed 20% to 30% among the companies surveyed.** In other words, 70% to 80% of these companies failed to achieve their expected results. An academic study from 2021 even claimed that failure rates could be as high **as 90%** (Ramesh & Delen, 2021). Moreover, a 2024 study conducted by DocuSign, which included 1800 decision-makers from European companies, revealed that despite increased investments in technology, **fewer than half of the respondents reported having a high level of digital maturity** (Fleureau, 2024).

Similarly to Lean, it appears that we still have a long way to go before fully harnessing the potential of digital technologies. In the next chapter, we will explore the necessary conditions for successful digital transformations, identify factors contributing to failure, discuss effective use cases, and define the limitations of digital initiatives.

5.2.5 *Best Use and Limitations*

We have just addressed deployment and organization. These aspects highlighted key conditions that are necessary but not sufficient for effective digital implementation. Let's look at the main points.

- As with Lean, an adapted framework and a Leadership approach that are meaningful:
  - Scope

    - "Why": what problem should be solved? What improvement is to be made?
    - Contribution to the "How":

      - **Digital technology as a means**, not an end
      - **Benefiting the customer,** their product and/or service, lining up experience with present and future expectations
      - **Benefiting the employees** to improve their capacity while respecting their integrity, lining up experience with present and future expectations
      - **Initial scope and target, defined and aligned**
  - Leadership

    - **Understanding the principles** that Leaders need
    - Clear strategy:

      - **Global maturity diagnosis** (organization, skills, available resources, available and necessary IT infrastructure)
      - **Understanding and identifying AV** (Added Value) and waste (Non-Added Value) **and applying this to the need, facilitate prioritization, and avoid digitizing waste**
      - **Doing what you can without investment first**
      - **Identifying when to Make or Buy**
      - **Human investment** (skills, trust, purpose, change management, cultural compatibility, adaptation, diversity and inclusion) as much as the technology

- Data at the heart of the system:
  - **Consistency of Data Management** from Suppliers to Customers
  - **Governance of Data Management** for Data Cleaning/Continuity/Reuse/Feedback
  - **Reinforced Cybersecurity**
  - **Centralization and Standardization** as Enablers

  - **ERP integration and upgrade** to facilitate data management

- At the hardware level:
  - IT architecture and information system ready for deployment
  - Ensure **continuous consistency between technological and human maturity**.
  - For **automation-related solutions, a logical approach** for teams
  - **Global consideration of impacts**: costs versus benefit and flexibility, environmental (GHG emissions, water, energy) and social impact versus benefits.

To complete this summary, I asked the Senior Executives interviewed to give the main factors of failure and the main conditions for success that they had been able to draw from their experiences. The number x shown (x) in Table 5.16 indicates the number of times the item was mentioned. At the academic level, there are still few studies on the subject but the failure factors cited are consistent with those addressed in the bibliographic summary of Oludapo et al., (2024) in Fig. 5.8.

Table 5.16 Main factors of failure and the main conditions for success from 19 interviewees

Failure factors	Conditions for success
• **Same as for Lean (7):** lack of vision, plan, monitoring, prioritization and focus, team empowering, pushed instead of pull, too much projects without consistencies…. • **Cultural aspects were not taken into account (3).** • **Wanting to do everything on your own** (2). • **Underestimated resistance to change** at startup or scale-up, or more difficult to manage than expected(3). • **Refocusing on VA tasks were more complex to implement** than expected (2). • **Disconnect between the maturity of the teams and the tools deployed** (2). • **Lack of collaboration** between teams and siloed deployment, deployment roadmaps that are not cross-functional (5). • **Focus on tools first** and foremost. • **Lack of resources or profitability** of the solutions deployed (3). • **Lack of centralization and standardization** of data (3). • **Wrong selected tools or solutions and/or unused solutions (2).** • **Return on investment required** from the test phase (4).	• **Customer needs and business focus** taken into account and maintained with digital deployment (4). • **Prioritization (and therefore strategic choices)** to ensure that the right resources are allocated to the right topic (4). • **Lean (Agile) first** to use the basics of animation, avoid digitizing waste that can be eliminated and activating solutions that are too complex and inflexible (6). • **Long Term support from Top leaders** who embody digital transformation (and understand it) (7). • A **complete preliminary impact assessment (2).** • **Operability and Sustainability defined at the beginning** (2). • **Creation of a long-term ecosystem** with the necessary external partnerships (4). • **Investment in change management and communication** just as important as in technology (3). • **Knowledge and Training Development** (4). • **Daily support and engagement from frontline teams** are secured **(4).** • **Right skills for right people (3).** • **Massive investment on Cybersecurity,** both on material and human aspects (3).

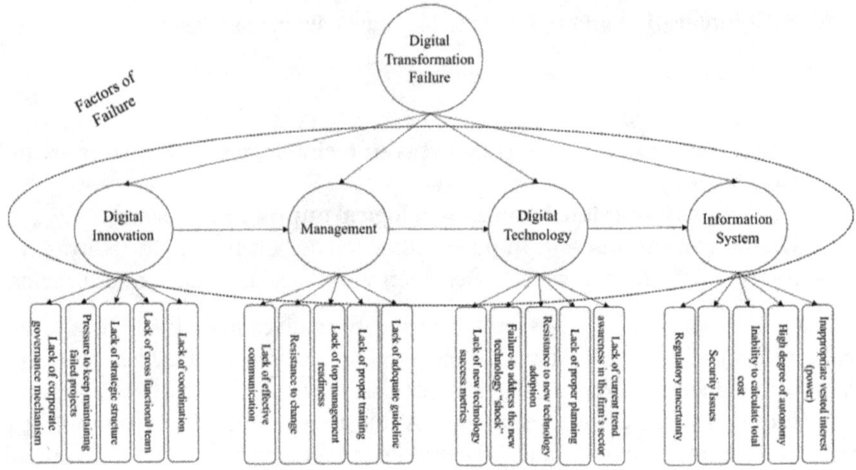

Fig. 5.8 Digital transformation failure model (*Source* Oludapo et al., 2024)

It is important to note that some studies conducted in emerging countries or those with fragile economic conditions indicate that this situation poses a significant challenge to implementing digital technologies, which require **investments and infrastructure that may not be readily available outside companies**. Existing studies reveal the **national specifics for a number of countries, especially the lack of favorable economic conditions** in the country as **a barrier to deploy Digital** (Inna & Murat, 2020). For instance, Digital transformation in the Russian manufacturing industry coincides with existing evidence on digital transition in developing countries and, in many respects, in **developed countries** (infrastructure restriction and lack of stable economic position) (Inna & Murat, 2020).

After these general observations, I will discuss some hardware solutions that we have previously examined. These deserve particular attention to understand their optimal use cases and limitations.

Let's first consider the case of 3D printing. The consolidation of the 3D printing industry is either already underway or on the horizon. For most industries, the technology is still in the "Show Me" phase (Petch, 2023). 3D printing is now **viable for both small-batch production and large-scale manufacturing**. However, its speed has not yet reached that of traditional production equipment or techniques. Additionally, it is well-suited for **producing spare parts, prototypes**, and **customized production tools**. One noteworthy advantage of 3D printing is the ease of use of the equipment. This enables employees to **make their own suggestions and customize their tools, which has a positive impact on engagement**.

In addition, 3D Printing is **not yet "plug and play" and not yet the speed for higher size of batch** (Petch, 2023). However, it is progressing. For example, MIT researchers have developed a new solution that can print rapidly with liquid metal, producing large-scale parts like table legs and chair frames in a matter of minutes, ten times faster than comparable metal Additive Manufacturing (Zewe, 2024).

As technological obstacles are removed, **we can expect significant advancements in the industrial Supply Chain**, pushing the limits of technology further.

A second material solution, the limits of which need to be addressed, is the exoskeleton. Its optimal use should allow for the following:

- **Relief from uncomfortable and painful efforts** in tasks that cannot be eliminated.
- **Gradual assistance** in the case of rehabilitation or **complete support for individuals with motor disabilities**.

However, some companies have been adding load to the exoskeleton, which contradicts its intended purpose. **When exoskeletons are used to carry more weight, they fail to alleviate the physical strain, thus increasing potential discomfort and pain**. Such practices do not promote ergonomic working conditions to achieve best performance over time.

The Digital Twin concept bridges the gap between material and immaterial solutions by creating a digital replica of a physical system to optimize it. While creating Digital Twins for everything wouldn't be economically viable, these solutions are particularly beneficial for modeling complex systems. Some key applications include:

- Providing training by simulating physical scenarios without endangering equipment or people.
- Addressing issues related to capacity, consumption, optimization of operations, or ergonomics.
- Fostering innovation by exploring new ideas that go beyond existing solutions.

However, when it comes to "Human Digital Twins", this model requires special considerations. The initial literature did not use the term "human digital twin," opting instead for phrases like "digital twin for workers" or "digital athlete" (Lin et al., 2024). Developing a more comprehensive model raises several social and ethical concerns, particularly regarding collecting, using, and protecting individuals' data. There are also technical challenges, as the impact of thoughts on the body influences its functioning, yet thoughts themselves cannot be modeled.

When considering intangible solutions, we have discussed the case of Cloud ERPs. The centralization and standardization of data facilitate the implementation of digital solutions. However, implementing an ERP can pose significant challenges, and if things go wrong, it can negatively affect both customers and the social climate within the company. Once integration is complete, issues can persist if the deployment is not thorough or maintained. This is why ERPs often attract criticism and raise concerns among companies. **The difficulties encountered during implementation can provide valuable lessons for the rollout of digital solutions and platforms. This can also require data centralization**. Throughout my career, I have been involved in ERP implementations multiple times. Here are some key challenges to keep in mind:

- **Preparation**: Implementing an ERP system is a meticulous *process* [89] that typically spans one to two years. Beyond just the data and the tool itself, all functions within the company need to prepare for adopting new habits. A holistic approach is needed to avoid creating unnecessary complexity.
- **Resources**: ERP vendors often promise productivity gains post-implementation, but these are rarely reflected in actual figures, leading to frustration among the teams involved. It's essential to recognize that nobody wants to work on a project that may lead to job elimination unless the new position is attractive. Therefore, if you decide to implement an ERP, focus on enhancing overall performance rather than solely aiming for labor productivity.
- **Management**: An ERP system does not assist operational teams in managing their daily tasks. These teams provide data to the ERP and extract information from it, but they require additional tools to utilize this data effectively. Awareness of this need is crucial, so preparing the necessary supporting tools is essential.
- **Data**: Not all data can be transferred directly from machines to the ERP or from operators and employees to the ERP system. It is vital to plan ways to minimize downtime for those involved and ensure data continuity. Selecting the right data for analysis can unlock significant value.
- **Training**: Training should occur before, during, and after the implementation. Often, companies invest heavily in training before the implementation, provide minimal training during, and invest almost nothing afterward. To address this, it's beneficial to integrate project resources dedicated to the ERP implementation within your operational teams.
- **Standards**: A smooth and efficient ERP implementation requires the company's *processes*[90] and associated standards to be clearly defined and free of significant waste before implementation begins. This essential step helps prevent the organization from adopting operational methods contrary to what employees are familiar with.
- **Integration**: Transitioning to a new ERP system necessitates transferring and integrating data from the old system. To avoid significant issues, the integrator you choose must thoroughly understand their tool and your company's existing systems.
- **Data and Management**: Even after installing a common ERP system, the integrated data must be reliable and consistent across different teams and locations. Establishing a culture of standards is crucial, yet many companies struggle to develop this. Daniel Hager shares these insights, emphasizing the importance of this aspect.

[89] A more precise definition is found in the Glossary (See No. 2).

[90] A more precise definition is found in the Glossary (See No. 2).

> *"Employees use the* ***same ERP system (SAP)*** *in our 24 factories,* ***but the data is identified differently. Local adaptations have resulted in system variants****, and a common standard has not yet been established."*
>
> From an interview of Daniel Hager with Carine Vinardi.

Although I haven't interviewed him personally, I would like to share the insights of the CEO of Nexans, an industrial company (€5 bn in revenues and 28,500 employees in 2023). He discusses the limitations of using ERPs and associated software, as well as the potential prospects (Guérin, 2023):

> *"We have deployed a wide range of software, including large-scale ERP systems. The irony, or perhaps the lesson here, is that most customers candidly admit—behind closed doors—that* ***these systems have not significantly improved their results****.*
> *Many of these tools, which have cost companies millions of euros, primarily serve as reporting and information process[91] ing instruments. Instead of simplifying operations, they have* ***ultimately added complexity*** *by siloing and fragmenting information into smaller pieces.*
> *The* ***holy grail of any company is found in what is untapped****: data.*
> *It is evident that industrial companies suffer from a* ***lack of analytical capability****."*

Digital deployment necessitates the centralization of data. **Transitioning towards centralized digital platforms will mean addressing the same challenges that arose during the implementation of ERP systems**, as briefly mentioned earlier.

Furthermore, ERP systems rely on data by definition, and while Data Management may seem straightforward in theory, it can be quite complex in practice. Therefore, it's crucial to understand the challenges that must be overcome in this context.

Cleaning **and reusing data** is essential for deployment and provides benefits in two ways: directly, by saving time for sharing, and indirectly, by helping to eliminate the **production of dark data** (Jackson et al., 2023c). **Ownership** is also a significant issue that cannot be resolved with a simple decision. Daniel Hager from Hager and Stéphanie Perchet from Adopin share their perspectives on these two subjects.

[91] A more precise definition is found in the Glossary (See No. 2).

"Each department ***processes***[92] ***its own data, and the harmonization among this data is not always present****, particularly in the existing Excel files.* ***Harmonizing all this data and thoroughly cleaning beforehand*** *is a substantial undertaking."*

From an interview of Daniel Hager with Carine Vinardi

"The ***key question is about data sovereignty: who owns the data,*** *and* ***who has the ability to leverage it?*** *This is a crucial consideration for any digital transformation."*

From an interview of Stéphanie Perchet with Carine Vinardi

Like any source of information, **reliable data can effectively address a business problem if it is well-defined and correctly understood** (Brown 2024b):

- JP Morgan sought to open new bank branches in various US cities. They facilitated this with a tool using Google Data to predict urbanization rates.
- Visa aimed to identify potential growth areas in the US by analyzing historical transaction data.
- Coca-Cola Southeast Beverages, on the other hand, was looking for a tool to streamline store inventory assessment and order creation. They used an algorithm incorporating demographic consumption trends and historical sales data to generate forecasting proposals.

This last example illustrates the connection to AI, which enables data exploitation through complex mathematical algorithms.

Before discussing the best uses of AI, let's start with some insights from two world-renowned French experts in the field: Luc Julia, whom I had the opportunity to interview for this book, and Yann Le Cun.

- "It takes two images for a human to recognize a cat with 100% accuracy, while an AI requires approximately 100,000 images to achieve a 95% accuracy rate" (Julia, 2019).
- "The large language models (LLMs) that power GenAI do not understand the real world and do not have associative memory. A rough calculation shows that a 4-year-old child has encountered at least 50 times more visual information than the largest LLM, which has processed nearly all of the texts available on the web." (Le Cun, 2024)

---

[92] A more precise definition is found in the Glossary (See No. 2).

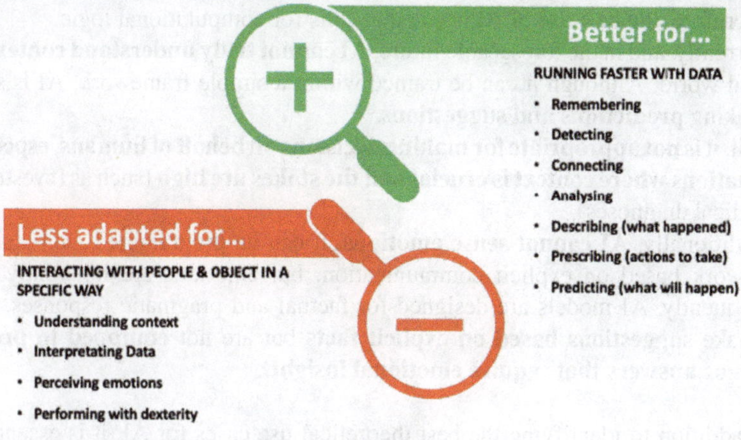

Fig. 5.9 Summary of AI best use and limits as of today

? What is AI capable of, and what are its limitations?

Before addressing this question, I'd like to share a press release published in Nature on November 7, 2019, entitled "*AI Copernicus "discovers" that Earth orbits the sun.*" Heliocentrism has been known since Nicolaus Copernicus's work in 1533, but what was groundbreaking was the discovery of physical concepts made without any prior knowledge. A neural network (DL), referred to as "SciNet," used experimental data and simple observations to determine that the Sun must be at the center of the solar system based on the movements of the Sun and the planet Mars as seen from Earth. Rather than merely confirming that the Earth revolves around the Sun, the AI inferred this conclusion from the data humans provided.

This underscores a fundamental principle that should not be overlooked: AI is created by humans, and such hypothetical discoveries would be more accurately made by people using advanced tools (Barraud, 2020).

This book does not aim to provide an exhaustive list of the best use cases for AI but to discuss how AI can be understood, its limitations, potential adverse impacts, essential conditions for success, and key factors to avoid failure.

To begin with, let's refer back to Fig. 5.9, mentioned in Chap. 2 and Sect. 5.2.3. This figure summarizes how to best approach AI given current technological constraints. It should serve as a starting point when considering its application.

It's important to remember that AI relies on mathematics; its performance depends on the volume of data it *processes*[93] es and the computing power of the available infrastructure.

[93] A more precise definition is found in the Glossary (See No. 2).

Therefore, the best use of AI is one that calls for computational logic.

Currently, and in the foreseeable future, **AI cannot truly understand context** and the real world. Although it can be trained within a simple framework, AI is suited for making **predictions and suggestions.**

Still, it is **not appropriate for making decisions on behalf of humans, especially in situations where context is crucial and the stakes are high** (such as investments or medical diagnoses).

Additionally, **AI cannot sense emotions**. It can be programmed with a simple framework based on explicit communication, but emotions are inherently tacit. Consequently, AI models are designed for factual and pragmatic responses. They can make **suggestions based on explicit facts** but are not **equipped to provide advice or answers that require emotional insight**.

In addition to identifying the best theoretical use cases for AI, it is essential to determine how we can leverage its full potential effectively. The goal should not be to **have AI perform tasks that interest us or that we are proficient in,** but **rather to focus on areas that we are less attracted to or where we are not as "strong" as AI**.

The tasks that interest us less tend to be **intellectually or physically demanding** and cannot be entirely eliminated. This aligns with the initial targets for automation we discussed. Conversely, the areas where we are less competent usually involve **computing power and speed**—these are value-added tasks that can be enhanced by algorithms.

Here are some appropriate applications:

- **Sorting: using large volumes of images, texts, or data** efficiently by using a specifically developed GenAI App.
- **Planning**: Effective planning is critical for all companies, especially when there are thousands of suppliers, materials, products, and services to consider, along with various production constraints. S&OP can be a valuable case study, allowing for a shared understanding and configuration to yield relevant proposals.
- **Production**: AI can significantly enhance production by identifying innovative assembly methods without necessitating equipment changes and integrating employee skills.

  This approach is exemplified by MIT's initiative, which focuses on optimizing mechanical part assembly, particularly for small production runs that often present economic challenges. **Their algorithm efficiently determines the order for multipart assembly, and then searches for a physically realistic motion path for each step**. Instead of one assembly line specifically designed for one specific product, if we can automatically figure out ways to sequence and move, we can use a fully adaptive setup. Impressively, the method can discover an assembly plan for a 50-part engine in a few minutes (Goardon, 2022).
- **Decision-Making**: **AI can assist in decision-making by providing predictions and proposals**. This is already evident in radiology, where studies show that the

error rate of AI alone is lower than that of humans. Furthermore, when AI assists a radiologist in focusing on the right areas, the overall error rate drops significantly, regardless of the radiologist's level of fatigue.

For these Best Use Cases to be relevant and adopted, the limitations of AI must not be forgotten and addressed:

- People must **understand the algorithms and training** *processes*[94] es to gain trust. Customers, suppliers, and employees who share their data, even voluntarily, should be able to **know and understand how their data is used**. While this is relatively straightforward with Machine Learning (ML), it becomes more complex with Deep Learning (DL), particularly in the context of GenAI, a subset of DL. Some researchers are actively working on AI models designed to explain their functioning. In the meantime, it is advisable to communicate what is known in simple terms and to avoid making unsubstantiated claims, especially those that require careful legal and regulatory justification.
- Another significant limitation is the potential for **biases and errors** that may arise from AI training. Despite their performance, these models can generate inaccuracies known as "hallucinations." Biases are introduced by humans who select the training data. However, there are methods to mitigate these risks:
  - **Diverse stakeholders should be involved in the training phases** to ensure that their interdisciplinary perspectives contribute to the model's development. Testing the model in various environments through a "test and learn" approach also helps meet ethical standards.
  - **Identify potential sources of bias in the results** to run safety checks.
  - **Introduce "frictions," or filters,** that temporarily slow down the model, enabling it to ask the right questions before proceeding quickly again—similar to how a human might pause to reflect during problem-solving.

    Medium level frictions resolve tradeoff between accuracy and time. Indeed, friction can help to capture better potential errors while not spending too much time more (Richardson Gosline, 2024)

    Introducing "targeted friction" into AI workflows can improve overall accuracy and reduce uncritical adoption. When you're talking about trillions of proliferation of potential error and bias, it's a motivator to take a hard look at what's happening. **People designing digital experiences are over-fixated on reducing friction to make it easier to do things**. However, the researchers found that friction should not be universally viewed as bad in the context of AI but rather can serve as a deliberate tool for promoting more responsible and successful generative AI use (Brown, 2024c).

- **Respect for confidentiality and data protection** continue to spark debate regarding the best approaches to address these issues. Companies have a choice:

[94] A more precise definition is found in the Glossary (See No. 2).

they can either wait until they are required to comply with regulations or proactively implement measures that build trust among employees, customers, and stakeholders. Here are a few key points to consider:

- **Data Feedback**: This proactive approach involves **reassuring customers and employees by providing targeted information** about the use of their data, including retention periods and options for action.
- **Enhanced Cybersecurity**: This is **essential for protecting sensitive data**. When individuals entrust their data to a company, they expect it to be safeguarded.
- **Multiple classifiers**: classifiers are algorithms that are trained to classify data into different categories based on a set of features. By selecting the right classifier, organizations can **effectively analyze data while safeguarding individual privacy**. Utilizing a systemic approach with multiple classifiers makes systems **harder to compromise**, which enhances overall security.
- **Normative Approach**: Europe has established the GDPR (General Data Protection Regulation), but it aims to go further with **the AI Act**, which is set to be gradually implemented starting in 2026 and will have an extraterritorial scope. Similar to CSRD, companies will be required to assess all AI projects and comply with the standards established by European authorities (based on risk assessment). These raise important questions: Do we need a new set of standards, along with the bureaucracy they entail, to take action? Is this approach effective? Can we expect the market to self-regulate when history shows it often fails to do so? As with environmental issues, these are questions that require further exploration. It is clear that **"safeguards" are necessary**, and **as AI continues to be deployed, the need for human expertise will grow accordingly.**

This can be illustrated by the case where the use of AI is appropriate but can also lead to problematic applications among the same actors. Let's consider the context of research. AI can be beneficial for tasks that may seem tedious and where the human brain operates more slowly than a computer.

- For example, AI can assist in searching for publications on a specific topic within a designated timeframe. While search engines have already performed this function on dedicated sites, the Gen AI tool offers a more comprehensive approach. Nevertheless, this task remains challenging and time-consuming. Although AI can provide an initial selection that saves time, researchers must still verify its relevance, locate the original publications, and retrieve the materials from official sources.
- Additionally, AI can propose research topics that have yet to be explored, but these suggestions must be analyzed critically.

However, researchers may not stop there. This is evidenced by the surge in the number of publications, which has coincided with the rise of mainstream GenAI models.

- Driven by a "race to publish," some researchers have started using GenAI to write posts. The problem is that writing with the context in mind is one of the researcher's added values, and AI is not the most suitable for this task. This discrepancy leads to another issue: proofreading publications.
- With an overwhelming number of publications and a shortage of reviewers to evaluate them, some individuals resort to using AI for this task, which does not represent the optimal use of AI in research.
- When outsourcing value-added tasks to AI, we often overlook that it searches for available written content, also known as explicit knowledge. However, as we will discuss further in Chap. 7, explicit knowledge constitutes only a tiny portion of overall knowledge. **Most human knowledge is tacit, acquired through experience and interaction rather than text alone** (Le Cun, 2024). If AI is involved in both the input and output *processes*[95] and relies solely on explicit knowledge, the outcome hasn't been achieved by the researcher but rather by AI. This reliance can only lead to **a decline in the quality of research.**

.

Some proposed methods for detecting AI-generated content involve using AI itself! However, models like ChatGPT operate based on probabilistic or randomized word associations. When a **detection tool analyzes a sequence of words, it may simply recalculate phrases that could have been generated by the AI's programming, but this does not mean that the model created them**. Furthermore, these models can copy and paste definitions and other online content, claiming them as their creations. This complex issue remains unresolved and may even create additional challenges.

Lastly, using AI in research brings significant issues, particularly concerning intellectual property. Here are some essential considerations that remain unresolved:

- **Inventors**: inventors are only recognized as real persons. This is still true in Europe and the USA, but some countries have recently begun to acknowledge AI as an inventor.
- **Status of AI-Assisted Inventions**: The legal status of inventions created with the assistance of AI is still unclear.
- **Training AI with Copyrighted Photos or Text**: The practice of training AI models using copyrighted material poses a growing challenge. This issue is becoming increasingly important and is expected to be central to ongoing debates. Personally, I have seen some of the complications that arise from excesses. I often use generative AI models (like ChatGPT, Deepseek, and Le Chat by Mistral AI) and always ask the models to provide sources, which I then verify. However, there have been instances where I couldn't validate a source because it was protected by copyright, but not enough, as the model managed to generate content.
- **Protecting Algorithms**: There are challenges surrounding the protection of algorithms. It must be disclosed to obtain patent protection for an algorithm; however, patents are only granted if the invention can be demonstrated. This presents a

[95] A more precise definition is found in the Glossary (See No. 2).

problem, as closed source applications do not share their algorithms. Some, therefore, offer AI applications available in Open source, but there would be no more patents possible, or the system would have to be rethought… And what would a profitable business model be?

AI continues to challenge researchers and developers alike. It's crucial to remember that AI should remain a tool and an assistant rather than a replacement for human intelligence. Thus, the **cases and conditions for using AI, rather than the technology itself, are the real issues that must be addressed**.

We discussed the use of AI in relation to its strengths and how to address the obstacles that need to be overcome. We have seen that AI can quickly switch from coherent cases to incoherent ones when performing the same activity.

Furthermore, AI is sometimes intentionally used for tasks perceived as most interesting or for which we are supposed to be stronger. This may seem counterintuitive, but it is a common practice with many applications of GenAI. For instance, when asking a public GenAI model to write a speech, it often merges content found online, sometimes disregarding copyright issues. In addition to existing biases and errors typical in classic applications, GenAI models are far from consistently stable.

Instead of using AI to generate entire speeches, seeking assistance on tips for structuring and writing a speech would be more beneficial. This approach positions AI in an assertive and supportive role, focusing on substance. Such support is already **available through traditional search engines that are significantly less energy-intensive and offer better privacy protection, like DuckDuckGo, for example**.

It is also important to note that *blockchain*[96] technology that I briefly introduced, poses limitations regarding personal data protection. While it allows transactions to be encrypted and safeguarded against cyberattacks—a critical aspect for crypto assets—it does not permit the modification or deletion of personal data. This limitation makes it incompatible with Europe's General Data Protection Regulation (GDPR), for example. Future advancements in quantum cryptography, which we will discuss in Chap. 9, could provide solutions that either improve *blockchain* [97] encryption or operate outside of it (Bonnecaze & Thiry-Atighehchi, 2025).

After discussing the implications of personal data, let's shift our focus to another limitation of digital technology: its environmental impact. The rapid adoption of modern digital technologies can pose significant challenges to sustainability, particularly due to the reliance on data centers, which are essential for cloud computing and artificial intelligence. I have previously highlighted some environmental impacts relating to digital technology, including greenhouse gas (GHG) emissions, water usage, and energy consumption. I will present key figures related to these impacts to clarify the limits and conditions needed for achieving (relative) success in this area.

[96] A more precise definition is found in the Glossary (See No. 8).

[97] A more precise definition is found in the Glossary (See No. 8).

- **Regarding GHG emissions**: by 2024, the digital industry is projected to account for **4% of global GHG emissions**, with an anticipated increase to **10% by 2030**. Data centers alone contribute approximately **2% to 2.2% of these emissions**, which is comparable to the global aviation industry's contribution of roughly 2.1% (Jackson et al.,).

  Companies such as Google and Microsoft report increasing emissions. In 2023, it represented +50% versus 2019 for Google and in 2024, + 30% versus 2020 for Microsoft (Kemen et al., 2024).
- **Electricity consumption:** the appetite of digital giants continues to grow inexorably.
  - In 2021, the **energy used by Bitcoin alone was estimated to be equivalent to that of Norway** (The Royal Society, 2021)
  - In 2022, **a single *blockchain* [98] transaction already consumed 350 times more electricity than a traditional credit card transaction** (Meunier et al., 2022). Furthermore, in 2022, data centers, cryptocurrencies, and artificial intelligence collectively accounted for nearly 2% of global electricity consumption (Ligozat & Vires, 2024).
  - **In 2023, data centers represented 2% of total global electricity consumption**, and it was projected that their usage would reach levels equivalent to Japan's consumption by 2026 (I.E.A, 2023).
  - In 2024, the entire **digital industry is expected to consume between 6 and 12% of the world's electricity** (Sternchuss, 2024). If ChatGPT powered all of Google's 9 billion daily searches, it would result in an additional 10 to 30 TWh of electricity consumption each year (Ligozat & Vries, 2024). Note that 30TWh is roughly the annual electricity consumption of Ireland.
  - Altogether, a new report by the Electric Power Research Institute found that **AI could comprise roughly 9% of the USA's total energy demand by the end of the decade** (Parsheley, 2024) Such an appetite for energy consumption seems unrelenting, as David Briggs from Naarea points out:

> *"As we are working to provide solutions to generate electricity, we can also witness the expected growth to come. We were recently consulted for a tender from GAFAM companies for an SMR to provide electricity* ***for 2040. Their expectation, only for their needs, was 100 GW (installed), which is the equivalent of Engie, who is the 2nd provider in France."***
>
> From an interview of David Briggs with Carine Vinardi

- **Water consumption** is a significant issue that is likely to lead to problems more quickly than other factors in the context of current climate change. However,

---

[98] A more precise definition is found in the Glossary (See No. 8).

obtaining comprehensive global figures on this topic is challenging. Estimates for OpenAI's GenAI model suggest that while we are currently on version 4 (which is more advanced), GPT-3 needed to "consume 1 L of water for every 20 to 100 responses", depending on the time and location of deployment. Just its training, GPT-3 used about **700,000 L of water**. (Li et al., 2023**).** Indeed, Data Centers use water for server cooling (scope 1), off-site water for electricity generation (scope 2), and supply-chain water for server manufacturing (scope 3) (Li et al., 2023).

Sharing these figures can be overwhelming and presents several paradoxes.

- By 2025, various industries, including aeronautics, will face increased pressure regarding their carbon impact, **while the digital industry has already exceeded these levels.**
- **Many industrial companies still do not fully grasp this issue, as the environmental impact of digital technology often represents only a small percentage** of their overall impact.
- **Transactional companies, many of which outsource their cloud and data center activities, frequently overlook Scope 3 emissions** and expect their service providers to take action.

However, it is important to note that this is not the case for all companies, as Bruno Echardour from La Poste points out:

> *"We are committed in a global Sustainable & Social approach (ESG) and we have a Group Roadmap leaded at the top level.* ***Digital is a small part of our Carbon footprint*** *(logistic is the biggest one), however,* ***Digital strategy has to comply with it and we have a Green Digital & IT approach:***
>
> - *"Green Digital": makes sure that our Digital Strategy has a positive impact on ESG.*
> - *"Digital for Green": selects projects where Digital can help to reduce our impact. For that we are partnering with Start-ups."*
>
> From an interview of Bruno Echardour with Carine Vinardi

- **Digital companies,** particularly the GAFAM, are **beginning to address environmental issues, but the figures I shared for two of them indicate that their environmental impact is not improving**. These companies face the same challenges as industrial firms, which are scrutinized regarding their business models.

We talk about digital technology extensively as a driver of growth**. However, when we mention its environmental impact, the conversation often shifts to topics like making data centers "greener" and increasing renewable electricity production.**

? In addition to avoiding unnecessary activation of digital solutions, what can we do to minimize environmental impacts and address our future needs?

It is possible to act on several levels; some refer to this approach as "Green Digital." However, given the figures mentioned above, I will not use this term.

- **Use common sense for basic needs:** Avoid using generative AI models (such as ChatGPT) when a simple search engine will suffice.
- Minimize energy and water use: **Implement methods like "power capping"** on the electricity supplied to processors and graphics processing units (GPUs). Power of the constraint with "power capping" for example, limiting the amount of power feeding processors and graphics processing units (Stackpole, 2025).
- **Reduce the impact of Data Centers**: There are approximately 8000 data centers worldwide, with around 33% located in the USA, 16% in Europe, and about 10% in China (Ligozat & Vries, 2024). To minimize their environmental impact, we can:
- **Address Dark Data**: Dark data is the majority of information stored in data centers. We can proactively eliminate unnecessary data center creation, which in turn reduces companies' storage needs and, consequently, the required data center capacities and their numbers.
- **Address existing Data Centers**: We should focus on investing in renewable energy sources and implementing measures to limit energy and water consumption for existing facilities.
- **Choose Locations for Future Data Centers**: New data centers should be strategically located in regions with plentiful carbon-free electricity resources, where water shortages are not a concern, and where the climate is generally temperate to facilitate natural cooling. This choice should align with operational needs, adding complexity to the overall planning *process* .[99]
- Rethink the use of materials for the digital industry
  - **Implementing a circular economy (see** Fig. 5.1)**:** In 2024, only 7.2% of the global digital economy was circular. There was still much to be done to minimize waste and increase recycling (Sternchuss, 2024). Eco-design could reduce the need for certain rare earths and use recycled materials without necessarily involving the high levels of purity needed to make magnets (222 T)
  - **New semiconductor:** new 3D chips could be an option if positive environmental impacts are confirmed. Instead of squeezing ever-smaller transistors onto a single surface, the industry is aiming to stack multiple surfaces of transistors and semiconducting elements — akin to turning a ranch house into a high-rise. Such multilayered chips could handle exponentially more data and carry out many more complex functions than today's electronics (Chu, 2024)

[99] A more precise definition is found in the Glossary (See No. 2).

- Review of AI models and their training
- **Use Simple AI Models: The ML** advancements that revived AI after its second winter can effectively solve many problems with a smaller environmental impact.
- **Use Specific AI Models**: These models can be custom-built or consumer models with limited functionality. An example of this is Deepseek, which was introduced in January 2025. Without knowing that it would be launched, I intended to highlight that **general AI models**, such as ChatGPT, do not always meet specific needs. For summarizing documents, some more modest models are smaller, have less computing power and energy, and have lower environmental costs.

  The Deepseek-V3 model was designed to be **more specific and efficient** than its American competitors. Its training took 55 days using 2,048 Nvidia H800 GPUs for $5.5 million, significantly less than a tenth of ChatGPT. It also had **8 to 10 times fewer GPU chips, which were less advanced, less energy-consuming, and cheaper**. Based on an open model, ongoing analyses confirm that it delivers respectable performance.

  Once again, the focus should not be solely on using advanced models when they are unnecessary, but rather on considering more sustainable alternatives based on our needs.
- **Rethink Model Training**: It is essential to consider that **AI models consume 40% of the electricity used during their training.** Use cheaper, less-robust AI models for training could be an option**.** For example, with **a training speed estimation tool that can predict end-state accuracy after only 20% of a computation** is complete, shaving off compute time without impacting model accuracy (Stackpole, 2025).
- **Make software AI- and carbon-aware:** Software can be designed to adjust for variations in carbon emissions and carbon impact throughout the day.

  Intelligent programs can shift workloads that MIT explained that the **carbon intensity** could be **reduced for different types of operations by about 80% to 90%** (Stackpole, 2025).

It is essential to recognize the limitations of the impact we aim for. The ideas outlined below should encourage us to limit our impact rather than increase it. **When adopting a technological solution, we must consider it as a whole.** While we can support using RFID technology for traceability and inventory management, we should avoid using disposable RFID tags (Bihouix, 2021).

To fully address environmental impacts, we must explore what digital technology can do for the environment. Some refer to this as "Digital for Green," while a more provocative term is "AI impact limitation by AI" (AI being the most significant contributor).

- Use of AI model to optimize energy consumption: The variability in renewable energy production often results in overproduction during peak times and underproduction during lulls, leading to wasteful energy consumption and grid instability. By analyzing vast datasets, from weather patterns to energy consumption trends, **AI can forecast energy production with remarkable accuracy** (Kemen et al., 2024).

- Use of Digital twins (with AI) could make possible to establish **feedback loops to monitor, understand, optimize and reduce GHG emissions** in many sectors (The Royal Society 2021):
  - In electricity grids, digital twinning can support power grids increasingly impacted by renewable energy, by managing numerous decentralized sources.
  - In planning and development, twins can help in siting assets.
  - In cities, digital twins modelling could help to predict and optimize.

Some estimate a reduction in GHG emission of – 20% by 2050 thanks to Digital (George et al., 2022). However, **these solutions will not sufficiently offset the increased consumption of water and electricity, nor the overall GHG emissions.**

And the goal should not be to innovate to counteract the effects of existing innovations. Indeed, some argue that the increase in electricity demand to satisfy digital demand could be offset and supplied by Renewable Energies. However, Renewable Energies such as wind and solar power for example, are not sufficient, do not provide continuous energy and are a source of instability for the power grid. So there is no logic to use Digital to solve grid problems…generated by Digital.

It's important to remember that information technology was expected to eliminate paper usage; however, despite a slight decline in consumption, we still rely on paper thirty years later (Bihouix, 2021).

Another **common misconception in environmental discussions is the focus on the ecology of supply, such as using recycled coffee cups, rather than developing an ecology of demand, as illustrated by the idea of having a reusable cup** (Bihouix, 2021).

Lastly, it is crucial to recognize the social limitations of digital transformation. It is already affecting our personal and professional lives, and this trend will continue to evolve. This impact is undeniable, as even farmers or craftsmen who work with their hands interact with digital tools when dealing with customers. Most jobs will adapt with the rise of digital technology—some will vanish, while new roles will emerge—but predicting the proportions and pace of these changes is challenging. Current studies tend to focus on artificial intelligence.

When it comes to AI for example, research is filled with dramatic forecasts. AI would affect almost 40% of jobs around the world, according to the International Monetary Fund. In 2023, it was expected to increase global GDP by $7 trillion—or 7%—over 10 years according to Goldman Sachs or to grow between $17.1 and $25.6 trillion annually according to McKinsey (Acemoglu, 2024).

We have never been able to predict the emergence of new jobs accurately, nor can we easily foresee which jobs will disappear. However, we can **identify two extreme categories**:

- **Jobs that require contextual reasoning, emotional intelligence related to strong social interactions, creativity, and manual dexterity that machines cannot reproduce** are unlikely to be replaced by machines for a long time.

- **Jobs based on routine tasks that can be easily automated**, such as repetitive activities requiring little skill and minimal social interaction, are at risk of being eliminated.

Between these two extremes are jobs that may be easily automated but involve strong social interactions, and jobs that are difficult to automate but have limited social interactions. Automation does not directly threaten these positions**; instead, they are likely to collaborate with AI assistants in the future**.

Several factors influence this scenario, especially for the professions concerned:

- **Feasibility, resource availability, and costs** (both initial and for scaling up). As has been evident with previous instances of automation, there is a **notable gap between what can be automated or digitized and what actually is** (Thompson, 2024).
- **Investment requirements can also pose challenges**, particularly as companies strive to mitigate uncertainties by shifting from **capital expenditure (CAPEX)** to **operational expenditure (OPEX)**. Stéphanie Perchet from Adoptin, who works in the IT/Digital, shares this perspective.

> *"My experience has shifted from selling Capex systems to* ***focusing on OPEX systems, which are now the standard.***"
>
> From an interview of Stéphanie Perchet with Carine Vinardi

- **Overall reliability** is a concern with partial applications of automation. Automating a single task or position within a larger system that is not fully automated can create more problems than it solves.

  The reliability of any given task significantly impacts the performance of the entire system (It is the "O Ring effect", referring to one of the booster rocket of Challenger space shuttle which did freeze and lead to the connected drama).

Daron Acemoğlu, a 2024 Nobel laureate in economic sciences, specifically highlights the impact of AI. In US, nearly **20% of all tasks in the labor market could be replaced or augmented by AI**. But **only about a quarter of those tasks—or 5% economy-wide—could be profitably performed.** In the other 75% of cases, costs for implementation may exceed the benefits. Then, it is estimated that the total **increase in AI-driven productivity over the next 10 years will be roughly 0.7%.** This would translate into a maximum growth in **GDP of about 1.8%; a more realistic finding under this scenario is about 1.1%.** But even that is optimistic (Acemoğlu, 2024).

There is a **mismatch between investment in AI, which is mostly taking place in large companies** in certain sectors, and the fact that many tasks that AI can perform or complement are undertaken in small-to-medium-sized enterprises (Acemoğlu, 2024).

Moreover, when it comes to Gen AI, it is has mostly been used for what he calls "**easy-to-learn tasks**," which are defined by two characteristics: There is a straight line between the action and the outcome, and a successful outcome is easy to measure. As AI is incorporated into the economy more broadly, however, it will be applied to a greater number of "hard tasks"—such as diagnosing a persistent cough—and productivity gains will likely be more limited, at least at first (Walsh, 2025).

To understand the subject comprehensively, **it is essential to cross-reference technological developments with demographic changes and identify locations where skilled workers are most available.**

The study by Korn Ferry, which I referenced in Sect. 4.3.3, indicated that, except for India, many parts of the world could face a shortage of human resources starting in 2030 (Binvel et al., 2019). Since 2019, COVID-19 has hit and may have shifted this timeline slightly, particularly with the rise of remote work for white-collar jobs. However, most of our job roles have not fundamentally changed in the past five years, and the survey results remain valid.

It's also important to note that digital technology has created its own demands for existing resources. With the rise of AI, the **proportion of computer engineers from emerging countries continues to grow** (The Economist, 2024). Consequently, the center of gravity in the computer coding industry, **once dominated by the United States, is shifting to India**. For instance, Wipro engineers in India played a significant role in developing Microsoft's Teams video conferencing service. **Between 2023 and 2029, the number of programmers in the Asia–Pacific region is projected to increase by 21%, while Latin America is expected to see a 17% increase**. In contrast, **North America will likely see only a 13% rise, and Europe, a 9% increase**. This imbalance continues to drive offshoring (The Economist, 2024).

In any case, companies that succeed in the long term are those that not only meet their customers' current needs but also anticipate future demands. This proactive approach is the best way to define added value and the associated skills needed for the organization. By addressing the needs for both blue-collar and white-collar digital skills, companies can position themselves more confidently for the future.

5.2.6 Compliance with Selected Criteria

As with Lean, let's refer to the selection criteria outlined in Sect. 4.4.

? What are the impacts of deploying digital technology on these criteria, provided it is implemented under the conditions for success?

Let's look at Table 5.17.

Without reservation at this stage, and despite our best efforts, criteria 3 and 4 stand out due to their negative impacts based on our previous discussions. This remains

Table 5.17 Assessment of a "State of the Art" digital deployment according to sustainability criteria

ASSESSMENT CRITERIA			STATUS
POSITIVE IMPACT ON PERFORMANCE (Customers, employees, shareholders, suppliers)	1	**Positive impact on the KPIs** by considering effectiveness, efficiency and *Benchmarking*[1]	++
	2	**Positive impact on KPIs confirmed over time over a period of at least 3 to 5 years** (Empirically, this is the time to change a company's performance culture (Vinardi 2013).	+
ENVIRONMENTAL ISSUES (Customers, employees, shareholders, suppliers)	3	Transformation from **linear to circular production**	-
	4	**Non-generation and/or elimination of waste** related to material, product, energy and water and the carbon footprint, more broadly.	-
SOCIETAL ISSUES (Customers, employees, shareholders, suppliers)	5	**Promotion of Diversity & Inclusion, Ethics, Resilience**	+
	6	**Protection of personal data**	++
	7	**Effective and efficient cybersecurity**	++
	8	**Need for greater trust** (versus data sharing, changes, etc.)	+
	9	**Need to act quickly** (action, reaction, decision)	++
SOCIAL & PSYCHOSOCIAL ISSUES (Customers, employees, shareholders)	10	**Elimination of waste related to human resources** (illogic, non-ergonomic tasks)	++
	11	**Safe working conditions** (limitation of physical risks)	++
	12	**Working conditions limiting psychological risks** (stress, lack of ethics, meaningless actions, etc.)	+
	13	**Economically decent working conditions**	+
	14	**Evolution of the workplace culture** being more collective, with more flexible time management, lower hierarchical distance, need for feedback and stronger meaning (versus Baby Boomers culture)	++
CHALLENGES OF EFFICIENCY AND DIGITAL MEANING (Customers, employees, shareholders, suppliers)	15	**Global data strategy and governance**	+
	16	**Balance between human and digital**, automation of interfaces when it is meaningful	++
	17	**Customization of digital interfaces, adaptable** according to needs and expectations	++
	18	**Fluidity and efficiency of interactions** between human and tech/digital	++
COHERENCE ISSUES FOR COMPANIES (Development in line with the various social, societal and environmental issues	19	**Coherent organization** (actions/words), **genuine, true** social and societal responsibilities	+
	20	**Organization maximizing the retention and development** of its human resources	+
	21	**Organization, Leaders, *process*[2], Supply Chain**	No impact
	22	**Development of remote work and autonomy**	++
	23	**Development of collaborative work** (including networking) to decide and act quickly	++
	25	***Simple and efficient S&OP process*[2]**	+
	26	Leaders: **role models, psychological safety basics, with a greater focus on soft skills**: change management, resilience, ability to create links, to give meaning	+
ALIGNMENT ISSUES FOR SHAREHOLDERS & GOVERNANCE	27	**Performance and Profitability**	++
	28	**Expertise for the development of corporate performance in line with** social, societal and environmental issues	+

a significant challenge in digital deployment. Here are a few remarks regarding the impacts indicated in yellow and marked with a single plus (+):

- Criterion 2: The lasting impact has not yet been demonstrated, so we must qualify it further.
- Criterion 5: The impact is indirect; allowing open communication of deviations encourages companies to take action.
- Criterion 8: The impact is also indirect; centralization compels companies to build trust to function effectively.
- Criterion 12: The impact is nuanced. Eliminating repetitive, intellectually demanding tasks can be very positive, but the overall impact depends on how we utilize the remaining time and how well the teams are prepared.
- Criterion 13: The assessment is nuanced here as well. Digital technologies, such as 3D printing, can lead to very local production, which could reduce global competition. Automation tends to decrease the impact of labor costs, reducing the risks associated with the relocation of blue-collar jobs. However, some white-collar jobs may still be relocated to low-cost countries for IT services, which can drive down salaries.
- Criterion 15: While digital technologies impact data governance, they do not address strategy through their deployment.
- Criterion 19: Easier and more instantaneous communication will drive companies toward coherence, but it does not guarantee authenticity.
- Criterion 20: Development is limited to the digital component.
- Criterion 25: The impact of digital can be very positive regarding efficiency, but it may have less influence on flexibility and simplicity, as it often involves more complex systems to achieve improvements.
- Criterion 26: The impact on the role model for digital application is limited; there should also be a greater focus on the soft skills required for digital implementation.
- Criterion 28: Expertise is clearly related to digital technologies.

References

Anand, B. N., Collis, D. J., & Hood, S. (2008). *Danaher corporation, case study*. Harvard Business School, https://store.hbr.org/product/danaher-corporation/708445

Acemoglu, D. (2024). *The simple macroeconomics of AI, MIT, shapping the future of work*. Economic Policy. https://shapingwork.mit.edu/wpcontent/uploads/2024/05/Acemoglu_Macroeconomics-of-AI_May-2024.pdf

Bajwa, F. (2024). *When you should (and shouldn't) have an AI leader in the C-suite, Russell Renolds insights*. https://www.russellreynolds.com

Barraud, B. (2020). *L'intelligence de l'intelligence artificielle*. hal-02327501v3, extract from L'intelligence artificielle—Dans toutes ses dimensions (L'Harmattan, 2019)

Bajwa, F., Jervis, T., & Richards, T. (2024). *Decoding the future: The RRA systems view on leading through AI transformation*. Russell Reynolds/Technology and Innovation. https://www.russellreynolds.com/en/insights/articles/the-rra-systems-view-on-leading-through-ai-transformation

Bihouix, P. (2021). *L'âge des Low Tech—Vers une civilisation techniquement soutenable* (Editions du Seuil, 2021).
Binvel, Y., Franzino, M., Guarino, A., Laouchez, J.-M., & Penk, W. (2019). *Future of work: The global talent crunch.* https://www.kornferry.com/insights/this-week-in-leadership/talent-crunch-future-of-work
Brown, S. (2024). *3 business problems data analytics can help solve analytics.* MIT Management Sloan School. https://mitsloan.mit.edu/ideas-made-to-matter/3-business-problems-data-analytics-can-help-solve
Brown, S. (2024). *To help improve the accuracy of generative AI, add speed bumps.* Artificial Intelligence/ MIT Management Sloan School. https://mitsloan.mit.edu/ideas-made-to-matter/to-help-improve-accuracy-generative-ai-add-speed-bumps
Bonnecaze, A., & Thiry-Atighehchi, K. (2025). Protéger et modifier les données: le casse-tête Blockchain, La Recherche Magazine, No. 580, January/Marcj 2025.
Christofi, M. (2024). The role of chief digital officer: Critical insights into an emerging field and road map for future research. *Journal of Business Research, 172*, 114390. ISSN 0148-2963. https://doi.org/10.1016/j.jbusres.2023.114390
Chu, J. (2024). *MIT engineers grow "high-rise" 3D chips.* MIT News. https://news.mit.edu/2024/mit-engineers-grow-high-rise-3d-chips-1218
Doguc, O. (2020). Robot process automation (RPA) and its future. In *Research anthology on cross-disciplinary designs and applications of automation* (pp. 35–58). https://doi.org/10.4018/978-1-6684-3694-3.ch003
Ehrenfeld, T. (2018). *Lean lessons from Tesla.* https://www.lean.org/the-lean-post/articles/lean-lessons-from-tesla/
Feng, C., & Ali, D. A. (2024). Levering digital transformation and ERP for enhanced operational efficiency in manufacturing enterprises. *Journal of Law and Sustainable Development.* https://doi.org/10.24857/rgsa.v18n3-101
Fleureau, G. (2024). *Étude: malgré des investissements croissants, les entreprises européennes peinent à atteindre la maturité numérique.* 12 septembre 2024. https://siecledigital.fr/2024/09/12/etude-malgre-des-investissements-croissants-les-entreprises-europeennes-peinent-a-atteindre-la-maturite-numerique/?
França Teles, N. C., Minoro Abe, J., Sestari do Nascimento, S., & Corrêa de Oliveira, C. (2022). The synergy of Lean Manufacturing methodology in the context of Industry 4.0: An integrative review (2022). *Research, Society and Development, 11*(13), e353111335449. CC BY 4.0. ISSN 2525-3409. https://doi.org/10.33448/rsd-v11i13.35449
George, M., O'Regan, K., & Holst, A. (2022) *Digital solutions can reduce global emissions by up to 20%. Here's how.* World Economic Forum. https://www.weforum.org/agenda/2022/05/how-digital-solutions-can-reduce-global-emissions/
Gordon, R. (2022). *An automated way to assemble thousands of objects.* MIT News. https://news.mit.edu/2022/automated-way-assemble-thousands-objects-1207
Guérin, C. (2023). *Pour aller dans le bon sens* (Le Cherche-Midi ed.).
Haffke, I., Kalgovas, B., & Benlian, A. (2016). The Role of the CIO and the CDO in an organization's digital transformation. In *Thirty seventh international conference on information systems, Dublin.* https://api.semanticscholar.org/CorpusID:33427304
International Energy Agency : I.E.A. (2023). *CO_2 emissions in 2023.* Annual report. www.iea.org
Inna, L., & Murat, B. (2020). *Digital transformation in manufacturing: Drivers, barriers, and benefits.* Higher School of Economics Research Paper No. WP BRP 107/STI/2020. SSRN: https://ssrn.com/abstract=3570342 or https://doi.org/10.2139/ssrn.3570342
Jackson, T., & Hodgkinson, I. R. (2022) *Dark data' is killing the planet—we need digital decarbonization.* Global Education Network. https://world.edu/dark-data-is-killing-the-planet-we-need-digital-decarbonisation/

Jackson, T., & Hodgkinson, I. R. (2022). *What is 'dark data' and how is it adding to all of our carbon footprints?* World Economic Forum. https://www.weforum.org/agenda/2022/10/dark-data-is-killing-the-planet-we-need-digital-decarbonisation/

Jackson, T. W., & Hodgkinson, I. R. (2022). Keeping a lower profile: How firms can reduce their digital carbon footprints. *Journal of Business Strategy, 44*(6), 363–370 (Emerald Publishing Limited, ISSN 0275-6668). https://doi.org/10.1108/JBS-03-2022-0048

Julia, L. (2019). *L'intelligence Artificielle n'existe pas* (1st ed.).

Kemen, E., Valkhof, B., & Tladi, T. (2024). *AI and energy: Will AI help reduce emissions or increase demand?* World Economic Forum, Energy Transition https://www.weforum.org/agenda/2024/07/generative-ai-energy-emissions/

Le Cun (2024). *Il manque aux machines le sens commun* (pp. 20 – 23). Interview for La Recherche Magazine, L'intelligence artificielle pour les sciences, No. 577

Li, P., Yang, J., Islam, M. A., & Ren, S. (2023). *Making AI less "thirsty": Uncovering and addressing the secret water footprint of AI models*, https://doi.org/10.48550/arXiv.2304.03271

Lin, Y., Chen, L., Ali, A., et al. (2024). Human digital twin: A survey. *Journal of Cloud Computing, 13*, 131. https://doi.org/10.1186/s13677-024-00691-z

Ligozat, A. L., & De Vries, A. (2024). *IA générative: La consommation énergétique explose.* Polytechnique insights. https://www.polytechnique-insights.com

Linner, T., & Book, T. (2023). *Automation, robotics, services: Evolution of large-scale mass customization in Japanese building industry.* Mass Customisation and Personalisation in Architecture and Construction, Routledge

Meunier, L., Vairet, F., Wong, C., & Lemarchand, J. (2022). Les sept questions qui fachent sur le métavers. *Journal Les Echos du Lundi.* https://start.lesechos.fr/innovations-startups/tech-futur/exemplaire-gratuit-de-notre-journal-metavers-the-new-place-to-be-1410104

Moshood, T. D., Rotimi, J. O. B., Shahzad, W., & Bamgbade, J. A. (2024). Infrastructure digital twin technology: A new paradigm for future construction industry. *Technology in Society, 77*, 102519. ISSN 0160-791X. https://doi.org/10.1016/j.techsoc.2024.102519

Oludapo, S., Carroll, N., & Helfert, M. (2024). Why do so many digital transformations fail? A bibliometric analysis and future research agenda. *Journal of Business Research, 174*, 114528. ISSN 0148-2963. https://doi.org/10.1016/j.jbusres.2024.114528

Petch, M. (2023). What is the future of 3D Printing? 80 Additive Manufacturing experts forecast 3D Printing trends for 2023. In: *The annual 3D printing industry executive survey.* https://3dprintingindustry.com/news/what-is-the-future-of-3d-printing-80-additive-manufacturing-experts-forecast-3d-printing-trends-for-2023-220261/#respond

Parsheley, L. (2024). *The hidden environmental impact of AI*. US/Science and Technology/Environment. https://jacobin.com/2024/06/ai-data-center-energy-usage-environment/

Precedence Research Website. (2024). *Robotic process automation market size, share, and trends 2024 to 2034.* https://www.precedenceresearch.com/robotic-process-automation-market

Ramesh, N., & Delen, D. (2021). Digital transformation: How to beat the 90% failure rate? *IEEE Engineering Management Review, 49*(3), 22–25 (1 thirdquarter). https://doi.org/10.1109/EMR.2021.3070139

Rayna, T., & Erwan Le Pennec, E. (2024). 4 Mythes sur l'IA générative, Polytechnique insights No. 8, L'IA & L'HUMAIN

Ribeiro, J., Lima, R., Eckhardt, T., & Paiva, S. (2021). Robotic process automation and artificial intelligence in industry 4.0—A literature review. *Procedia Computer Science, 181*, 51–58. ISSN 1877-0509. https://doi.org/10.1016/j.procs.2021.01.104

Richardson Gosline, R. (2024). *Human-first AI.* MIT Initiative on Digital Economy—MIT IDE Annual Conference Ep. 3. https://www.youtube.com/watch?v=y8i92aJfD3Q

Schreech, T. (1996). The Western Scientific Gaze and popular imagery in Later Edo Japan. The Lens within the Heart Cambridge University Press. *The Australian Journal of Art, 14*(1), 151–161 (page 67). https://doi.org/10.4324/9781315029313

Somers, M. (2024). *The risk of letting junior professionals teach AI to senior colleagues.* MIT. https://mitsloan.mit.edu/ideas-made-to-matter/risk-letting-junior-professionals-teach-ai-to-senior-colleagues

Stackpole, B. (2025). *AI has high data center energy costs but there are solutions.* MIT News/Sustainability. https://mitsloan.mit.edu/ideas-made-to-matter/ai-has-high-data-center-energy-costs-there-are-solutions?utm_source=mitsloanlinkedin_tf&utm_medium=social&utm_campaign=2025_01_15

Sternchuss, A. (2024). *Les chiffres 2023–2024 du marché de l'IA dans le monde, AI and globalisation are shaking up software developers' world.* The Economist, September 29, 2024

The Royal Society. (2021). *Computing for net zero: How digital technology can create a 'control loop for the protection of the planet'*. Climate Change/Science and Solutions/Briefing 2 based on the Royal Society's report Digital technology and the planet: Harnessing computing for net zero (see https://royalsociety.org/-/media/policy/projects/digital-technology-and-the-planet/digital-technology-and-the-planet-report.pdf). Accessed 23 Mar 2021

Thompson, N. (2024). *Artificial intelligence, quantum and beyond.* MIT initiative on Digital Economy—MIT IDE Annual Conference Ep. 6. https://www.youtube.com/watch?v=JYb6Fl0A6nA

Tounkara, T. (2020). *Automatisation Robotisée des Processus (RPA): Quels défis pour la transformation numérique des entreprises?* (pages 95 à 118/Approches Théoriques en Information-Communication (ATIC) 2020/1, Éditions DICEN IdF, 2020). https://doi.org/10.3917/atic.001.0095

Vinardi, C. (2013). *Le Lean : atouts, impacts et limites* (Vuibert ed.).

Vinardi, C. (2016). *Les équipes multiculturelles en entreprise : Les comprendre, les gérer, les développer*. Maxima Editions.

Vinardi, C. (2019). PhD Thesis on Industrial Engineering : *"Les défis du Lean à l'ère de la mondialisation et de l'industrie 4.0*". Université de Technologie de Compiègne. Français (NNT : 2019COMP2500) (tel-03201462)

Vinardi, C., Eynard, B., Fanio, E., Poulier, J. M. (2019). Amélioration de la performance opérationnelle de PME/TPE de l'industrie aéronautique française dans le cadre d'un projet national d'accompagnement. *Génie industriel et productique, 2*(2). https://doi.org/10.21494/ISTE.OP.2019.0424 (hal-04294268)

Vinardi, C. (2023). *Business strategy with Hoshin Kanri* SpringerBriefs in Applied Sciences and Technology. https://doi.org/10.1007/978-3-031-20963-5_1

Walsh, D. (2025). *A new look at the economics of AI.* MIT Management Sloan School, Artificial Intelligence. https://mitsloan.mit.edu/ideas-made-to-matter/a-new-look-economics-ai

Yokota, Y. (2009). A historical overview of Japanese clocks and Karakuri. In H. S. Yan & M. Ceccarelli (Eds.), *International symposium on history of machines and mechanisms.* Springer, Dordrecht. https://doi.org/10.1007/978-1-4020-9485-9_13

Yuji, Y., & Sandström, K. (2018) Karakuri IoT—The concept and the result of pre-study. In P. Thorvald & K. Case (Eds.), *Advances in manufacturing technology* (Vol. XXXII, p. 311). https://doi.org/10.3233/978-1-61499-902-7-311

L. Zhang, J. Liu, J., & Zhuang, C. (2024). Digital twin modeling enabled machine tool intelligence: A review. *Chinese Journal of Mechanical Engineering, 37*, 47. https://doi.org/10.1186/s10033-024-01036-2

Zewe, A. (2024). *Researchers demonstrate rapid 3D printing with liquid metal.* MIT News. https://design.mit.edu/news/researchers-demonstrate-rapid-3d-printing-with-liquid-metal

Chapter 6
Making Performing Combinations

Abstract In the previous chapter, we explored the key points of addressing Lean and Digital deployments separately in a meaningful way, against their selected performance criteria. In this chapter, we will first discuss the chronology of these two approaches: does it make more sense to implement Lean before or after Digital? Additionally, we will consider how to tackle both together. The second part of the chapter will focus on the organization and skills necessary for a combined deployment, concluding with the changes that this may require from companies.

Chapter 5 concluded with Digital's compliance with the doubly sustainable performance criteria I presented to you. By cross-referencing Tables 5.10 and 5.17, we could identify combinations that could optimize compliance and maximize impact. By highlighting the criteria that are below the target (indicated by: ++), we can focus on:

- Table 5.10, nine criteria: 5, 6, 7, 13, 14, 15, 17, 22 and 28.
- Table 5.17, fourteen criteria: 2, 3, 4, 5, 8, 12, 13, 15, 19, 20, 21, 25, 26, 28

If synergies can be created, theoretically **only four criteria remain**: 5, 13, 15 and 28.

? How can we transition from theory to practice and activate potential synergies? Is there a specific deployment order to consider?

6.1 Combinations and Their Impact on Performance

6.1.1 Lean Before Digital

As mentioned in Sect. 5.2.2, the **deployment of digital technology is strategic rather than technological.** To be effective, the digital strategy must consider the needs and maturity of customers and users while being integrated into the overall

C. Vinardi, *Sustainable Performance in the Digital Age*,
https://doi.org/10.1007/978-3-032-01252-4_6

company strategy. Consequently, it is better to adopt a dual approach that incorporates both *Bottom Up*[1] and *Top Down*[2] perspectives within a learning framework that should be flexible, easily adaptable, and aligned with other company objectives such as Sales and Growth, Safety, Environment, Quality, Delivery, Costs, and Management.

The *Hoshin kanri*[3] approach, which I include in the Lean System, addresses this need for a strategic approach that is both efficient, based on the criteria above, and flexible. It facilitates alignment and coordination among all levels of the organization, from Top Managers down to the Front Lines, and vice versa.

Therefore, implementing *Hoshin kanri*[4] as part of the Lean System before adopting Digital technology is an essential advantage.

Implementing a digital strategy requires active facilitation, and as we saw in Chap. 2, the Lean System provides the key managerial principles that apply to Digital deployment. Here are a few examples:

- Establishing and promoting digital standards—whether concerning services, products, or *processes*[5]—through Visual Standard Management and quick, iterative animation methods like Action Work Out.
- Enhancing digital standards using Continuous or Breakthrough Improvement.
- Addressing issues related to implementation using Problem Solving and Escalation techniques.
- Learning from the implementation *process*[6] by replicating successful outcomes and avoiding repeated mistakes through the PDCA cycle.

When the fundamental principles are not implemented, it can be a significant obstacle, as Eric Ciampi from Olivier Wyman points out. Conversely, when these principles are applied, results often exceed those achieved by industry leaders, according to David Briggs from Naarea. Although the contribution of Lean and their core concepts is well recognized, they are often insufficiently deployed into the Lean System format I have described. It can be tempting to take shortcuts and rely solely on the tools available. This observation is also noted by Olivier Dréan from Dassault System:

*"**Lean before Digital by the Leaders directly** (myself as COO). We have started to implement the basics of management that are the baseline of any Lean*

[1] A more precise definition is found in the Glossary (See No. 6).
[2] A more precise definition is found in the Glossary (See No. 7).
[3] A more precise definition is found in the Glossary (See No. 3).
[4] A more precise definition is found in the Glossary (See No. 3).
[5] A more precise definition is found in the Glossary (See No. 2).
[6] A more precise definition is found in the Glossary (See No. 2).

system: Performance Management, Standard & Visual management, Problem Solving Management"

From an interview of David Briggs with Carine Vinardi

"Lean is often a blocking point for Digital *when Lean basics are not deployed before."*

From an interview of Eric Ciampi with Carine Vinardi

*"**Lean Basics are key to support Digital deployment**. However, despite everything published for the last 30 years, it is not **sustainably deployed so well**, even in Industrial Companies. There is still **confusion between tools and lean principles (philosophy)**."*

From an interview of Olivier Drean with Carine Vinardi

In addition to management, there is a common starting point, as Digital transformation can influence products and services, as well as ***processes***.[7] To implement this effectively, start from the customer and then address market expectations and challenges appropriately while creating value. This approach is aligned with the principles of Lean, as outlined in Chap. 2. By beginning with the customer and moving up the value chain to suppliers, and concentrating on the needs of employees who deliver value, Lean helps identify activities that add value (AV) and those that do not (waste). **By eliminating non-value-adding activities, a Lean approach ensures organizations avoid digitalizing wastes, thereby avoiding unnecessary investments**.

The importance of implementing Lean is emphasized by Richard Vinardi from L'Occitane En Provence, Eric Ciampi from Oliver Wyman, Eric Le Bourdon from Labeyrie, and Vincent Di Lorenzo from Baumert. Luc Julia from Renault, an international expert in AI and a founder of several start-ups, also highlights this point. When Lean methodologies are not applied beforehand, it can lead to difficulties, particularly in transactional activities where waste may not be immediately apparent. Without identifying waste, organizations may struggle to fully understand their levels of performance and non-performance. Eric Ciampi from Oliver Wyman share this perspective as well:

[7] A more precise definition is found in the Glossary (See No. 2).

> *"My perspective is that we should not introduce a digital approach* ***without first evaluating or re-evaluating organizations and activities of Lean expectations****. Digitalization* ***could unnecessarily reinforce routines that are not optimized or need to be reinvented****. In this context, relying on digital technology may hinder productivity growth and the system's overall performance. Additionally, since the Lean approach encourages everyone to propose and implement improvements, if digital solutions are* ***too restrictive, they could require more financial and human resources than necessary****, ultimately affecting implementation timelines."*
>
> From an interview of Richard Vinardi with Carine Vinardi

> *"Lean before digital makes sense with basics such as 5S, safety standards, and macro-diagnostic of key processes.* ***Lean does not require investment first****, out of human time, whereas digital requires time and money." "Lean is* ***a prerequisite."***
>
> From an interview of Eric Le Bourdon with Carine Vinardi

> ***"Even when you develop a start-up, money is still crucial****, and you* ***must avoid waste****. Therefore, the basic but mandatory drivers to* ***implement first for performance are mainly Lean and a focused mindset****."*
>
> From an interview of Luc Julia with Carine Vinardi

> *"Especially in Transactional activities: Impatience can push to digital without having deployed Lean basics, which can lead to* ***digitalize wastes and can require investing more time in deployment****.*
>
> *It can be challenging* ***to understand what part of the savings could have been made without investing in digital."***
>
> From an interview of Eric Ciampi with Carine Vinardi

This observation holds true for both industrial and transactional activities.

Beyond the unnecessary expenditure, the Lean approach emphasizes maximizing the use of human resources before committing material resources. This strategy aims to avoid unnecessary spending. It applies to automation by encouraging the consideration of Low Cost/Tech Automation approaches (such as *karakuri*) before moving on to High-Tech options (both traditional and digital). It also suggests utilizing small, flexible equipment instead of large, versatile but restrictive machines. This approach **aligns with achieving sustainable performance, and in this regard, Lean principles are also beneficial for digital initiatives.**

From a cultural perspective, the Lean approach is a practical method that allows you to start small and learn before considering a full-scale deployment. This "learning by doing first" strategy, often called "Test and Learn" in digital contexts, will be explored in more detail in Chap. 7, where we discuss how deployment and its sustainability are handled similarly.

Test safely and securely, test small et learn. Moreover, like digital initiatives, the Lean approach emphasizes strong collective orientation, required cooperation, flexible time management, and a low hierarchical distance. It also highlights the importance of strong feedback. **Thus, committing to a Lean approach in advance will help facilitate the cultural adaptation required for digital initiatives.**

Finally, the deployment of digital has much to learn from the deployment of Lean, especially in its implementation difficulties. The failure factors highlighted for Lean are common in many ways to Digital, whether for the Leadership part (lack of *vision*,[8] prioritization and focus, lack of Team Empowering, etc.) or the Tactical part (focus on tools, lack of standardization, the disconnect between the maturity of the teams and the tools deployed, etc.). The fears expressed about the deployment of Digital, such as the fear of losing one's job, not having the skills, or changing one's habits, are reminiscent of those expressed for Lean. The same goes for the conditions for success.

When a Lean approach is properly implemented before adopting digital solutions, it enhances the learning cycle and facilitates sharing experiences. This was empirically confirmed by at least seven people I interviewed (see Table 5.16). **Integrating Lean beforehand can be crucial, as it generates an acceleration effect that can lead to significant benefits.**

Adopting a Lean approach before implementing a digital strategy has significant advantages in terms of strategic management and facilitation. It emphasizes a customer-centric focus and adds value by optimizing expenses, facilitating cultural adaptation, and fostering a learning environment.

To truly benefit from Lean, it must be effectively implemented. However, since it is often viewed as less "trendy" and can suffer from negative perceptions, prioritizing its impact during digital transformation can be challenging.

On the surface, initiating a digital transformation may seem easier, as noted by Eric Ciampi from Oliver Wyman and Stéphanie Perchet from Adoptin, who shared insights in their recent testimonial.

[8] A more precise definition is found in the Glossary (See No. 4).

"By experience, it is ***easier to engage people in Digital than Lean transformation****:*

- *Is it because* ***Digital is already part of people's daily lives****, including operators'?*
- *Is it because, contrary to Lean,* ***no big labor productivity / no "famous" restructuring plan*** *can be directly connected to Digital deployment (yet)?"*

From an interview of Eric Ciampi with Carine Vinardi

"From my experience, in many transactional businesses, ***Lean and Agile principles are not deeply embedded.*** *This is often because top leaders either* ***fail to recognize inefficiencies and the associated impact on internal performance*** *or* ***feel they have 'enough money', choosing to hire additional staff to address problems rather than tackling the root causes*** *. (...)* ***Transformation Team / Office is trendy for Digital.*** *However, if we look at content, it is very close to Lean."*

From an interview of Stéphanie Perchet with Carine Vinardi

Nevertheless, there are a few reasons for optimism. Industrial companies that have already committed to Lean practices can benefit by reevaluating their approach to adopt a more systemic methodology if they haven't done so already. The same applies to transactional companies that have implemented only a few key elements of the Agile approach; they must integrate these into a cohesive, comprehensive system. Additionally, as Stéphanie Perchet from Adoptin pointed out, some Lean principles are now integrated into other frameworks related to digital transformation.

It is important to note that at the academic level, publications on the combined Lean principles and Digital technologies are relatively recent, although this topic is gaining momentum (Scriven et al., 2024). Several studies highlight the role of Lean in implementing Digital strategies, emphasizing its significance prior to Digital adoption:

- Lean is a **required foundation** for Digital implementation (Chiarini et al., 2021), **particularly through enhancing the understanding of internal *processes***.[9] For 75% of companies, **understanding internal *processes***[10] helps organizational efficiency and is essential to start (Rodrigues Terra et al., 2023).
- Lean is a **facilitator for implementing** digital solutions, making them **more accessible** (Naciri et al., 2022), (Scriven et al., 2024).
- Lean provides a **competitive advantage** in the Digital era (Gatell & Avella, 2024).

[9] A more precise definition is found in the Glossary (See No. 2).

[10] A more precise definition is found in the Glossary (See No. 2).

However, it is noteworthy that, similarly to the combined deployment of these approaches, the existing **publications do not give clear implementation guidelines.**

6.1.2 *Digital Before Lean*

Some companies begin deploying digital solutions without first considering Lean methodologies (or similar approaches).

? Does this make sense? If so, in which cases?

This situation often applies to companies that develop digital products or services, particularly start-ups. But which are these companies? Start-ups typically, even if there are no specific figures on their share in the development of solutions. Start-ups have three key characteristics (Vinardi, 2019):

- They are relatively young, usually existing for only a few years.
- They bet they make on innovation and its market potential.
- Their innovations are primarily related to new technologies.

A start-up does not necessarily aim to be the best or the most profitable; its primary goal is to be the first in a market. Consequently, start-ups often lack a performance culture that encourages the implementation of Lean. You can argue that it is possible to perform without Lean training by developing the key management and facilitation skills. My grandmother, a shopkeeper and a savvy businesswoman, was a personal example of this principle.

However, if you want to secure funding and successfully penetrate the market, you will need to promote a culture of performance. This entails learning the principles of Lean on your own or applying internally acquired insights that others have already discovered. Notably, entrepreneurs like Luc Julia from Renault have emphasized that adopting Lean practices from the outset is crucial for performance, even in the context of a start-up.

We have seen that the fundamentals of Lean leadership are needed for effectively deploying digital initiatives in both transactional and industrial companies. To engage in a sustainable performance approach, it is ideal to start implementing Lean practices—particularly those related to team facilitation and management—before introducing digital tools, or at the very least, to do so concurrently. If Lean is introduced simultaneously with digital initiatives, there may be a risk of performance loss due to the challenges of simultaneous learning.

In addition to the leadership aspect, there is also the tactical aspect. In my interviews, I shared insights with David Briggs about the experience of Naarea, a deep tech company. It is important to note that this company aims to construct its first Nuclear SMRs (Small Modular Reactors) by 2030, so it is not yet in production. The fundamentals of Lean Leadership were established before introducing Digital initiatives, aligning with what I have just shared.

However, early on, Naarea formed a partnership with Dassault System to leverage its digital platform and develop a Digital Twin model. Given that small nuclear reactors are complex and specialized products, the company aimed to use digital capabilities to enhance the speed and quality of development processes, facilitate the sharing of explicit knowledge, and be ready to improve production efficiency. This situation illustrates that **Digital initiatives were initiated alongside or even before Lean, particularly in the tactical aspects, although some methodological elements are still being implemented.**

Let's shift our focus from the product or service being developed to examining the performance of the company's internal *process*[11] and consider the situations where it may be beneficial to implement digital solutions before applying Lean methodologies.

We've highlighted the advantages of digital tools, subject to selecting the right *processes*[12] for digitalization. To do this effectively, it's necessary to conduct an assessment—even a simple one if there is only one item to evaluate. This ensures that we do not digitize waste that can be eliminated.

? Are digital tools be beneficial for upstream diagnosis?

Often, conducting a Value Stream Mapping (VSM) before its digital version @VSM is more economical and efficient, and even more so, for transactional activities. However, certain situations can be challenging to analyze, such as large industrial sites or longstanding structures with complex and automated *processes*[13] in environments that restrict staff access. In these cases, connected equipment can provide valuable data for analysis. **Therefore, it may be worthwhile to consider digital diagnosis tools— partial or complete—before implementing Lean.**

Let's go further upstream and look at training, particularly focusing on **sharing explicit knowledge—knowledge that is written, formalized, and can be digitized**. Before the 2000s and the rise of the internet, theoretical training was done with books and paper publications obtained from bookstores and libraries. The content was chosen beforehand, and since publishing has a cost, it results in potentially higher quality material. However, access to this theoretical knowledge was limited, often restricted by geographical disparities, as not all cities and villages had access to the same books. Consequently, disseminating theoretical knowledge was a physical and directive *process*,[14] flowing from the author to the reader.

With the advent of the internet, the distribution of books broadened significantly. While books are not free, @book versions have become increasingly available, enabling physical and virtual access and so reducing geographical disparities. Additionally, the first interactive and engaging training courses—inspired by the

[11] A more precise definition is found in the Glossary (See No. 2).

[12] A more precise definition is found in the Glossary (See No. 2).

[13] A more precise definition is found in the Glossary (See No. 2).

[14] A more precise definition is found in the Glossary (See No. 2).

world of video games—were developed on web platforms. However, they were often constrained by the capabilities of existing networks.

With the advent of digital technology and advanced network infrastructures, smartphone apps have opened up far greater possibilities than during the earlier days of the internet. The ability to acquire explicit knowledge through interactive, easily accessible, and asynchronous formats has become much simpler. Now, with a few clicks, one can quickly find content to meet our needs, without the hassle of carrying a physical book or using a computer to access a virtual module. In this context, I developed and tested digital apps starting in 2016, focusing on theoretical training in the company's Lean System principles. These modules incorporated gaming elements, making the learning *process*[15] both interactive and enjoyable. This approach proved highly successful, as it catered to the need for personalized training of younger generations while also appealing to older employees.

An *omnichannel*[16] content was also available in traditional formats on the intranet, using conventional media such as PowerPoint. This format has also been employed for other training purposes.

When training is universal, meaning it is not specific to a particular company, such as the principles of negotiation or ergonomics, it is possible to have pay Apps on dedicated training platforms that have developed these resources. For more personalized training needs, it is essential to create materials with **the help of experts who are skilled in the relevant field**, proficient in training, and experienced in producing digital media. Additionally, one must find a suitable platform to host the training, which is what I have done.

Consequently, **theoretical training in a company's Lean System is a use case that benefits from digital solutions before implementing Lean practices at ground level**. However, this discussion only addresses the explicit aspects of training; actual implementation requires more than just theoretical knowledge.

It is important to note that some studies have explored the potential of digital tools as enablers of Lean practices. However, they have not clearly outlined the specific reasons or methods (Scriven et al., 2024).

6.1.3 Lean and Digital Together

The matching relationship between Lean practices and digital technologies is an increasingly discussed. Firstly, because many industrial companies have already adopted a Lean approach, although with varying degrees of success, they are now looking for effective ways to integrate Lean with digital solutions. Secondly, some industrial and transactional companies have not yet implemented Lean but are interested in understanding the prerequisites and conditions necessary for successfully deploying digital technologies. Indeed, automating/digitalizing a low- productivity

---

[15] A more precise definition is found in the Glossary (See No. 2).

[16] A more precise definition is found in the Glossary (See No. 10).

task can simply speed up poor performance: leaders may discover that human workers can compensate for variability and instability, whereas sensors and software can't.

We have seen that the basic principles of Lean management significantly enhance the implementation of Digital strategies. Lean encourages us to start from the customer's needs and then identify value-added activities and areas of waste. This approach improves our ability to determine what and how to digitize (product or service) or digitalize [*process*].[17] Following this analysis, Digital technology can effectively address waste, as Eric Le Bourdon from Labeyrie explains.

"After implementing Lean basics, Digital can help ***identify and manage waste****."*

From an interview of Eric Le Bourdon with Carine Vinardi

Finally, this approach allows us to address better the cultural changes associated with digital transformation while minimizing unnecessary expenses without activation.

By concentrating on the customer and Added Value, Lean can help identify the key questions regarding how digital tools can effectively meet people's needs (Xu et al., 2021)

- What unique new value could this technology create that otherwise would not be possible?
- How could this technology enhance our organization's overall performance?
- Will this technology augment our people and enable them to engage in a safer, more productive, more satisfying way?

Lean also encourages asking questions, which is especially valuable when crafting a request for a Gen AI engine, such as Mistral AI's Chat, OpenAI's ChatGPT, or DeepSeek. Various books and prompt training courses introduce the acronym "CRAFTIK," which stands for Context, Role, Action, Format, Tone, Information, Kaizen. The notion of "Kaizen," a Japanese concept, has recently been incorporated and refers to the philosophy of Continuous (Incremental) Improvement, discussed in Chap. 2 (without using the Japanese term, which is sometimes misconstrued). "Kaizen" suggests rephrasing your prompt if the answer is incomplete (Continuous Improvement) or encouraging the AI to ask you to clarify questions to enhance its responses.

It is **necessary to align their maturity levels to ensure that Lean and Digital can sustainably support each other**. This approach has already been demonstrated with the Lean System, which emphasizes the importance of coherence in management, behavior, and methodology (we will revisit this in Chap. 7). For Lean to effectively facilitate its implementation, the management and behavior must take precedence over methodology (Vinardi, 2013, 2019). **Similarly, the basics of Lean must be**

[17] A more precise definition is found in the Glossary (See No. 2).

established before deploying Digital solutions to provide support. This strategy helps prevent the implementation of overly complex solutions that may confuse users, as Stéphanie Perchet from Adoptin rightly points out:

> *"We must be cautious* ***not to let Digital/AI tools become a Ferrari handed to a team with the maturity of a young driver*** *—unable to fully leverage its potential."*
>
> From an interview of Stéphanie Perchet with Carine Vinardi

The importance of the combination was highlighted by Daniel Hager from Hager and Olivier Drean from Dassault System:

> *"I believe that Lean and Digital technologies complement each other well."*
>
> From an interview of Daniel Hager with Carine Vinardi

> *"The performance potential is also* ***huge in each case: with Lean Deployment alone, with Digital Deployment alone, and in combination.****"*
>
> From an interview of Olivier Drean with Carine Vinardi

Olivier Drean's recent testimonial raises an important issue that often arises when Digital is implemented before Lean, or at least before Lean has fully reached its potential. This issue is performance attribution. If we neglect to establish the fundamentals of Lean before introducing Digital, any performance improvements may incorrectly be credited to Digital. **However, if we first implement the fundamentals of Lean, it will be easier to identify the performance gains brought about by Digital and how it can work with Lean.**

Similarly, Delphine Asseraf from Harvest and Luc Julia from Renault emphasized the importance of balancing Lean and Digital approaches in a pragmatic and coherent manner:

> *"With* ***Digitalization and Lean is always a question of finding the right balance*** *between: the top Company Objectives, the industrial maturity, the technological drivers and their maturity, the culture of the organization, the*

> *skills of the people, the speed and deepness of the change introduced and it's adoption and the operability and sustainability of the new model."*
>
> From an interview of Delphine Asseraf with Carine Vinardi

> *"I believe in* ***hybrid solutions with a pragmatic approach****. It makes no sense to "use a sledgehammer to crack a nut". We should not plan to do everything manually or digitally."*
>
> From an interview of Luc Julia with Carine Vinardi

Digital tools can enhance Lean practices in a positive cycle. Let's examine some key examples:

- **Digital Value Stream Mapping (VSM):** I talked about Value Stream Mapping as a powerful technique used to identify added value (AV) and various types of waste while organizing an action plan with stakeholders. VSM is typically conducted in an Action Work Out, where "pencils and paper" are the only materials needed. It is done by going to ground level with the people involved in the *processes*[18] at least with recognized individuals when the scope is large, along with leaders with the authority to make decisions, and one or more facilitators (Vinardi, 2013). This straightforward approach helps to concentrate on the right areas and align teams on the necessary actions by prioritizing the quickest and most logical tasks before addressing the more complicated ones, which can often have a greater impact (Vinardi, 2013). When **flows and *processes*[19] are complex or the most challenging tasks cannot be tackled using conventional Value Stream Mapping (VSM)**, digitizing **VSM** in a later phase can be beneficial. This will lead to data collection for further analysis, which can then be used to define a simulation model based on algorithms, similar to what some Digital Twin models have.
- **(Smart) *Process*[20] Control & Inspection:** Delivering high-quality products and services is a fundamental objective for any company. Achieving this goal must also be cost-effective for the company and its customers. This means the appropriate amount of human and material resources must be used. In this context, as discussed in Chap. 2 regarding Jidoka, it is essential to prevent defects from occurring by filtering them. If a defect is produced, it should be addressed as soon as possible at source, and if necessary, a final inspection should be conducted before delivering the product to customers.

---

[18] A more precise definition is found in the Glossary (See No. 2).

[19] A more precise definition is found in the Glossary (See No. 2).

[20] A more precise definition is found in the Glossary (See No. 2).

When designing filters, consider them at the Design stage and implement as many simple and cost-effective solutions as possible. This could involve manual or visual control of key parameters, or the activation of error proofing systems based on a Karakuri. However, these approaches may have limitations, particularly when dealing with complex or multi-parameter systems.

Once you have maximized results with minimal resources and gained a firm understanding of your equipment, it may be worthwhile to integrate or activate any available sensors or actuators to enable data collection, facilitate automated control, and allow for prompt reactions to any deviations that occur, helping to prevent or at least mitigate potential problems.

The same approach will apply to the final inspection, which should be designed in the simplest way possible for the initial implementation. This inspection can be completed using sensors and actuators that facilitate data collection through digital inspection tools, such as visual and dimensional recognition. These tools will also aid decision-making by isolating potentially non-compliant products or services.

- **(Smart) Predictive maintenance** is for companies that rely on numerous machines or high-capacity equipment, expensive, specialized, and hard to replicate, that must operate reliably 24/7. Implementing autonomous, preventive, and corrective maintenance (as described in Chap. 2) aims to reduce the frequency and duration of downtime. Establishing is crucial for operational efficiency.

  However, the frequency of preventive maintenance operations can be reduced, particularly to what is just necessary, if we predict when components need to be replaced based on their actual wear rather than following a set schedule. This is the aim of **predictive maintenance**, which requires an organization to **have mature traditional maintenance *Processes*.**[21] When an organization reaches this level of maturity, it is feasible to implement predictive maintenance. Sensors will collect data to determine optimal timing for performing preventive maintenance tasks.

- **Digital Problem Solving**: I am tackling a sensitive topic here, as it involves a managerial action that humans, rather than machines, should fundamentally understand.

  The aim is not to digitalize Problem-Solving facilitation—at least not until a substantial level of maturity on the subject has been achieved—but rather to support it. This is a common misconception, and I will explain why.

  Digital screens are now flourishing in companies, replacing traditional whiteboards. Whiteboards were designed to be wiped clean when information was managerial, for example, and did not necessarily need to be kept. Conversely, when whiteboard data, such as production quantities, needed to be kept, it had

[21] A more precise definition is found in the Glossary (See No. 2).

to be recorded manually before being entered into a computer. **Automating data collection makes sense here, but only if you clearly understand what needs to be automated and why [which requires maturity in the *process*].**[22]

Today, we've reached the opposite extreme by collecting almost all data, which we then store in the Cloud in case we might use it later. Unfortunately, much of this data ends up as "dark data". Moreover, the digitization of management has made it harder to quickly and visually understand what is happening within the organization and identify problems. Therefore, **facilitation of problem-solving** *processes*[23] **and other managerial routines should ONLY be digitized after they have been adequately implemented and understood "physically".**

On the other hand, **when the collected data is relevant and thoroughly analyzed, digital solutions can assist in:**

- **Logging Previous Issues**: Similar past problems can help with conclusions and tests. This is sharing explicit knowledge, as humans cannot remember everything, as in the use cases of AI illustrated in Fig. 5.9. Additionally, potential actions can be suggested, allowing people the final say in interpreting results and deciding on next steps, depending on the issues.
- **Identifying new problems** that can be hard to detect, such as those arising from rapid change or the complexity of numerous parameters. Furthermore, these solutions can lead to possible causes and solutions to be tested, giving people the final say once again.

The same applies to all managerial rhythms, including the most strategic ones like *Hoshin kanri*[24]. **Digital solutions should assist in areas where automation is worse or not as good.** For the facilitation, it is better to digitize *Processes*[25] only after you have established or acquired sufficient knowledge and maturity in the subject. This will help identify what is worthwhile to digitize. By doing so, digitalization will be better understood and more readily accepted, and it will unlock its full potential. Other examples of this approach include Digital Kanban and Digital 5S.

Finally, and as I discussed in Sect. 6.1.2, the link between Lean practices and digital transformation is more prevalent in research on industrial companies than on transactional ones. This is not surprising, as Lean methodologies are more commonly implemented in industrial settings, where specific cases of Industry 4.0 and Industry 5.0 have been thoroughly analyzed. This has enabled an academic exploration of the prerequisites for integrating digital technologies with Lean practices. The combination of Lean and digital approaches is also examined from other perspectives:

---

[22] A more precise definition is found in the Glossary (See No. 2).

[23] A more precise definition is found in the Glossary (See No. 2).

[24] A more precise definition is found in the Glossary (See No. 3).

[25] A more precise definition is found in the Glossary (See No. 2).

- **The car industry as a reference for the combination**
  - The automotive industry is determined as the best strategic alignment between Lean and I4.0 technologies (Ozbiltekin-Pala et al., 2024).
  - Automotive is one of the sectors where Ind4.0 has been adopted the fastest linked to alignment between education, skills and effective planning of investment, linked to maturity of Lean (Ozbiltekin-Pala et al., 2024).
- **The need for coherent maturity between Lean et Digital**
  - Reach a higher level of **maturity by combining** (Dos Santos et al., 2020)
- **Lack of academic understanding on the combined implementation**
  - While the combined potential is understood, there is still poor understanding of "how" companies should embrace digital transformation to make it successful and sustainable with Lean and "how" the coexistence can be achieved (Rossini et al., 2021), (Rodrigues Terra et al., 2023).
  - Since Lean and Industry 4.0 are not fully applied in every industry, t**he number of experts who are establishing the relationship between both concepts is very low** (Ozbiltekin-Pala et al., 2024), (Buer et al., 2021)
- **The role of Lean practices and the time needed for its deployment**
  - Several studies have proposed **Lean as prerequisite** (Buer et al., 2021)
  - Lean can support with two patterns (Rossini et al., 2021):

    Sustaining digital transformation is typical for companies that show high commitment to Lean: deployment and consolidation of a series of **small and incremental projects.**

    Disruptive digital transformation emerges from companies positively committed to Lean: **few and large digital steps that imply a disruptive and radical change** in the company system. It requires bigger investment organized in complex and integrated plans.
  - Lean System is not something that leaders can whip up in time for the next quarterly report. It may take years—even decades—to unlock all the latent talent in an organization and deploy it to its fullest possible effect (Xu et al., 2021). If leaders take shortcuts and decide that "the tools alone will get us where we need to go," **the impact will likely evaporate in months** (Xu et al., 2021).
- **How Digital changes Lean Systems** by modernizing it to make is more performant (Xu et al., 2021; Naciri et al., 2022; Teles et al., 2022)

- **Combined results**
  - To achieve the greatest benefit, **Lean and digital should be used concurrently.** Recent examples of companies which have **cancelled their Lean Program to focus on Digital technologies. Studies show that it is absolutely not recommended** (Buer et al., 2021)

    New technology trends extend traditional performance measures, combined with traditional approach (Lean), enables a step change in performance (Dos Santos et al., 2020; Roland Berger, 2024).
    Whereas most of the existing studies says that the highest performance comes from Lean (Dinis-Carvalho et al., 2023; Xu et al., 2021); others say that digital can double the gains obtained by Lean (Zhang et al., 2022; Roland Berger, 2024).

- **Changes to take into account**
  - The companies are **moving resources and investments, initially dedicated to Ind4.0, back to Operations, trying to increase the circularity of their** *processes*[26] **and reduce the impact on planet.** "Digital Lean Green" is anticipated as a novel concept fusing Ind4.0, Lean and sustainability principles (Gatell & Avella, 2024).

    This is precisely the subject of the book.

6.2 Organization and Competencies for Performance

6.2.1 Organization to Design, Deploy and Support

I propose approaching this chapter similarly to how I addressed Lean (in Sect. 5.1.4) and Digital (in Sect. 5.2.4) in previous sections. First, we will consider the general factors necessary for a combined deployment. Next, we will discuss the needs of the Front Lines, and finally, we will examine the organization at the Top Leadership level.

The combination of Lean and Digital involves at least three key themes to maximize a sustainable performance that encompasses social, societal, and environmental dimensions (beyond mere sustainability over time): Lean, Digital, and CSR (Corporate Social Responsibilities):

- Lean implementation requires Leadership commitment and specific resources to facilitate rapid change. As mentioned in Sect. 5.1.4, these resources, which I refer to as Lean Leaders for simplicity, can be **in-house, outsourced, or hybrid, depending on the company's size. Internal resources can be fully dedicated to Lean initiatives or shared with other functions** such as Safety, Environment, Quality, or Supply Chain.

[26] A more precise definition is found in the Glossary (See No. 2).

- The digital functions include not only traditional IT roles—such IT risks and mitigation, Help Desk, Material and Software Infrastructure Managers —but also specific roles tailored to digital needs, such as Data Engineer, Data Analysts, Digital Lean Leaders, Development Managers, Integration Managers, Digital Designers, I4.0/Digital Product Managers, Digital Help Desk, as outlined in Sect. 5.2.4. Depending on the size of the company, IT and digital resources can be **in-house, outsourced, or a hybrid, and they may operate as dedicated teams or be combined with IT or Lean resources**.
- **CSR is a topic we've outlined as an objective (see** Sects. 4.3.4 and 4.4**),** complete with KPIs to measure progress. While CSR is a critical focus, **Digital and Lean primarily serve as tools to achieve these objectives**. CSR encompasses various components, including environmental impact (the effects of the company on the external ecosystem), social responsibilities (internal to the company), and societal aspects (external to the company).

  Before 2015, CSR was managed by leaders, particularly the Top Leaders, who drove the initiative within their respective functions:

  - The combined Safety and Environment function has existed for decades, especially in industrial companies. The primary responsibility for managing environmental objectives fell under Operations, which included internal sites as well as suppliers and service providers.
  - Human Resources generally oversaw the facilitation of social and societal objectives, while the operational objectives were implemented by the relevant Top Leaders.

  Since 2015, with the establishment of the UN's sustainable development goals aimed at eradicating poverty and combating climate change, the theme of CSR has gained significant traction. This momentum has been further fueled by the CSRD, adopted by the European Union in 2022, along with anticipated regulatory requirements set to take effect in 2024. I have personally pursued specific training in this area as part of my role as an independent director. In conjunction with an increasingly aware European public (relating to economic balance), these regulatory constraints have prompted many companies to establish specific or combined CSR functions. These initiatives include **conducting CSR audits of suppliers and assessing GHG emissions**. However, as this book is being written (end of 2024 and into 2025), the urgency of these issues appears to have diminished among national leaders due to the prevailing geopolitical climate. Consequently, the implementation of CSRD directives seems to be on hold at the EU level in favor of prioritizing security concerns. As previously discussed, it is not the case in China with CSDS.

  Regarding the operational teams, also known as the Front Lines, it's important to note that they are not responsible for their company's CSR objectives. However, they do apply the standards that directly impact their daily lives. Additionally, they play a crucial role in the advancement of these standards because they are directly affected when issues arise that impact their physical health, such as pollution, as well as their psychological well-being.

In the Lean and Digital context, I combined Tables 5.7 and 5.15 into Table 6.1, emphasizing the needs of the Front Line Teams:

- The "Management and Facilitation" section shows a very strong convergence between Lean and Digital approaches.
- Similarly, the "Organization" section also reflects some convergence.
- However, the "Methods and Tools" section is more specific, with data primarily driving digital-related needs.

Similar to the initial deployment of digital, the support functions interacting with the Front Lines make it difficult to successfully combine the operational deployment of Lean and Digital if there is a proliferation of different stakeholders acting in succession without coherence. Moreover, it is better not to add more CSR representatives.

To support the Front Lines effectively, it is often necessary to make connections between the various functions at that level:

- Safety and Environment functions, typically, even within SMEs, can include CSR initiatives.
- Quality function is often present independently, but can sometimes be combined with Safety and Environment or Lean initiatives.
- If the Lean function is independent, it can be incorporated with the Digital function for better leverage. However, if Lean is already combined with another function, the integration is more difficult.
- The Digital function can also be paired with IT, provided that individuals adopt a business, people, and *process*[27] oriented approach.

It is essential to consider the specific constraints and priorities of the company, as well as the skills and interests of the available people, just as with Top Management.

The key is to align the objectives and strategic actions with *Hoshin kanri*[28] and to create versatile points of contact (or relays) to interface with the Front Lines. Ideally, these relays include Leaders and/or relevant functions that make sense for your organization. Organizational decisions are the responsibility of Top Leaders, who must determine whether to make or buy and how this choice changes over time. Additionally, it is worth noting that how functions are organized and versatile can be an attractive argument for Talent Retention among Leaders, as it allows them to expand their role within the bounds of their capabilities and interests.

This concept also applies at the highest level. As Daniel Hager from Hager emphasizes, the role of the CIO does make sense, even if, as Olivier Drean from Dassault System points out, this is not always clear in every organization:

[27] A more precise definition is found in the Glossary (See No. 2).

[28] A more precise definition is found in the Glossary (See No. 3).

Table 6.1 Needs of front line teams regarding Lean and Digital

<table>
<tr><th></th><th>LEAN/front line needs</th><th>Digital/ front line needs</th></tr>
<tr><td rowspan="7">Management and facilitation</td><td colspan="2">Understand contributions to Strategic Objectives[29] and make proposals</td></tr>
<tr><td colspan="2">To take on, formalize, correct, improve, and learn</td></tr>
<tr><td colspan="2">Participate in the definition, maintenance and development of processes[30] and standards (including Digital ones)</td></tr>
<tr><td colspan="2">Correct and refer standards (including Digital ones)</td></tr>
<tr><td colspan="2">Improve standards (including Digital ones)</td></tr>
<tr><td colspan="2">Participate in projects that make sense within the scope</td></tr>
<tr><td colspan="2">Understand and participate in relevant changes</td></tr>
<tr><td rowspan="5">Organization</td><td>Understand the Management line</td><td>Understand and set aside local needs when they don't align with collective needs</td></tr>
<tr><td>Be inspired on what is done elsewhere</td><td>–</td></tr>
<tr><td colspan="2">Having a trusting environment and daring to share data and problems with Management</td></tr>
<tr><td colspan="2">Be in control of the expectations of your position and scope (with digital solutions)</td></tr>
<tr><td>Meet with managers, talk to them about problems, whether solved or not, and their ideas</td><td>–</td></tr>
<tr><td rowspan="8">Methods and tools (including digital ones)</td><td>Understand your environment and the priorities chosen using the tools and methods</td><td>Understand the motivation for data management</td></tr>
<tr><td rowspan="7">Understand and participate in the implementation of tools and methods that make sense</td><td>Follow the rules of storage and cleaning</td></tr>
<tr><td>Provide the data and understand how it is used</td></tr>
<tr><td>Understand the main principles of development</td></tr>
<tr><td>Understand the main principles of integration</td></tr>
<tr><td>Contribute to feedback as a user</td></tr>
<tr><td>Understand and participate in the implementation of appropriate tools and methods</td></tr>
<tr><td>Follow the rules of storage and cleaning</td></tr>
</table>

(continued)

[29] A more precise definition is found in the Glossary (See No. 5).

[30] A more precise definition is found in the Glossary (See No. 2).

Table 6.1 (continued)

	LEAN/front line needs	Digital/ front line needs
		Ask for and receive help with problems about digital solutions
		Request and receive help with network, server, and storage issues

*"**Our CIO, who is an effective educator, provides digital training** and has done an excellent job of teaching within the group. However, management also has a responsibility to follow through."*

From an interview of Daniel Hager with Carine Vinardi

*"Digital transformation is often **considered an IT responsibility,** which **is the main reason for the failure of so many projects.**"*

From an interview of Olivier Drean with Carine Vinardi

? What organizational structure is precisely most effective at the highest level of a company to support the combined Lean and Digital deployment?

Similar to considerations at the front lines, this requires a **common-sense approach that considers the current structure versus the desired outcome, as well as the importance of teamwork**. Marie-Ange Giuseppin from Inetum also highlights this point:

*"What is important is **to collectively find a good balance** between **all the constraints** expressed by all the roles in a company."*

From an interview of Marie-Ange Giuseppin with Carine Vinardi

It is the responsibility of the CEO or the head of the organization to foster a culture of common sense and collaboration from the Group Executive Committee down. This maximizes the strategy performance with *Hoshin kanri*3

The CEO and the Board must define the optimal organizational structure to implement the strategy while considering the needs of the front-line employees to enhance their performance. This organizational structure **may vary with changes in strategy**.

However, in reality, many companies, including those at the highest levels, **continue to operate in silos**, as noted by Olivier Drean from Dassault System. Additionally, even with strong motivation, a lack of prior alignment can lead to divergent interests among functions, as highlighted by Eric Ciampi from Oliver Wyman and Marie-Ange Giuseppin from Inetum:

"As it is very often ***managed in silo*** *between IT, Operations and Lean,* ***significant losses of values happen at the interface of work.*** *Any changes need a strong support from top leadership. Without executive sponsorship transformation is set to fail."*

From an interview of Olivier Drean with Carine Vinardi

"Digital requests are very centralized, but their sponsors are different:

- ***CEOs*** *can have strong convictions.*
- ***CIOs and COOs are the main other sponsors, but they are difficult to align.*** *CIOs are focused on standardization, speed, and costs, while COOs focus on pragmatism, engagement, and impact. Saying who is accountable for savings is not easy."*

From an interview of Eric Ciampi with Carine Vinardi

"For our customers, several stakeholders are often involved in the digital strategy with different objectives and constraints resulting in a power game from which a fragile balance emerges: *the innovation director or CTO pushes for new technologies, not always mature and stable enough, the CIO pushes for rationalizing and securing the Information System thus slowing down innovation and the COO pushes for best of breed solution that can solve their operational problems as quickly as possible, so against harmonization and rationalization."*

From an interview of Marie-Ange Giuseppin with Carine Vinardi

Let's examine the functions, apart from the CEO, that can leverage the deployment of Lean and Digital practices to enhance a dual sustainability performance approach, including CSR, as previously mentioned.

This discussion will use the titles outlined in Sects. 5.1.4 and 5.2.4, as well as those encountered in my experience related to CSR. The focus will be on two common options: a C-Level role or reporting directly to a C-Level role. Of course, there are other possibilities.

For the Lean function

- A C-Level function on the Group Executive Committee
  - Chief Performance/Competitiveness Officer or
  - Company Business Excellence Officer
- A function reporting to a C-Level: COO (Operations) or CFO (Finance)

For the Digital function

- A C-Level function on the Group Executive Committee
  - Chief Data/Digital/AI/ Officer or
  - Chief Transformation Officer
- A function reporting to a C-Level: mainly CIO (IT) or COO (Operations)

Moreover, there is an increasing presence of **CSR functions** at the highest levels of European companies

- A C-Level function
  - Chief Sustainability Officer or
  - Chief CSR Officer
- A function reporting to a C-Level: CHRO (HR for Human Resources) or Group General Counsel (Legal) or COO (Operations) for Sustainability versus Environment.

Choosing an organization even before the advent of Digital and CSR was already a subtle art, but the *process*[31] has become increasingly complicated. The proliferation of functions has made it challenging, and **Group Executive Committees cannot be expanded infinitely; a maximum of ten reports to the CEO is manageable** (Vinardi, 2016). Additionally, a simple and clear organizational structure and effective coordination among team members make for motivated teams. The more individuals involved, the more complex the coordination becomes.

While adding new functions at the Company Executive Committee level can **show the company's ambitions in the market, it only makes sense if it operates effectively within the organization**. Various combinations at this high level can be

[31] A more precise definition is found in the Glossary (See No. 2).

viable. Reporting to C-level executives often involves these combinations. A common approach at the C-level is adding Lean and Digital roles, such as:

- A Chief Performance and Digital Officer or Company Business Excellence and Digital Officer
- Or a Chief Transformation Officer, who oversees Lean, Digital, and other transformations

Eric Ciampi from Oliver Wyman, also supports this perspective as well as Stéphanie Perchet from Adoptin:

> *"At the beginning of Digital deployment, Lean/opex resources are often been used to support digital actions, but as the needs increase, separate teams are created. **They can work together under the same head.**"*
>
> From an interview of Eric Ciampi with Carine Vinardi

> *"Transformation Team / Office is trendy. However, if we look at the content, **it is very close to Lean**"*
>
> From an interview of Stéphanie Perchet with Carine Vinardi

Finally, Digital technology can be divided into two main approaches: an industrial one, often called Industry 4.0, and a transactional one, particularly concerning customer interfaces. Coordination between the two digital approaches is crucial.

Once the organization's structure is defined, it is important to ensure that team training and development align with both their daily needs and the company's *Strategic Objective*.[32] To achieve this, it is necessary to identify the relevant Skills and Competencies.

6.2.2 Skills and Competencies

To **train effectively, you must understand what you need to learn, why it is needed, and the specific gap you aim to fill in your knowledge** (where you currently are and where you want to go).

- The lowest starting level is "I don't know anything," which means you possess neither theoretical knowledge nor practical experience in the subject.

[32] A more precise definition is found in the Glossary (See No. 5).

- On the other hand, the highest level represents advanced expertise, characterized by in-depth theoretical knowledge, extensive practical experience, and the ability to teach the subject. Some experts suggest "10,000 hours" of training is the minimum needed to achieve excellence.

Between the two boundaries, development should occur gradually, encompassing at least two additional levels: "beginner" and "intermediate." Both levels should address theoretical understanding (grasping the mechanisms of operations) and practical application (consistent practice across various scenarios).

In the previous chapter, I discussed a specific aspect of CSR in the context of sustainable performance that underlines this book. Industrial companies have long had teams trained to address environmental constraints relevant to developed countries.

Issues such as air, water, and soil pollution have been considered for some time. CSR led to the emergence of **new roles, particularly related to the time to fulfill these responsibilities. However, not all of these roles need new skills**. For instance, CSR auditors ensure suppliers comply with the company's established requirements. In contrast, some roles, like those assessing GHG emissions, require new skills, as they are relatively recent.

Lean, which emerged several decades ago, has what is known as the Skills and Competencies Matrix. Some standard skills are found everywhere, often aligning with the key elements of Leadership and Tactics defined in Chap. 2. However, there are variations, particularly in the tools and methods, and between transactional and industrial activities.

It is important to note that companies frequently use external assistance from consultants to start their Lean journey. These consultants should help to identify and define the necessary skills, which influence recruitment decisions following organizational choices.

Currently, there is no clear list of "Skills and Competencies" for Digital technologies. In almost every interview, AI and generative AI came up as significant topics.

However, what does it truly mean to train for AI? Does it involve defining a need, developing a model and an algorithm, or integrating it?

I completed a training course at MIT on "Business Strategy with AI" for several weeks, but I do not know how to define or integrate an algorithm.

This highlights a key challenge with **digital technology: implementing it means business stakeholders need diverse skills. The person who develops the algorithm may not understand how to define the need or manage transformations.**

This is why I emphasized the importance of **creating an ecosystem** to initiate the *process*.[33] Seeking external help is often more critical than it is for Lean methodologies. In many cases, it is better to involve several stakeholders.

[33] A more precise definition is found in the Glossary (See No. 2).

As Eric Ciampi from Oliver Wyman explained, this way will help companies determine which skills should be developed internally and which should be acquired externally:

> ***"Our customers do not yet have the skills**. They **train internally** after hiring, often with the help of consultants.*
>
> *While consultants can initiate digital transformation, **it is rarely sustainable**. Companies recognize that the critical and strategic aspect of such transformations **is the data utilized**. They **want to retain ownership of this data**. To truly maximize its value, you need to ensure a strong connection between data analysts and industry experts."*
>
> From an interview of Eric Ciampi with Carine Vinardi

It is important to note that implementing concrete physical solutions, while not simple, can be easier to understand in terms of expertise at the I4.0 level. This is because solutions such as exoskeletons, cobots, and automated guided vehicles (AGVs) can be initially sorted by solution.

References

Buer, S.-V., Semini, M., Strandhagen, J. O., & Sgarbossa, F. (2021). The complementary effect of lean and digitalisation on operational performance. *International Journal of Production Research, 59*(7), 1976–1992. https://doi.org/10.1080/00207

Chiarini, A., & Kumar, M. (2021). Lean six sigma and industry 4.0 integration for operational excellence: Evidence from Italian manufacturing companies. *Production Planning and Control, 32*(13), 1084–1101. https://doi.org/10.1080/09537287.2020.1784485

Dinis-Carvalho, J., Sousa, R. M., Moniz, I., Macedo, H., & Lima, R. M. (2023). Improving the performance of a SME in the cutlery sector using lean thinking and digital transformation. *Sustainability, 15*, 8302. https://doi.org/10.3390/su15108302

Dos Santos, V. A., Pierin Ramos, L. F., de Freitas Rocha Loures, E., & Deschamps, F. (2020). Lean & Industry 4.0 Integration: Proposed roadmap to assess maturity and select new technologies. In *Proceedings of the international annual conference of the american society for engineering management; Huntsville*

Gatell, I. S., & Avella, L. (2024). Impact of Industry 4.0 and circular economy on lean culture and leadership: Assessing digital green lean as a new concept. *European Research on Management and Business Economics, 30*(1), 100232. ISSN 2444-8834. https://doi.org/10.1016/j.iedeen.2023.100232

Naciri, L., Gallab, M., Soulhi, A., Merzouk, S., El Bhiri, B., & Di Nardo, M. (2022). Risks and opportunities within lean 4.0 systems: SWOT analysis. In *Proceedings of the 5th European international conference on industrial engineering and operations management, Rome, Italy, July 26–28, 2022*. https://doi.org/10.46254/EU05.20220204

Ozbiltekin-Pala, M., Kazancoglu, Y., Kumar, Y. A., Garza-Reyes, J. A., & Luthra, S. (2024). Analyzing critical factors of strategic alignment between operational excellence and Industry

4.0 technologies in smart manufacturing. *The TQM Journal, 36*(1), 161–177. https://doi.org/10.1108/TQM-07-2022-0232

Rodrigues Terra, J. D., Tobal Berssaneti, F., Quintanilha, J. A. (2023). Challenges and barriers in integrating Industry 4.0 and continuous improvement into an operational excellence plan. *Production & Manufacturing Research, 11*(1), 2288175. https://doi.org/10.1080/21693277.2023.2288175

Roland Berger, T. (2024). Digital—and AI—enabled plant performance Improvement.

Rossini, M., Cifone, F. D., Kassem, B., Costa, F., & Portioli-Staudacher, A. (2021). Being lean: How to shape digital transformation in the manufacturing sector. *Journal of Manufacturing Technology Management, 32*(9), 239–259. https://doi.org/10.1108/JMTM-12-2020-0467

Scriven, P., Ledwith, A., & Nagle, T. (2024). Towards a lean digital transformation research framework: A literature review. *Journal of Decision Systems*. https://doi.org/10.1080/12460125.2024.2354608

Teles, N. C. F., Minoro Abe, J., Sestari do Nascimento, S., Corrêa de Oliveira, C. (2022). The synergy of lean manufacturing methodology in the context of Industry 4.0: An integrative review. *Research, Society and Development, 11*(13), e353111335449 (CC BY 4.0, ISSN 2525-3409). https://doi.org/10.33448/rsd-v11i13.35449

Vinardi, C. (2013). *Le Lean: atouts, impacts et limites* (Vuibert ed.).

Vinardi, C. (2016). *Les équipes multiculturelles en entreprise : Les comprendre, les gérer, les développer*. Maxima Editions.

Vinardi, C. (2019). Ph.D., Thesis on Industrial Engineering, *Les défis du Lean à l'ère de la mondialisation et de l'industrie 4.0* (Université de Technologie de Compiègne, Français, 2019) NNT :2019COMP2500, tel-03201462.

Xu, X., Lu, Y., Vogel-Heuser, B., & Wang, L. (2021). Industry 4.0 and Industry 5.0—Inception, conception and perception. *Journal of Manufacturing Systems, 61*, 530–535. ISSN 0278-6125. https://doi.org/10.1016/j.jmsy.2021.10.006

Zhang, T., Shi, Z. Z., Shi, Y. R., & Chen, N. J. (2022). Enterprise digital transformation and production efficiency: Mechanism analysis and empirical research. *Economic Research-Ekonomska Istraživanja, 35*(1), 2781–2792. https://doi.org/10.1080/1331677X.2021.1980731

Chapter 7
Key Steps from the Gap to Fill to a Performing Action Plan

Abstract After discussing key elements and the advantages of integrating Lean principles with digital technologies for enhanced sustainability, this chapter will focus on the practical implementation of this integration. We will begin with the diagnostic stage and provide a detailed, step-by-step guide on implementing one or more pathways, up to the definition of a concrete and actionable action plan.

7.1 Capture the Gap to Fill

The first step in implementing Lean and Digital together is to understand the starting point of each and where you want to go, just as with any strategic initiative. In more formal terms, this means you need to:

- **Conduct an objective initial assessment** that closely reflects reality. This can be challenging because it requires the organization to recognize its current situation. Therefore, the people involved in this *process*[1] are crucial. Bringing in external help or new team members can provide valuable perspectives.
- **Define your future state** in terms of both *vision*[2] (what you aspire to be) and performance (what you aim to achieve).

These two states are needed because **the gap between your current state and your desired state helps to determine the resources** (human, material, and financial) required for the transition, as well as the time needed to achieve it.

[1] A more precise definition is found in the Glossary (See No. 2).

[2] A more precise definition is found in the Glossary (See No. 4).

C. Vinardi, *Sustainable Performance in the Digital Age*,
https://doi.org/10.1007/978-3-032-01252-4_7

7.1.1 *Using Roadmaps, Masterplans and Action Plans*

To conduct an initial assessment, a reference system is needed. Thus, we will start from a **Lean and Digital reference framework** incorporating the required **prerequisites and expectations.** The simplest way involves using a questionnaire to collect data. The questions must be specific, verifiable (with deliverables), and as clear as possible to facilitate comparison of the responses. To ensure a consistent facilitation routine, it is essential to request reports and their frequency (Vinardi, 2019).

An alternative option is to use an assessment tool to show both the initial and desired future states, assisting in the stages of progression. This "3-in-1" format is a diagnostic roadmap to guide deployment, the Path, considering the change cycle, coverage, depth, and cultural aspects (Vinardi, 2019).

Several steps are required to transition from the diagnosis to a coherent action plan for the teams involved (Vinardi, 2023):

(1) WHAT: Identify the gap between the initial state (where you are currently) and the future state (where you aim to be), particularly focusing on the selected Key Performance Indicators (KPIs) (Fig. 7.1).
(2) HOW: Establish priorities and choose a Path for progress. Multiple paths may exist, and adjustments might be necessary along the way. Various options were mentioned during the interviews, especially concerning digital transitions.
(3) HOW: Define the Key Steps needed for this path (Fig. 7.2).
(4) HOW (actionable): Integrate the Key Steps into a detailed Action Plan (Fig. 7.3).

We will look at the structure of the roadmaps in Sect. 7.2.2.

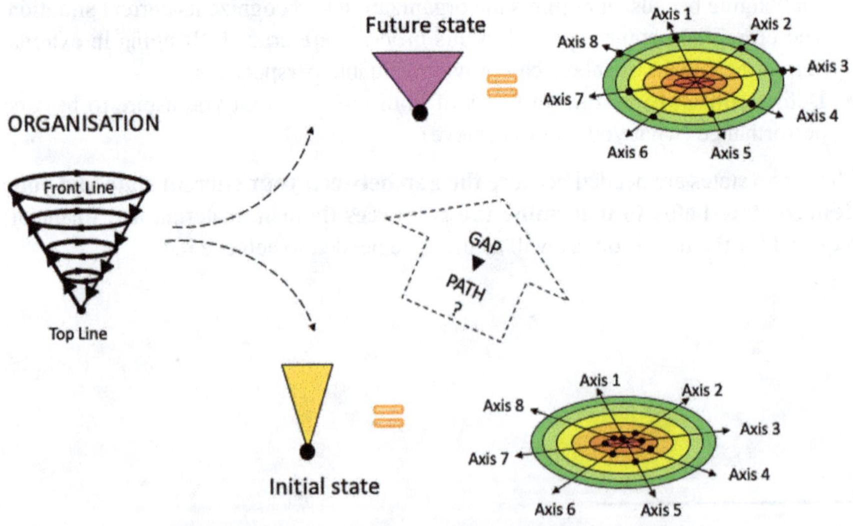

Fig. 7.1 Clarify the gap and the assessments (example with 8 axes)

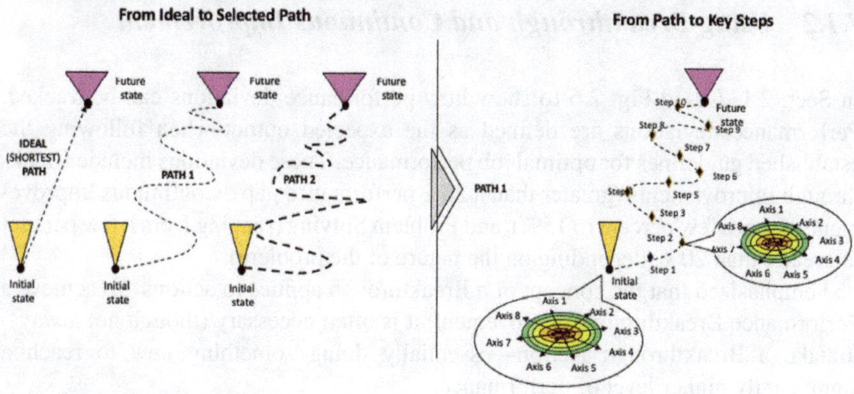

Fig. 7.2 From the path to key steps

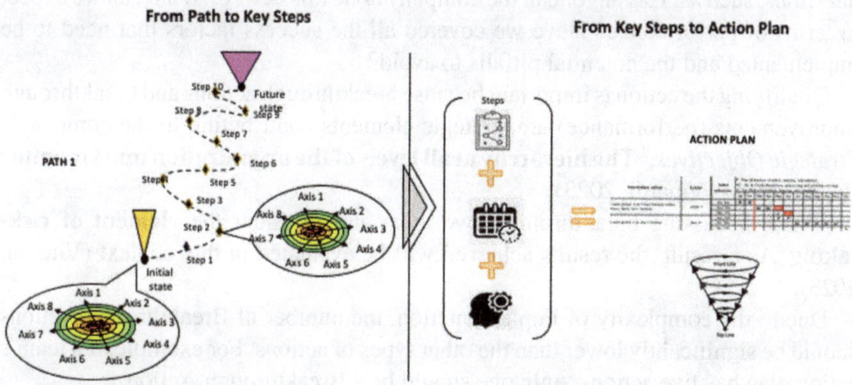

Fig. 7.3 From key steps to action plan

Note that this approach aims to provide detailed insight into implementing solutions using the content outlined in Chap. 5.

The action plan must include, at a minimum, the specific actions, the people responsible for the execution, a schedule that follows the PDCA cycle, and the expected results (Vinardi, 2023).

Moreover, each team's action plan should be on a single page. This concise format facilitates clear communication and ensures the list is manageable and easy to update.

It is important to note that it is okay to track reservations (using a priority matrix, for example, see Fig. 2.5). These can be added if progress is faster than anticipated.

7.1.2 Using Breakthrough and Continuous Improvement

In Sect. 2.1, I used Fig. 2.6 to show how performance deviations can be tracked. Performance deviations are defined as the expected output when following the established guidelines for optimal job performance. These deviations include Breakthrough Improvements (greater than a 20% performance gap), Continuous Improvements (from a few percent to 15%), and Problem Solving (ranging from a few percent to greater than 20%, depending on the nature of the problem).

I emphasized that the concept of a Breakthrough applies to actions. To achieve a Performance Breakthrough Improvement, it is often necessary (though not always) to take a Breakthrough Action—essentially doing something new to reach a significantly higher level of performance.

When teams make their action plans, **the leader and the team must understand the nature of the actions** they want to pursue. They should ask themselves simple questions, such as: Has anyone in the company done this before? What can we expect in terms of performance? Have we covered all the success factors that need to be implemented and the potential pitfalls to avoid?

Qualifying the action is important because breakthrough actions and breakthrough improvements (performance) are strategic elements contributing to the company's *Strategic Objectives.*[3] **The hierarchy at all levels of the organization must monitor these actions** (Vinardi, 2023).

When discussing breakthroughs, we must also consider the **element of risk-taking**. As a result, the results achieved will be evaluated in this context (Vinardi, 2023).

Due to the complexity of implementation, the number of Breakthrough Actions should be significantly lower than the other types of actions. For example, if a team's action plan has five actions, only one should be a Breakthrough Action.

> A good guideline for teams without experience is to aim for a "**one in five ratio**". As the team gains more experience and matures, its capacity to handle these actions may increase. **However, based on experience, this ratio rarely exceeds two out of five, even for the most experienced teams.**

Breakthrough Actions have the potential for the most significant rewards, but they also come with **increased complexity and a high failure rate**. Consequently, they may not always be the most motivating for a team. This is why it's important to balance these actions with **easier, less risky tasks, known as Quick Wins, that can still contribute to the overall goals.**

---

[3] A more precise definition is found in the Glossary (See No. 5).

7.1.3 Getting Started with Quick Wins

Quick Wins are evaluated using a priority matrix. These actions are crucial for maintaining motivation. Indeed, one effective way to motivate oneself is to monitor progress, and Quick Wins are designed to foster that sense of accomplishment. Even if the results are small, they can give the drive to tackle more challenging tasks.

Some people refer to these as low-hanging fruit, with its image of easy harvest.

? What do Quick Wins look like in the context of Lean deployment?

Implementing the performance basics outlined in Sect. 2.1.1—such as Standard Visual Management, Daily Performance Management, PDCA Cycle Management, and Problem Solving Management—can serve as Quick Wins. Even if these implementations are not perfect, they still yield fast initial results.

It is indeed possible to identify Quick Wins for nearly all the components of Lean, whether in Leadership or Tactics. This can be achieved in a couple of ways:

- By implementing Lean practices on a small scale, such as on a workstation or on a piece of equipment. A notable example is the 5S approach, which I briefly described in Sect. 2.1.2. This method is quick and straightforward to implement in a small area, and it proves to be highly effective when not considered merely a large cleaning action.
- By initiating the first stages of implementation. This approach applies to Daily Performance Management when it is developed collaboratively with the teams involved. It also helps to design roadmaps, which I will discuss further in Sect. 7.2.1.

? What do Quick Wins look like in the context of Digital deployment?

Deploying digital technologies requires investments and resource allocation. As such, discussing "Quick Wins" is not as easy as it is for Lean methodologies.

We have seen that prior implementation of Lean principles, particularly its fundamental aspects, facilitates the deployment of digital solutions, assuming these fundamentals are already in place, similar to those in IT.

Exoskeletons are an excellent example of quick wins regarding physical solutions. In cases where strenuous tasks cannot be eliminated or further optimized but still require human intervention, implementing an exoskeleton can be a practical solution to alleviate the burden on operators. For a few hundred euros and taking only a few weeks, mainly due to the ordering and delivery timeline, it is feasible to deploy exoskeletons in targeted positions.

Setting up collaborative robots (cobots) is also relatively straightforward. In cases where tedious tasks cannot be eliminated or optimized, implementing cobots can be a coherent solution, as outlined in the conditions of Table 5.13. With a relatively small investment in resources, typically for a few tens of thousands of euros, cobots can often provide significant returns.

Regarding **intangible solutions, such as artificial intelligence, ML can be seen as a "quick win," relatively speaking**. Implementing ML allows teams to become familiar with modeling and algorithm principles while achieving tangible results. Requiring an investment of a few tens of thousands of euros and only a few weeks of development using limited resources, it is easy to define and integrate an ML model for various activities. Here is a real-life example:

> We all have team members with strong expertise, and we want them to focus on the most complex problems and innovations to be tackled. This focus often motivates experts. However, these experts are sometimes asked to work on simpler problems that operational teams could manage independently if they had the right support. This is where ML comes into play. **A system can provide operational teams with a model by modeling part of the expertise captured from experts' existing data.** This model helps make informed suggestions based on their inputs, which can be verified. This approach frees up valuable time for experts—time that can be easily quantified—and helps operational teams develop greater autonomy in their problem-solving capabilities.

Machine Learning models can be quite sophisticated, and we will explore business cases further in Chap. 8. However, setting aside Quick Wins, I want to emphasize the often-overlooked power of ML. For instance, companies like Netflix depend heavily on ML. They state that their algorithms suggest 80% of the content watched by users.

When it comes to developing Deep Learning (DL) models, which represent the most complex form of AI, it is challenging to achieve Quick Wins. While some of the free templates provided by Generative AI (GenAI) are currently in use, not all applications yield Quick Wins. Simply using these text and image generation models for free is not enough; the results must justify the investment.

I'll let you evaluate whether you have seen tangible benefits in relation to the time and effort you have invested....

7.2 Succeeding and Learning with Execution

7.2.1 Starting and Scaling Up

It is better to start on a (safely) small, controlled scope before expanding to a larger scale. This approach has at least two advantages:

- First, it **helps ensure feasibility and enables team members to ask questions, reducing the stress** associated with unexpected changes. To facilitate this, it's essential to select a representative **pilot test to gain confidence in the organization.**
- Second, starting small enables you to **identify and refine the success conditions** as well as the **criteria needed to scale-up.**

In the context of Lean, this strategy is often referred to as using Models or Pilots, which can be a specific entity, department, factory, call center, or production line, for example.

For Digital initiatives, we can think of it as a *process*,[4] a product, or a service. These models or pilot programs serve similar functions as those in Lean methodology and can encompass multiple activities, as long as support is available.

As mentioned in the introduction to Chap. 5, the solutions tested require financial investments, which is not typically the case with Lean approaches. While Digital Quick Wins can yield rapid returns on investment with minimal effort and cost, many digital solutions may not achieve quick results.

Therefore, a crucial step for Digital initiatives is to assess whether the investment is worthwhile. **The concept of a Pilot or Model is divided into two parts,** often using specific terminology: POC (Proof of Concept) and either POV or POS (Proof Of Value or Solution).

Both the Proof of Concept and the Proof of Value follow the PDCA cycle:

- The **POC** is a quick and cost-effective method to assess whether the technology aligns with the business, the IT system, and its needs.
- In contrast, the **POV** is more detailed and builds on the POC, aiming to determine whether the technology delivers value, produces the expected results, and under what conditions.

This *process*[5] requires companies to invest in a number of prioritized solutions that may suit their needs. The POC and POV steps help clarify whether these solutions make sense for scaling up, as illustrated in Fig. 7.4. **Note that the POC and POV phases progress more quickly when the Technology Readiness Level (TRL) is high** (see Chap. 4):

- Proven solutions that are bought are already at TRL 9, which allows for faster POC and POV phases. In these cases, testing is conducted internally without further developing the TRL.
- For solutions being developed—whether in-house or by external parties such as startups—the POC phase begins when the TRL reaches 4.
- Once a solution has been validated at TRL 8 (the end of the POV phase), deployment can start (at TRL 9).

---

[4] A more precise definition is found in the Glossary (See No. 2).

[5] A more precise definition is found in the Glossary (See No. 2).

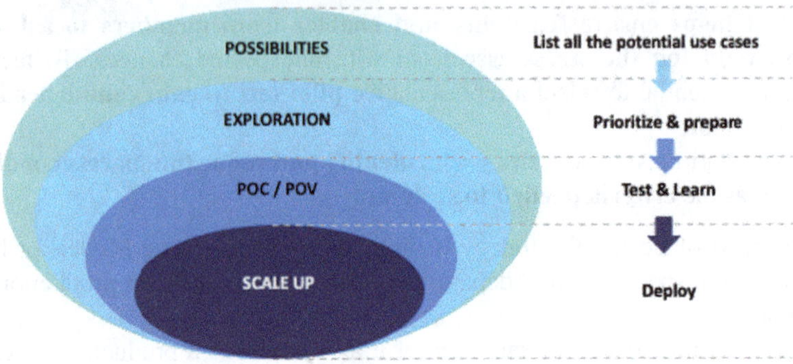

Fig. 7.4 From possible to scalable

Ideally, organizations should learn to invest in a "Test and Learn" phase without expecting an immediate return on investment, as the outcomes are still uncertain. Philippe Roger from Orange Consulting also emphasizes this approach:

> ***"I don't believe it is necessary to evaluate the Return on Investment for each individual project*** *at the beginning.*
> *The initial investment can be determined at the ExCom level based on a rough estimation of the expected tangible and intangible benefits.*
> ***However, once the use cases have been tested and validated, managing the ROI becomes essential for their large-scale deployment with industrial and sustainable solutions."***
>
> From an interview of Philippe Roger with Carine Vinardi

AI development can occur either internally or through collaboration with service providers or universities that focus on algorithm development. An integrator may also be called to connect the solution to the existing IT infrastructure.

Figure 7.5 is a visual representation of AI development, illustrating the relationships between TRL, POC, and POV phases.

Once a solution is validated at TRL 8, deployment can begin at TRL 9.

A rule of thumb estimates that data preprocessing accounts for approximately 80% of a successful AI project, while modeling contributes around 20% (Torrents, 2025).

The Pilot, Model, or POC/POV phases define the scale-up phase. As Olivier Drean from Dassault System and François Moallic from Cerratizit emphasize, it is good to act quickly but not rush the *process*[6]. This applies to both Lean and Digital initiatives:

[6] A more precise definition is found in the Glossary (See No. 2).

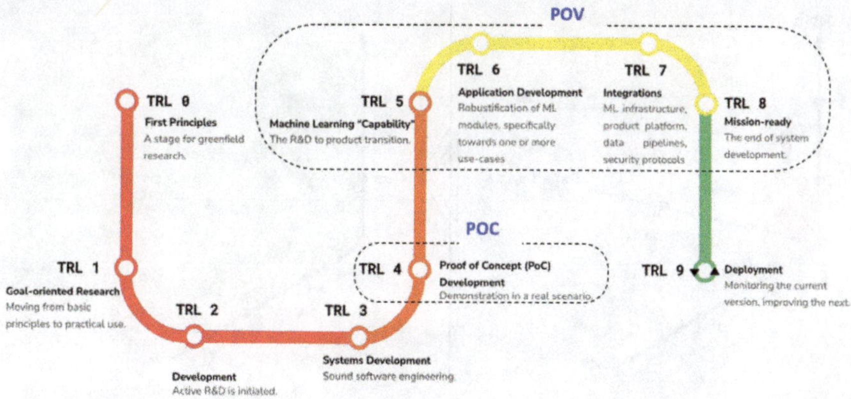

Fig. 7.5 POC and POV in relation to TRL—example with machine learning (*Source* inspired from Lavin et al., 2022)

> ***"Scaling up is not easy*** *if people are not convinced there is resistance, if the company moves* ***too quickly,*** *or if it is* ***not organized for it."***
>
> From an interview of Olivier Drean with Carine Vinardi
>
> ***"It is all too easy to develop digital tools quickly,*** *but it is* ***not easy at all to make those tools robust, flexible, and sustainable."***
>
> From an interview of François Moallic with Carine Vinardi

The deployment *process*[7], which includes the Pilot, POC, POV and scale-up phases, must **include duration and notions of the organization**. It has two dimensions (Vinardi, 2019):

- **Depth** (of deployment): This covers all phases of skill development on a topic, ranging from getting initial training to completely mastering all main scenarios.
- **Cover** (of deployment): This reflects how a topic gradually influences all teams and equipment within the company, aiming for 100% coverage.

Figure 7.6 summarizes the deployment scope, across three dimensions. While Cover may be enough to evaluate the deployment from Top Management to Front Line Teams, some choose to add a third dimension to consider the Departments involved (such as Sales, R&D, Operations, etc.).

Depth and Cover are guides for defining roadmaps and setting deliverables for deployment. Achieving integral depth (Mastering) and 100% coverage for Lean and/or Digital initiatives requires embedding these practices within the company's culture.

---

[7] A more precise definition is found in the Glossary (See No. 2).

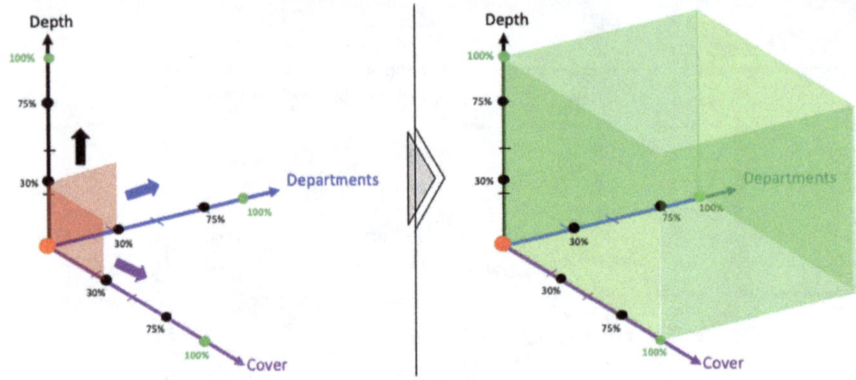

Fig. 7.6 Summary with a 3D representation for the deployment scope

For the Depth aspect, Mastering entails both theoretical understanding and practical application across a range of possible scenarios. Cover has three key stages prior to full deployment (Vinardi, 2019):

- The **first threshold is at 30% (or 1/3)**. Reaching this level allows teams to begin exerting influence.
- The **second threshold is at 50% (or 1/2)**. Although significant progress has not yet been made, this is a critical tipping point.
- The **third threshold is at 75% (or 2/3)**. At this stage, a majority is reached, which enables teams to persuade the hesitant.

It is **only beyond the 75% threshold that we can consider the deployment to be more sustainable and secure.**

7.2.2 *Progressing and Sustaining*

With the **constant inflow and outflow of personnel**, along with various **physical and material changes**, the deployment *process*[8] of a company is never truly finished. It requires **ongoing maintenance to ensure it remains 100% effective over time**.

Additionally, the deployment must align with the **cycle of change** (see Fig. 3.1) to effectively integrate Lean and Digital principles into the company's culture for the long term. This change cycle can also be incorporated into the different levels of the roadmap. In Table 6.2, you will find a proposed structure for the roadmap, which includes both the Cover and Depth components and the change cycle associated with the five-level approach (Vinardi, 2019). Note that:

[8] A more precise definition is found in the Glossary (See No. 2).

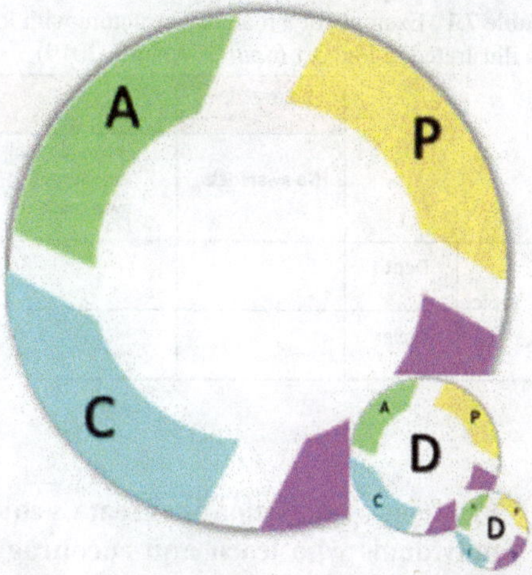

Fig. 7.7 PDCA cycles can fit inside each other

- The Roadmaps include a list of topics (or components) of the Lean and Digital methodologies. These components are deployed consistently both horizontally and vertically across different levels.
- Ideally, the Lean Leadership components precede the Tactical components, targeting one level above Level 4. More generally, the Lean components should be prioritized over the Digital components.
- The deployment of Lean and Digital follows the PDCA learning cycle, just like all associated actions.
- The overall deployment is divided into stages, and at the end of each stage and key action, feedback of experience (the Act of PDCA) must **be gathered to capitalize on the insights and initiate a new cycle**. This *process*[9] occurs at the end of a deployment cycle, at the end of each year, or after completing each action in the plan, for example. The PDCA cycles fit together like nesting dolls (see Fig. 7.7).

Capitalizing and learning are required to avoid repeating past mistakes, replicating successes, and sustaining good results.

Implementing a **learning organization is crucial** and must be included in our deployment strategy. I previously discussed this in Table 5.8 concerning the key success factors of Lean, but let's explore the concept in more detail (Table 7.1).

? What is a learning organization?

The origin dates back to the 1940s, but Peter Senge popularized the concept in the 1990s (Vinardi, 2013). There are several definitions of a learning organization, but I suggest the following version:

[9] A more precise definition is found in the Glossary (See No. 2).

Table 7.1 Example of a roadmap structure with its progress levels following the Cycle of Change as illustrated in Fig. 3.1 (*Source* Vinardi, 2019).

		Level 1	Level 2	Level 3	Level 4	Level 5
		No awareness	Factual and emotional awareness	Change of way of thinking and behaving	New habits implementation	Established culture
Topic	Depth					
	Cover					

> A learning organization can **create, acquire, and transfer knowledge** through **individuals who learn and encourage themselves both individually and collectively**.

To achieve this, it is essential to **organize knowledge and the methods of transmission** by addressing the two main types of knowledge mentioned in Chap. 2:

- **Tacit knowledge** is learned and passed on primarily through **observation, social interactions, and personal experience.**
- **Explicit Knowledge** is typically learned and shared through **written or codified forms, such as documents**.

Tacit and explicit knowledge play complementary roles in learning and knowledge transfer. Although it is challenging to assign fixed percentages universally, several studies provide valuable insights, including the work of Nonaka and Takeuchi (1995):

- **Tacit Knowledge**: It is estimated that tacit knowledge is a significant portion of the knowledge in organizations, ranging from **70 to 90%**, particularly in environments that rely heavily on practical skills and expertise.
- **Explicit Knowledge**: In contrast, explicit knowledge tends to account for a smaller share, typically between **10 and 30%** of the organization's total knowledge. This knowledge can be easily transferred through formal and digital media.

Vigilance is needed in deploying (Gen)**AI, which will look for explicit knowledge** but **cannot access tacit knowledge. Nevertheless, tacit knowledge is a significant part of overall knowledge** (as mentioned in Sect. 5.2.5 regarding the limitations of AI).

Nonaka and Takeuchi (1995) proposed a model of knowledge conversion known as SECI (Socialization, Externalization, Combination, Internalization). To maximize the transfer of knowledge and facilitate learning, organizations should consider the following strategies:

- **Create collaborative spaces** for exchanges focused on tacit knowledge. Leaders often learn more effectively with their peers, so these spaces should cater to their interests and preferences.
- **Organize and disseminate** both tacit and explicit knowledge through regular, well-structured activities such as **workshops, Action Work outs, visits or meetings.**
- **Invest in user-friendly and straightforward management systems for collecting and sharing** explicit knowledge, including standards, best practices, Lessons Learned, and Feedback of Experience, etc.

Digital solutions can help share explicit knowledge through easy-to-access applications, as noted in Sect. 6.1.2 regarding theoretical training. Additionally, as Olivier Drean from Dassault System mentioned in Sect. 5.2, specific platforms can further support this initiative:

"In our company, we also use our digital platform for Digital Twins (the one we commercialize) to:

- ***Share explicit knowledge** (documents, good practices,* etc.*),*
- ***Share tacit knowledge** through a network supported by AI."*

From an interview of Olivier Drean with Carine Vinardi

Apart from Apps that are simple solutions, **Custom GenAI templates can collect and consolidate significant explicit knowledge effectively**, similar to the approach MIT used for Pfizer (Brown, 2024):

- The Biopharmaceutical company Pfizer aimed to reduce the time required to get new drugs to market. One way was to speed up the knowledge transfer *process*[10], which is when relevant research and development and other documents are transferred to the manufacturing department.
- Using manual methods, it could take nine months per molecule to classify thousands of documents for a drug, and observations can be lost, hidden, or siloed off from the broader enterprise.
- The MIT Sloan team used more than 33,000 documents to build a suite of products that makes scientists' work easily available by quickly recognizing and retrieving information. This is already accelerating the speed of discovery and the speed to market for therapies.

It's essential to choose a **data management system that facilitates easy information retrieval rather than relying solely on Generative AI,** as this can lead to being

[10] A more precise definition is found in the Glossary (See No. 2).

overwhelmed by a vast amount of data. For instance, Pfizer has addressed this issue with Generative AI but it's crucial to find effective ways to update the newly created database.

Don't expect digital technology and machines to learn for you in the future. As Yann LeCun, a leading expert in AI, states (Le Cun, 2023):

> *Machines, as sophisticated and powerful as they are, remain* ***quite specialized****. Although their* ***computational capacities*** *greatly exceed those of humans,* ***they learn much less efficiently than humans and animals****.*

In addition to how to promote sharing, it is essential to consider the factors that encourage it and how we can influence or be influenced by others (Colotla et al., 2024). Here are some key aspects:

- **Reciprocity**: We must first have something to give in return to influence others.
- **Affection**: It is easier to say yes to people we like.
- **Social Proof**: The perceptions of others become more significant when we are in doubt.
- **Scarcity**: We tend to assign higher value to things that are hard to obtain.
- **Authority**: We are more inclined to listen to experts and those in positions of power.
- **Coherence**: Keeping our promises is crucial.
- **Simplicity**: We are more likely to implement ideas we understand clearly.

Furthermore, we should **not overlook the obstacles that can hinder effective sharing**, particularly **two factors that can lead to failure** (Norton et al., 2012):

- ***"Sunk cost effects":*** Leaders can be tempted to continue to devote resources to failing projects in which they have previously invested.
- ***"Not invented here" syndrome***: leaders can be tempted to refuse to use perfectly good ideas developed elsewhere in favor of their – sometimes inferior – internally-developed ideas.

These two challenges are related to what science refers to as the "Ikea Effect." (Norton et al., 2012): **When people imbue products with their own labor**, their **effort can increase their valuation**. There is also an "ironic link" between the arduous, unpleasant nature of tasks and their simultaneously rewarding properties. It seems that t**he more effort people put into some pursuit, the more they come to value it.**

? How do we overcome the challenges to implementing a learning organization?

To overcome the challenges, we need to address and acknowledge them and find ways to neutralize their effects:

- To counter the ***sunk cost effect*** top leaders must value those who question decisions, as this should not be seen as a failure.

  This attitude must begin with the leaders themselves, as their willingness to admit mistakes can have a powerful impact.

  I recall a time when Roland Vardanega was the Number two at PSA Peugeot Citroën (currently Stellantis). He publicly acknowledged before senior executives at an important convention that he had made a mistake by not adopting Toyota's Production System (TPS) sooner. Toyota and PSA had recently formed a joint venture. Instead of diminishing his credibility, publicly admitting this error earned him deep respect from his team. Additionally, it inspired others to follow suit and encouraged a stronger commitment to the Lean.
- To address the "***Not Invented Here***" syndrome, leaders should be encouraged to include **personal touches when adapting practices** from outside their organization. In my experience, when team members **can contribute their own input, they are more receptive to accepting new ideas**. This aligns with the concept of reciprocity mentioned by Colotla et al., 2024, as it establishes a social contract linked to the principle of gift and counter-gift, which I discussed in Sect. 4.2.2 (Mauss, 1925). This collaborative approach also helps enhance good practices.
- Moreover, Top Leaders must recognize the importance of collective learning, **value it, and engage actively** in it. One effective way to achieve this is by **incorporating time for knowledge sharing into the team's workload** and **participating in Feedback of Experience** sessions. Learning together is a valuable **human investment that also has indirect financial benefits**.

7.3 Main Failure Factors to Avoid and Conditions for Success to Repeat

To achieve this, I will address the elements outlined in Tables 5.8, 5.9, and 5.16, along with Chaps. 5, 6, and 7, to emphasize the most significant aspects of the sustainable and integrated implementation of Lean and Digital (Table 7.2).

Table 7.2 Macro-analysis of the main failure factors and the key conditions for success in the combined deployment of Lean and Digital

	Failure factors	Conditions for success
Strategy	• **Failure to take into account customer needs** • **No vision or prioritization** • **Lack of preparation** • Deploy Lean **only in Operations or in production and at ground level** • **Focus on tools first** • **Focus on costs** as a primary objective • **No integration of Hoshin kanri** • **No centralization and standardization** of data	• **Understand customer needs and deliver on promises** • **Ensuring a customer experience** in line with the expectations of different generations • **Prioritize (and therefore make strategic choices)** to ensure those limited resources are allocated relevantly • **Integrate Lean and Digital strategies** into the overall strategy **with Hoshin Kanri** • **Onboard vendors as early as possible** into the Lean & Digital deployment as quickly as possible • **Embody Lean & Digital transformation** at the highest level and over time (ensure this with regular feedback loops) • **Implement Data Management and Data** • **Governance** from the collection to the Feedback and Recycling phase of the data
Execution	• **Lack of monitoring** of implementation to ensure sustainability and **learning cycles** • **Lack of coordination of Feedback on Experience** • **Pushed instead of pull deployment** • **Too many projects** without consistency… • **Return on investment required** as of the (POC/POV) test phase for Digital • **Wanting to do everything on your own** for Digital • **No choice of progressive use case** (ML for AI, for example) • **No logical approach in the deployment** (focus on Added Value or waste without prioritization of non-eliminable and difficult waste first) • **Lack of meaning and acceptance** (including in the dispossession of Data)	• **Keeping business & people focused on the entire value chain** in implementing Lean and Digital • **Develop problem-solving efficiency** • **Massively invest in Cybersecurity,** both in material and human aspects • **Use the basics of Lean coordination** in the deployment of Digital • **First, use a classic VSM approach,** and if that is enough, carry out macro-diagnostics and align customer/ employee/company/supplier needs • **Align the maturity** of the organization, *processes*[11] and tools • **Integrate ERP** into the Lean and Digital deployment system • **Automate by giving meaning, and with Karakuri when possible**

(continued)

[11] A more precise definition is found in the Glossary (See No. 2).

Table 7.2 (continued)

	Failure factors	Conditions for success
System	• **Excessive internal performance** • **Exercises without an overall coherence** (Lean, SixSigma, Agile, Commercial Excellence, etc.) • **Focus on tools first** and foremost • **Lack of resources or profitability** of the solutions deployed • **Wrongly selected tools or solutions and/or solutions not used**	• **Integrating Hoshin Kanri** into the Lean & Digital system • **Implement a Lean**& Digital systemic approach in overall coherence (meaningful to everyone) • **Consider deploying all cultural dimensions** (national, company, activity, business) of managers and local teams • **Maintaining a durable ecosystem** with external partnerships
Human resources	• **Lack of collaboration** between teams. Siloed deployment. Little cross-functional roadmap deployment • **Maturity of the teams is not adjusted** to the tools deployed • **Too many interfaces** between Front Line Teams and Lean and Digital support teams • **Underestimated resistance to change** at startup and during scale-up, or more difficult to manage than expected • Lack of team empowering, • Recruitment of **Lean and Digital teams to implement improvements,** which "drip feeds" the organization with these resources • A **lack of data to understand current** and on-the-ground maturity and/or a lack of feedback on what is really happening at ground level	• **Checking the overall consistency of the Lean & Digital deployment with the needs** (using a regular feedback loop) • **Creating conditions so that (Top) Managers are ready to train and are always connected with the ground level** • **Gradually engage** all Leaders and **ensure their Role Model** (using a regular Feedback loop) • **Ensure that teams master the basics of their business** before developing Lean and Digital skills • **Use specific Lean & Digital teams** to support **(train, advise, mentor, coach**) (top)managers to develop their skills to create a new culture • **Gain buy-in** from Front Line teams daily • Establish a **learning organization**

References

Brown, S. (2024). *How companies use generative AI to execute with speed*. MIT Management Sloan School. https://mitsloan.mit.edu/ideas-made-to-matter/how-companies-use-generative-ai-to-execute-speed

Colotla, I., Fookes, W., Iverson, T., Schaefer, E., Sellschop, R., & Wijpkema, J. (2024). *Today's good to great: Next-generation operational excellence*. McKinsey Company/ Operations Practice. https://www.mckinsey.com/capabilities/operations/our-insights/todays-good-to-great-next-generation-operational-excellence

Cun, Y. L. (2024). *Il manque aux machines le sens commun* (2024) Interview for La Recherche Magazine. *L'intelligence artificielle pour les sciences, 577*, 20–23

Lavin, A., Ciarán, M., Gilligan-Lee, M., Visnjic, A., Ganju, S., Newman, D., Ganguly, S., Lange, D., Baydin, A.G., Sharma, A., Gibson, A., Zheng, S., Xing, E. P., Mattmann, C., Parr, J., & Gal, Y. (2022). Technology readiness levels for machine learning systems. *Nature Communications, 13*, 6039. https://doi.org/10.1038/s41467-022-33128-9

Mauss, M. (1925). *Essai sur le don. Forme et raison de l'échange dans les sociétés archaïques.* Année sociologique. https://gallica.bnf.fr/ark:/12148/bpt6k93922b/f36.item

Nonaka, I., & Takeuchi, H. (1995). *The knowledge-creating company.* Oxford University Press, New York, NY. https://doi.org/10.1093/oso/9780195092691.001.0001

Norton, M. I., Mochon, D., & Ariely, D. (2012). The IKEA effect: When labor leads to love. *Journal of Consumer Psychology, 22*(3), 453–460. ISSN 1057-7408. https://doi.org/10.1016/j.jcps.2011.08.002

Torrents, J. P. (2025). *Pourquoi la révolution de l'IA médicale n'aura-t-elle peut-être jamais lieu ?* Polytechnique Insights. https://www.polytechnique-insights.com

Vinardi, C. (n.d.). *Business strategy with Hoshin Kanri.* SpringerBriefs in Applied Sciences and Technology. https://doi.org/10.1007/978-3-031-20963-5_1

Vinardi, C. (2013). *Le Lean: atouts, impacts et limites* (Vuibert ed.).

Vinardi, C. (2019). PhD Thesis on Industrial Engineering: "*Les défis du Lean à l'ère de la mondialisation et de l'industrie 4.0*". Université de Technologie de Compiègne, Français (NNT : 2019COMP2500) (tel-03201462)

Chapter 8
Specific Company Cases and Prospects

Abstract This chapter presents several notable cases of companies engaged in transactional or industrial activities. These companies have either begun with a Digital objective and then implemented Lean or started their operations with a Lean system before incorporating Digital components. We will analyze the reports provided by these companies to offer a quick assessment. Additionally, we will reexamine some of the key criteria outlined in Chap. 4.4 to evaluate double sustainable performance and identify potential future developments.

As we discussed in Chap. 4.4, many of the criteria we have focused on show the benefits of combining Lean and Digital approaches (see Chap. 6). In this chapter, I will examine the environmental, social, societal, and governance impacts of Lean and Digital practices, as detailed in the published reports. However, because this set of criteria is broad, the analysis is an incomplete view of the double sustainable performance. To gain a better understanding of the specific contributions of Lean and/or Digital methods, the details of their implementation need to be examined.

8.1 Transactional Companies

8.1.1 Starting with Digital then Combining with Lean

As mentioned in Chap. 6.1.2, **applying Digital before Lean in a sustainably is possible, but there are few such cases.** The same applies to company deployment.

It is important to note that **transactional companies primarily rely on information flows**, although there may also be physical flows. These companies offer various types of services, including purely intellectual services (such as specialized support and training), financial services (like insurance and banking), and manual services (such as those provided by electricians and plumbers). For manual service companies, often small SMEs, expertise primarily comes from professional training and/

C. Vinardi, *Sustainable Performance in the Digital Age*,
https://doi.org/10.1007/978-3-032-01252-4_8

or skillful dexterity. Digital tools have emerged mainly for payment processing and increasingly for customer relationship management. This represents a Digitalization of *process*[1] rather than mere Digitization of products or services.

These small organizations, typically the most efficient ones, implement certain basic Lean principles, often based on common sense, to enhance performance without deploying a complete Lean system. More recent companies adopted digital tools before establishing basic Lean practices, but they soon need to integrate both. This trend also applies to very small businesses that offer intellectual services.

In contrast, companies with several tens of employees, particularly those in intellectual or financial services, have various scenarios. With the advent of the internet, information flows have become predominantly digital, allowing these organizations to be among the first to undertake digital business transformations.

Let's first examine companies founded during and after the internet boom of the 2000s, specifically focusing on start-ups with purely digital business models. These start-ups often face a high failure rate because they need to find and keep a viable market for their business model. Many of them also lack a strong performance culture.

Although it is possible to develop a digital product or service (in the sense of *digitization*) without using Lean, using Lean can be decisive for effective development, particularly during the Test and Learn phase. Eric Ries addressed this when he introduced the concept of the "Lean Start-Up" in 2011. While the idea is rooted in the typical Lean principles, it gained traction because of its buzzword and its specific emphasis on the needs of these companies.

The Lean Start-Up approach applies Lean principles during the business model definition phase. Instead of creating a comprehensive business plan and developing complete prototypes without engaging with customers, this method **rests on proposing and testing hypotheses** that are the foundation of the business model. It involves gathering early and frequent customer feedback and showing "minimum viable products" to prospects. This Lean approach has also been adopted by companies other than start-ups, particularly when defining new business models. When aiming to be sustainable and transition into SMEs, **Start-ups must foster a performance-oriented culture from the outset and throughout the execution of their business model**. Luc Julia highlighted the importance of Lean's very early introduction (in an interview with him in Chap. 4.1.1). In Sect. 8.1.2, we will explore examples of some start-ups that have successfully evolved into well-known companies.

Start-ups may be tempted to begin with a Digital approach, even though adopting Lean can provide advantages from the outset. **During my investigative research for this book, I found no compelling examples of transactional companies set up between 1995 and 2000, when the Internet emerged.** Moreover, none of these companies were recognized for their long-term performance or engaged in business transformation with a Digital focus without first embracing a Lean in its broadest sense.

[1] A more precise definition is found in the Glossary (See No. 2).

8.1.2 Combining Lean and Digital Early

Lean and Digital have often been implemented together from the outset in the operations of modern transactional companies, particularly in certain start-ups. Stéphanie Perchet from Adoptin and Delphine Asseraf from Harvest both shared insights reflecting this trend, even if Lean had a different name and wasn't applied as comprehensively as I previously discussed, its principles remain relevant.

Additionally, I have chosen three transactional companies well-known for their digital initiatives and that have demonstrated their commitment to implementing Lean practices.

8.1.2.1 The Case of Airbnb

Airbnb was founded in 2008, inspired by a difficulty faced by its creators. Let's use this example of how constraints can drive innovation. Then, shift the focus back to implementing Lean and Digital practices. The story goes that two friends, Brian Chesky and Joe Gebbia, lived together in a shared apartment in San Francisco.

In the fall of 2007, they faced financial difficulties due to a rent increase. At that time, San Francisco was preparing to host a major conference on industrial design, and all the hotels were fully booked. Recognizing an opportunity, Brian, who worked for a design agency, and Joe decided to rent out their modest apartment to help earn some extra money. They offered guests a place to sleep for one night on an air mattress, complete with breakfast. Although the concept was simple, it proved to be successful. Just a few weeks later, they launched their website, Airbedandbreakfast.com.

One of their clients suggested refining the concept and enhancing their website. Open to this idea, Brian and Joe reached out to Nathan Blecharczyk, Brian's former roommate, who had notable expertise in the field. Enthusiastic about the project, Nathan joined them, leading to the official creation of Airbnb in August 2008.

Back to Lean, it seems that Lean principles guided Airbnb's approach to validation and scaling. The business model was validated incrementally by analyzing local demographics and establishing key performance indicators like booking rate and user engagement. Therefore, Airbnb gauged the viability of their platform and iteratively expanded into new markets. It enabled Airbnb to scale efficiently without overextending resources prematurely. This case was not part of an academic study, but their application of Lean is similar to what Eric Ries later made popular as Lean Start-Ups.

In relation to the Digital deployment, Airbnb's communication focuses primarily on AI and GenAI. Airbnb has started to use AI to access a dynamic pricing strategy tool that adapts conversion rates in real-time to market demand and competitive rates, or to predict future occupancy rates and your users' rules based algorithms own preferences and booking patterns with impressive accuracy. At the end of 2024, Airbnb announced the introduction of a virtual assistant that uses AI to manage user interactions. This includes tasks such as answering frequently asked questions and canceling and rescheduling bookings (Fleureau, 2024).

Regarding Environmental, Social/Societal, and Governance (ESG) performance, Airbnb publishes its ESG report on its website, Airbnb.com since 2022. The 2023 report serves as a benchmark because the 2024 report is not available at the time of writing. As a result, the available insights are limited to just two years. Table 8.1 builds on items from Table 4.7 and highlights several key points:

- The Digital impact does not seem to have been considered, as it is outsourced, leading to contradictory results regarding environmental issues.
- No measures are given to evaluate the actions taken from a social and societal perspective.
- Data protection and cybersecurity are not specifically addressed.
- Governance issues are not explicitly stated either, despite being identified as a significant risk by several rating agencies.

The cases of Netflix and Amazon that I discuss below were not born of the Digital era like Airbnb; instead, they emerged with the advent of the Internet. The opportunities for developing a business in a "virtual" environment enabled these two companies to thrive. Both Netflix and Amazon successfully executed their Digital Business Transformations while simultaneously structuring themselves and integrating Lean principles. This progress represents a combined development in their growth.

Table 8.1 General overview of Airbnb environmental societal, social and governance issues

Assessment criteria	
Environmental issues	• The commitment is to achieve Net Zero Impact by 2030, with a goal of providing electricity solely from renewable energy sources without further details on the roadmap However, the results do not seem to align with these ambitions. Emission reductions were announced between 2019 and 2022, yet the MTCO2e emissions across scopes 1, 2, and 3 have continued to increase, particularly from 2021 to 2022 • Suppliers account for 96% of emissions. In 2022, a commitment was made involving 70 suppliers, who account for 61% of emissions, as part of a program lacking a clear definition and uncertainty regarding whether it includes data centers and the impact of Digital operations
Societal issues	• Employees: There is a diversity ratio focusing primarily on gender and underrepresented minorities (for the US) • D&I actions are listed on without any information on their impact • There is no specific information on data protection and cybersecurity
Social and psychosocial issues	• Several actions are listed for employees but no KPIs to evaluate them
governance issues	• There is a charter but it does not state any specific actions

8.1.2.2 The Case of Netflix

In 1997, Netflix was a DVD rental service. The company aimed to bring customer-centricity to the video rental. Without physical shops, it was a "mail-order business model". At the time, the SWOT was the following (Kishwar et al., 2023).

- Strengths: Mailing DVDs to service customers was disruptive versus competition with physical shops (less costs)
- Challenges: Logistic and transportation in addition to financial costs as customers had no limit for holding DVD, no fee subscription and DVD were only postal sent.
- Threats: Growing rise of internet, increase of shipping cost for DVDs.
- Opportunities: Shift to focus on streaming services with wider audience and more global expansion.

In 2000s, a change in the business model started to take place with a web subscription-based content and a dedicated streaming platform launched in 2007. The platform enabled viewing movies and TV show. Then Netflix partnered with different electronics companies to expand its user base. Therefore, Netflix grew from a **Transactional company (DVD rental) into a Transactional Digital company with a dedicated platform** that produces and streams its original licensed content. (Kishwar et al., 2023).

In the **2010s, Netflix began adopting Lean methodologies, integrating them with its Digital approach from the outset.** Nowadays, thanks to the Cloud, Netflix use data analytics and AI to improve decision making with a step by step market/country approach. Today Netflix faces the threat of an increasing competition and one of its challenges is the pricing strategy that they have addressed with partners including competitors (Kishwar et al., 2023).

Regarding Environmental, Social/Societal, and Governance (ESG) performance, Netflix publishes its ESG report on its website, netflix.com since 2020. The 2023 report serves as a benchmark because the 2024 report is not available at the time of writing. Table 8.2 builds on items from Table 4.7.

It highlights several key points:

- The Digital impact is not explicitly mentioned; the results align with the company's promises.
- Actions related to societal issues, including cybersecurity, are mentioned, though there are no KPIs to measure their impact.
- For the social component, actions are being taken with KPIs that show progress, but the challenge in achieving these goals is not assessed.

Specific actions regarding governance are highlighted, addressing key criteria such as competence, independence, and diversity within the organization.

Table 8.2 General overview of Netflix environmental societal, social and governance issues

Assessment criteria	
Environmental issues	• Ambition to reduce emissions by roughly half by 2030 versus 2019 • Overall carbon footprint is production-related (35%), therefore actions are highlighted to modernize production operations • Lowering emissions and global electricity consumption since 2022 following a rise up to that year • There are no KPIs on water consumption • Impacts of Digital practices are not noted
Societal issues	• Employees: There is a diversity ratio focusing primarily on gender and underrepresented minorities (for the US) • D&I actions are listed with KPIs • Running Actions on suppliers, in Cybersecurity but without KPIs • No specific information on the protection of data
Social and psychosocial issues	Specific actions on: • Recruitment and retention, with KPIs to measure the impact • Actions on pay equity with KPIs • Actions on benefits and well-being without KPIs
Governance issues	Several actions are noted: • The adoption of Corporate Governance Guidelines • The adoption of a Human Rights Statement that sets forth our ongoing • The commitment to respecting internationally recognized human rights' • An update to the Netflix Code of Ethics • The skill of Directors versus the needs (Innovation, Content, International) • Representation: nine of the top 10 shareholders are Independent Directors, women made up 40% of our independent directors and 31% of our overall board

8.1.2.3 The Case of Amazon

Jeff Bezos launched an online bookstore Amazon.com in 1994, and everyone thought it could not survive. Bezos' initial goal in starting his **own business was to create a small-scale bookstore**, and his hope was that the bookstore he presented could be **purchased by consumers throughout the world using an online system.** The e-business continued to expand, selling music, pet supplies, electronics, home improvements, pharmaceuticals and the world witnessed its evolution to a massive marketplace for third-party sellers (Onggowidjojo & Ellitan, 2023).

Although the Lean is not clearly featured in Amazon's public material, it has long been an integral part of the development. **Amazon** has integrated Lean principles throughout its operations, particularly under the leadership of Jeff Wilke, who helped incorporate it in the late 1990s in operations. The company **uses Lean to optimize** ***processes***[2] such as inventory management and logistics, and has implemented tools like the Andon system, which empowers employees to stop production. Amazon Web Services (AWS) has also used Lean Six Sigma to enhance customer service, improve operational efficiency, and reduce costs (Onggowidjojo & Ellitan, 2023). Amazon

[2] A more precise definition is found in the Glossary (See No. 2).

has succeeded so far to create a site that is **customer friendly** and **linked to the needs** of each type or characteristic of customers. With an efficient supply chain, including automated warehouse to decrease price and minimize mistakes (Onggowidjojo & Ellitan, 2023). According to Jeff Wilke (1999 Annual Report of Amazon.com):

> *To us, it implies two things: delivery continuous improvement in customer experience and driving productivity, margin, efficiency, and asset velocity across all over businesses.*

Amazon has **updated its marketplace with Digital technology** using AI of course but also **drone technology,** which can speed up package delivery compared to other marketplaces. The delivery drone is only capable of carrying packages weighing up to 3 kg to customers in half an hour. This tool can fly as far as 24 km (Onggowidjojo & Ellitan, 2023).

Amazon is investing heavily in **artificial intelligence (Artificial Intelligence)** which **helps drones navigate very safely to their destination and deliver packages comfortably and safely** (Onggowidjojo & Ellitan, 2023).

Regarding Environmental, Social/Societal, and Governance (ESG) performance, Amazon publishes its ESG report on its website, amazon.com since 2019. The 2023 report serves as a benchmark because the 2024 report is not available at the time of writing. Table 8.3 builds on items from Table 4.7 and highlights several key points:

- Several actions and indicators were highlighted for the environmental part, but there is no reference to emissions or water consumption, for example.
- Actions of developing electricity with renewable sources without referring to consumption.

  The impacts from Digital and Data centers are abundant in the report but are not explicitly highlighted.
- Actions are highlighted for the societal and societal part but they are very local (many concerning the US). A particular focus is on training and improving employee safety (linked to the "industrial" part of the activity).

Despite the significant role that Digital technology plays in the operations of Airbnb, Netflix, and Amazon, these organizations have not emphasized its contribution to environmental issues in their external communications. Additionally, there is **no mention of internal improvements related to their activities, such as the potential impacts of Lean practices.**

Table 8.3 General overview of Amazon environmental societal, social and governance issues

Assessment criteria	
Environmental issues	There are several actions and KPIs with a clear status update • Reach net-zero carbon emissions by 2040 • Match 100% of the electricity consumed by our global operations with renewable energy by 2025—five years ahead of our original target of 2030 • Water positive (return more water to communities than it uses in its direct operations) by 2030 • Make Amazon device packaging 100% recyclable by 2023 (versus 90% achieved) There was no increase in the proportion of renewable energy sources between 2022 and 2023, without stating whether the actions compensate for the increase in digital and data center activity that Amazon develops
Societal issues	• Diversity indicators and actions are given • Inclusion indicators are very local and focus on the US • Actions are announced about human rights in some developing countries without giving KPIs
Social and psychosocial issues	• Employees: – Training and upskilling actions and indicators – Security actions and indicators are improving • Inclusion indicators are very local and focus on the US • Supplier actions are stated without putting them into perspective
Governance issues	Diversity indicator on gender and race (specific to the US), but no precise actions are given

8.1.3 Starting with Lean then Combining with Digital

Many transactional companies that began operating before the internet age have since started their digital transformation. To improve their performance, most of these companies have adopted a Lean/Agile approach in their transformation efforts, as is the case for banks and insurance companies.

In my interviews, the testimonies of Nicolas Siegler from Maif and Bruno Echardour from La Poste illustrate this. The companies they represent are a large mutual insurance group (Maif) and a private organization from the French public sector (La Poste), which originally focused on mail delivery but has since expanded into banking, insurance, and online storage of sensitive documents.

Maif was established in 1934, while La Poste traces its origins back to around 1476 during the reign of Louis XI. Both companies have undergone numerous transformations before embracing Digital innovation. They have experimented with the Lean approach to varying degrees of success, alongside a more recent Agile methodology (as mentioned in Sect. 2.2; the Lean system has a broader scope).

In the 2000s, many banks and insurance companies began adopting the Lean approach. This was when the internet was just emerging, and Lean was experiencing its peak popularity. While implementing Lean in service activities has seen some successes, it has often led to disappointment and failure. As mentioned in Chap. 2, these outcomes were primarily linked to a misunderstanding of **key success factors, which were not effectively replicated, and a failure to adapt the approach to the**

specific cultural context of the organization. This was compounded by insufficient competent resources supporting the *process*[3]—a situation I observed and experienced firsthand. Many Lean-trained professionals coming from the automotive industry believed they could replicate the model without considering the importance of culture.

8.2 Industrial Companies

8.2.1 Starting with Lean then Combining with Digital

Before the advent of the internet, many industrial companies, even in developing countries, sought to optimize their performance using a Lean System, whether formally or informally. This point was discussed in the first part of this book and applies to all the industrialists interviewed. As these companies have been the subject of academic studies, here are two examples—Toyota and Danaher—illustrating how they have successfully enhanced their performance over time. Both companies achieved this by engaging their teams, suppliers, and partners. While the studies primarily focused on implementing Lean, both Toyota and Danaher have also incorporated Digital strategies to further optimize their operations.

8.2.1.1 The Case of Toyota

I do not claim to have detailed knowledge of Toyota's approach to integrating Digital technology with its Toyota Production System (TPS). However, it is clear that the implementation of Digital tools is aligned with TPS and the company's core values:

- **Human-Centric Approach**: **A key principle of TPS is its focus on people.** Employees are the ones who define, innovate, and improve *processes.*[4] Technology and Digital solutions assist people, rather than the other way around.
- **Cautious Adoption of Technology on large production series**:
  - Toyota is recognized for its pioneering use of robotics in production and for launching the first hybrid-electric vehicle in 1997. Each innovation is introduced only when its maturity has been thoroughly assessed, following the philosophy similar to that of the successful Millau Viaduct mentioned in Chap. 2.
  - Toyota takes a pragmatic approach to digitalization that seems to prevail today. Rather than rushing, it carefully implements technologies that have demonstrated clear value in specific situations. This includes using AI only when it contributes effectively to production goals.
  - The aim is to avoid unnecessary complexity.

[3] A more precise definition is found in the Glossary (See No. 2).

[4] A more precise definition is found in the Glossary (See No. 2).

- **Additionally, a frugal mindset extends to environmental considerations**: Digital solutions are evaluated for their environmental impact and potential to reduce waste.

Regarding Environmental, Social/Societal, and Governance (ESG) performance, Toyota publishes its ESG report on its website, toyota.com, titled "Toyota Sustainability Data Book" and been published since 2021. Unlike the previous mentioned companies, the 2024 report is available at the time of writing. **This report includes strategic objectives and a 2025 action plan, demonstrating an effective internal organization that will have the necessary information ready by January 2025**, thanks to the *Hoshin kanri*.[5]

Table 8.4 builds on items from Table 4.7.

Table 8.4 General overview of Toyota environmental societal, social and governance issues

Assessment criteria	
Environmental issues	• The ambitions are clearly aligned with TPS and the values of the Toyota Way. A defined roadmap for 2030 and 2050 is included, along with a detailed plan and objectives for 2025 • The engagement actions of all stakeholders—customers, suppliers, employees, and shareholders—are clearly detailed • The structure is consistent for each theme addressed, outlining aims and initiatives and detailing KPIs and roadmaps • Key focus areas include life cycle management (circular model, energy, and transportation) • Actions targeting electric vehicles and their batteries are primarily concentrated in Japan • The report includes results for the main selected criteria. It also mentions the involvement of an external service provider, PwC, in validating the data
Societal issues	• The ambitions are clearly aligned with TPS and Toyota values (Toyota Way). They outline a defined path for achieving goals by 2030 and 2050, along with a detailed plan and objectives for 2025 • The engagement actions of all stakeholders—customers, suppliers, employees, and shareholders—are clearly detailed • The structure is consistent for each theme addressed. It focuses on aims and initiatives, followed by detailed KPIs and roadmaps • The main themes include respect for human rights, the treatment of immigrant workers, harassment, etc • A specific emphasis is placed on integrating suppliers and customers through dealers as part of Value Chain Collaboration • Societal issues, such as customer security and data protection, are addressed with detailed actions, including the organization of feedback mechanisms • There is a particular focus on Diversity and Inclusion, emphasizing gender diversity from the perspective of societal integration (especially in nursing care) rather than just internal equality. The inclusion of people with disabilities and the LGBTQ + community is also included

(continued)

Table 8.4 (continued)

Assessment criteria	
Social and psychosocial issues	• The ambitions are clearly aligned with TPS and Toyota values (Toyota Way). They outline a defined path for achieving goals by 2030 and 2050, along with a detailed plan and objectives for 2025 • The engagement actions with all stakeholders—customers, suppliers, employees, and shareholders—are thoroughly explained • The structure is consistent for each theme addressed. It focuses on aims and initiatives, followed by detailed KPIs and roadmaps • For employees, recruitment, development, retention, and feedback collection are detailed in particular • There is a focus on mental health and physical security • Engagement surveys, including results and actions, are included
Governance issues	• As for the other themes: • The ambitions are clearly aligned with TPS and Toyota values (Toyota Way). A defined roadmap for 2030 and 2050 is included, along with a detailed plan and objectives for 2025 – The engagement actions of all stakeholders—customers, suppliers, employees, and shareholders—are clearly detailed – The structure is consistent for each theme addressed, outlining aims and initiatives and detailing KPIs and roadmaps • The skills and the work of the Board, including compliance, are given in detail

It highlights several key points:

- The report is highly detailed across all sections and is structured in a standard way that makes it easy to read.
- Many of the indicators mentioned in Chap. 4.4 (Tables 4.3, 4.4, 4.5, and 4.6) are included.

Both the TPS and Digital initiatives are clearly identified as means to achieve the company's objectives. However, the report does not specify how these approaches are combined or their individual contributions.

8.2.1.2 The Case of Danaher

I mentioned the case of Danaher, which was successfully inspired by the Toyota model. Additionally, Danaher has combined its DBS (Danaher Business System) with Digital integration, as mentioned by Martin Stumpe, appointed in 2024 as the Chief Data and AI Officer for Danaher Health and Life Science (Danaher.com):

> *We have the devices, the global install base and the assays, plus the world-class operational excellence (…). AI cannot solve problems in a vacuum. In order to develop AI and impact in the real world, many factors need to be considered.*

Regarding Environmental, Social/Societal, and Governance (ESG) performance, Danaher publishes its ESG report (Danaher Sustainability report) on its website, danaher.com, published since 2021. The 2024 report is available as for Toyota, at the time of writing.

Table 8.5 builds on items from Table 4.7 and highlights several key points:

- Danaher's report is less comprehensive than Toyota's but provides more detail than the reports from the transactional companies I presented previously.
- KPIs are given with some key actions for environmental issues, but there is no clearly stated roadmap.
- Although the DBS is referenced many times, the report does not discuss Digital deployment in detail, which is addressed separately. Like Toyota, Danaher approaches Digital technology as a means to an end (AI is one example).
- The governance section of the report does not have any specifics.

Table 8.5 General overview of Danaher environmental societal, social and governance issues

Assessment criteria	
Environmental issues	• Environmental protection is one of the three main priorities, along with innovation and building the best team. The ambition is linked to DBS, with a long-term goal of achieving Net Zero by 2050 at the latest. A medium-term objective is to reduce GHG emissions by 50% by 2032, using 2022 data as the baseline, though a detailed roadmap is not provided. Actions are currently being identified • KPIs for GHG emissions are tracked according to scopes 1, 2, and 3. The actions taken in this area are mentioned, but do not detail energy consumption, recycling, or water usage. Again, no roadmap is provided
Societal issues	• A few KPIs are highlighted, but there is no detailed roadmap • A specific focus on diversity and inclusion is given, emphasizing gender and racial diversity, particularly for the USA, as well as equality within the organization. The integration of people with disabilities and the LGBTQ + community is also addressed
Social and psychosocial issues	• Building the best team remains one of the three priorities, alongside innovation and environmental protection, and is in line with the DBS ambition • Three KPIs related to talent recruitment, internal fill rates, and training/coaching initiatives are highlighted, particularly related to actions to promote diversity • The commitment level is indicated by the annual engagement index, which has been monitored since 2019. However, the results have been declining since 2020, but no countermeasures have been shared
Governance issues	• The Danaher Sustainability Council is highlighted • Aside from legal and customary details, the report contains no additional information on this subject

8.2.2 Combining Lean and Digital Early

In Chap. 6, we discussed that having a sustainable performance, means implementing the fundamentals of Lean either beforehand (particularly in terms of engagement actions and management) or, at the latest, when Digital practices are started.

Modern industrial companies that are created now could greatly benefit from the combined integration of Digital with Lean deployment for the Tactics part, as illustrated by Naarea (as noted in David Briggs' testimonial). A company that wants to remain sustainable over the long time, must maintain a consistent level of maturity between Lean tools and methods and digital solutions.

To exemplify a promising combined deployment, I would like to highlight Tesla. This company is beginning to attract academic attention, although little research is still available.

8.2.2.1 The Case of Tesla

Elon Musk became chairman of the board of directors in 2004, at which point he began to play a significant role in the project of Tesla Motors (now Tesla Inc.), a company founded just a year earlier.

Tesla was viewed as a **black horse** in the auto-industry. Compared with the incumbent auto companies who has decades-experience in making and selling cars, Tesla was a new entrant with less inert as other incumbent automakers for business model innovation (Chen & Perez, 2018).

Free from the limitations of existing factories and prior operational methods, Tesla focused on a specific segment: electric vehicles (EVs). The company developed a disruptive model that caught the attention of multiple manufacturers. Indeed, Tesla quickly embraced an unconventional approach in many respects (https://ir.tesla.com/press), (Chen & Perez, 2018; Fabric Editorial Team, 2024; Thornhill, 2021):

- **Customer focus and relationship**: Tesla defined a direct sales model rather than through franchised dealerships. Moreover, the drivers connect to the car only with a big Tablet and scroll as in the personal life, rather than using a traditional dashboard. It also allows Tesla to control the customer experience and pricing more directly.
- **Revenues**: The idea was to generate income from software services, charging, maintenance and insurance as well as sales of powertrains, batteries and carbon credits to other manufacturers.
- **Vertical Integration**: Tesla decided to produce many components in-house, such as batteries, and owning its sales and service centers.
- **Infrastructure**:
  - IT: Tesla owns IT and In-House e-commerce platform (as no external existing solution were available)

  - ERP: All sales on line with a specific ERP software (all the *processes*[6] needed to run the company are connected) vs competition with Dealer point model.
  - Recharging Station: Tesla had to invest in a network of superchargers that were not available at the time, and charging was free for buyers of the first models.
- **Innovations in Manufacturing**: Tesla was noted for its "Giga Presses", to produce larger parts of a car in a single piece, reducing production steps and increasing the structural integrity of its vehicles.

At the same time, Tesla learned valuable lessons from the past experiences of other manufacturers:

- **Customer segment**: Tesla initially targeted the High-end market, which is the most profitable segment, before moving to the mass market.
- **Value configuration**: Tesla began by offering a minimal choice of free options on the base model. The company was criticized for this approach, noting that even the slightest option, such as metallic paint, was significantly more expensive than competitors.
- **Toyota experience for Manufacturing**: Tesla's first production facility was set up at the former NUMMI plant (New United Motor Manufacturing Inc.) in Fremont, California. This joint venture, operated by General Motors and Toyota, ran from 1984 to 2010. Tesla began producing the Model S at this factory in 2012, and for the general public, this sedan truly put Tesla on the map.

Therefore, Tesla's initial approach for production was broadly consistent with lean thinking and E. Musk initially tapped into proven lean knowledge in his early launch (Ehrenfeld, 2018). The Model X, launched in 2015 as an SUV based on the Model S, was produced at the same Fremont plant.

However, it was the **Model 3 that captured the founder of Tesla's intent to challenge traditional automotive industry paradigms with a futuristic vision.** The **extensive experience accumulated by leading automakers over the decades, including Lean principles, was no longer really considered relevant.** The message to teams was that production is fundamentally a technical problem for engineers to solve, and the key solution was automation. Additionally, it is important to note that Elon Musk had built his fortune on software, making him more receptive to the Agile approach, which he aimed to apply to automotive manufacturing. Musk shared his ambitions during a conference in 2016 (https://www.youtube.com/watch?v=f9uveu-c5us):

> *"We are pushing* ***robots to their operational speed limits*** *andasking* ***our suppliers to develop much faster robots.****Our suppliers are surprised by our demands, as no one has attempted this before. (...)We aim to create* ***machines***

[6] A more precise definition is found in the Glossary (See No. 2).

> ***that manufacture other machines, and we envision factories designed as products".***

In his dream of a "lights-out factory" E. Musk was not merely harkening back to previous dreamers, but **he was under-valuing the role of people as part of this socio-technical system** (Ehrenfeld, 2018).

In 2017, the launch of the Model 3—a compact sedan designed to make EVs more accessible to the general public—was on a different production line than the Model S. This decision did not take into account past Lean experiences, and led to more disruptions than Elon Musk had anticipated. Significant production issues and unfulfilled deliveries marked the launch, as the volume of orders exceeded manufacturing capacity, causing the planning system to fall behind. **Musk had even referred to this as "production hell."**

What Elon Musk was missing is exactly the point that has made Toyota so successful. **Toyota's living system approach is exactly what has been missing from Musk's mechanistic view** and needs to be at the center of his vision, not as an occasional response to a crisis. **Toyota is a learning organization with a long memory** (Ehrenfeld, 2018). It worsts to accelerate production by leveraging your people rather than your robots.

As I mentioned in Chap. 2.2, **Agile, which is based on Lean in a simplified version, does not incorporate the project aspect alongside Lean Product Development.** This could have help to test the hypotheses of Elon Musk. He later acknowledged in a tweet that he was mistaken:

> ***"(...) excessive automation at Tesla was a mistake.*** *To be precise, my mistake. Humans are underrated".*

E. Musk has gone from thinking that excessive automation was the answer to his productivity challenges to recently admitting that there are **areas that are best left to manual human labor.** Elon even **showed more admiration for low tech "dumb robots"** that **use simple sensors and magnetic strips** to move material to various workstations within the factory (Randall, 2018). It is important to note that the Low Tech robots mentioned by Elon Musk are improved versions of Karakuri robots (described above), enhanced with sensors and are already in use on Toyota assembly lines.

It is instructive to note that the Tesla Model 3 in Fremont California was the site of the NUMMI factory. The **productivity of the Model S line was much higher than Model 3 line, despite the fact that it was less automated**. NUMMI factory was able to produce so many cars using **less manpower and automation** mostly because of employing lean principles (Randall, 2018).

Additionally, it is worth mentioning that until 2019, Tesla was not profitable, struggling for more than ten years since its inception, and even coming close to bankruptcy in 2018. It wasn't until 2020 that the company began to generate profit. By 2023, Tesla's profitability per vehicle was compared to that of Toyota's.

However, while acknowledging the efforts of Tesla's teams to address their challenges, Tesla must yet sustain its sales and performance over time. Suitable comparisons are difficult to make: Toyota offers a much broader range of vehicles, allowing it to benefit from volume-based margins on lower-end models. Tesla's sales volume is still a small percentage of Toyota's total sales. While Toyota's approach has consistently yielded excellent results over time, Toyota's "frugal and incremental" strategy and Tesla's more "technological and radical" approach are frequently debated. **I believe we should not match them head-to-head, especially since they learn from one another.** Elon Musk was able to reevaluate his purely technological model, opting for simpler and more flexible aspects. Additionally, the Giga Press method, which simplifies assembly, comes at very high costs and also raises environmental concerns. In the event of repairs, large blocks need to be changed instead of replacing small ones.

Tesla was **launched by integrating its innovations with Lean principles**, and it corrected its errors after initially relying heavily on advanced automation. The company will continue towards double sustainability, not only through Breakthrough Improvements (BI) but also via Continuous Improvement (CI), which includes its BI efforts (Vinardi, 2023). This approach is chosen by its competitors, including Toyota, who draw inspiration from Tesla's digital model for the inside of the vehicles and its profitability model for high-end EVs.

Perhaps one of the most important underlying themes of the unfolding Tesla story, has to do with the idea of learning versus optimization (Ehrenfeld, 2018). Tesla will grow by **becoming a learning company and a high-performing one for the planet**.

Specifically, regarding Environmental, Social, and Governance (ESG) performance, Tesla publishes its ESG report, known as the "Impact Report," on its website, tesla.com since 2021. The 2023 report serves as a benchmark because the 2024 report is not available at the time of writing.

Table 8.6 builds on items from Table 4.7.

It highlights several key points:

- There appears to be an Environmental, Health, Safety and Security Vice President coordinating initiatives, although roadmaps do not have more details. The report does mention the distribution of emissions ranked by importance: (N°1: Other (?), N°2: Battery). The highlighted actions and KPIs predominantly focus on batteries, especially regarding their recycling and reparability, reflecting their impact.
- Societal issues are mainly addressed from the standpoint of customer safety, emphasizing vehicle technology and Tesla's appeal to younger generations. No additional societal concerns are noted.

Table 8.6 General overview of Tesla environmental societal, social and governance issues

Assessment criteria	
Environmental issues	• Protecting the environment is one of Tesla's stated priorities • The company conducts annual assessments but has surprisingly states: **"Current emissions frameworks weren't Built for a Company like Tesla."** However, the framework applies to everyone equally • A common issue is the frequent comparisons made with combustion engine vehicles without considering their whole life cycle • The report specifically addresses water and electricity consumption in production plants, giving KPIs • A large section focuses on suppliers due to their contribution to emissions (scope 3). The report highlights some actions and KPIs related to battery production and material extraction
Societal issues	• While a few actions and KPIs are presented, no clear roadmap is given • Diversity and inclusion are mentioned with some local initiatives, yet there is no clear corporate vision or KPIs • The actions mainly focus on inspiring new generations and developing professions related to sustainable energy, and some KPIs are given • Customer safety is prominently featured, with actions and KPIs concerning Driver-Assist and Automatic Emergency Braking Technologies
Social and psychosocial issues	• Employee safety is also addressed with relevant actions and KPIs • Well-being and harassment prevention are discussed, but without a clear roadmap • No specific actions or KPIs are outlined for cybersecurity and data protection, despite the topics being mentioned • The company emphasizes internal promotions, noting that two-thirds of its leaders are promoted from within, and highlights employee commitment. However the required participation rate for assessing commitment is not specified • A significant section is dedicated to suppliers, including actions to combat Forced Labor. However, progress in this area cannot yet be measured
Governance issues	• Oversee ESG risk • The report provides minimal beyond the usual legal and traditional disclosures

- The section on social issues primarily emphasizes the safety of employees and those working for suppliers.
- The governance section of the report does not have specific details.

It is important to note that, despite its type of activity, the comparisons by various rating agencies regarding Tesla's ESG record show that its performance is worse than several competitors, many of whom produce significantly more vehicles, predominantly combustion vehicles. One possible explanation for this could be that social and societal issues and the risks associated with governance are not fully considered. Additionally, the overall carbon footprint of electric vehicles (EVs) must be considered comprehensively. For instance:

- Manufacturing EVs requires more energy than combustion vehicles, mainly due to the impact of batteries.
- The carbon emissions during usage, particularly from vehicle charging, are influenced by the source of electricity generation, which varies by location.
- The EVs' carbon footprint is better than that of internal combustion vehicles after approximately 70,000 km, assuming electricity is mainly sourced from renewable energy. Tesla estimates the battery lifespan to be between 240,000 and 320,000 km.
- While Tesla appears to have considered the recycling of its end-of-life vehicles, the company's cars and their components seem to be more complicated to recycle, especially when they are damaged during their lifecycle. According to CarVertical (www.CarVertical.com), which provides automotive data for France, the average distance traveled by a vehicle before an accident is 159,758 km for 2023, and this decreases over time.

Despite its ongoing efforts, Tesla still has considerable work ahead to achieve double sustainable performance.

8.2.3 Starting with Digital then Combining with Lean

The last category of industrial companies to examine is those that began their Digital transformation before adopting Lean. This applies to companies established after the Internet and the widespread adoption of Digital technologies, specifically from the 2010s onward.

Aside from start-ups, which benefit from Lean principles right from the start of their operations, my research for this book did not find any substantial examples of industrial companies that have excelled over time that pursued a Digital transformation before adopting Lean practices (in the broadest sense) or by doing so after.

Achieving double sustainable performance in both Lean and Digital initiatives, means considering various challenges and ensure that Lean and Digital strategies align with the company's overall objectives in the long term. A smooth integration of all the aspects, progressing with maturity and capitalizing on learned experiences, is needed.

As shown with the two notable cases of Toyota and Tesla, **the goal is not to pit these approaches against each other but to create interaction. Strengths should not be compared; they must be combined, and different lessons should complement one another.**

As we conclude this chapter, it is important to note that the ESG reports I have referenced provide initial insight but are not absolute evidence. Reports are not standardized across companies, making comparisons difficult. **They primarily serve as a way for companies to share their progress, but they do not always reflect the true challenges. Although these communications are official, it would be best to measure the information presented with observations in the field**. Furthermore, since 2010, numerous studies have been published analyzing the consideration and communication of Corporate Social Responsibilities (CSRs) and the motivations and facts related to Corporate Social iRresponsibilities (CSiRs). Indeed, CSiR attributions have been shown to increase financial risks for firms, incur stakeholder sanctions and impair reputation, particularly when such perceptions receive significant and widespread attention. Despite many efforts around CSR, firms tend to continue to act irresponsibly and inflict harm on stakeholders with different levels of intensity (Cho et al., 2024).

References

Chen, Y.,] Perez, Y. (2018). *Business model design: lessons learned from Tesla motors. Towards a sustainable economy. Paradoxes and Trends in Energy and Transportation* (Springer, 978-3-319-79060-2. hal-01655959).

Cho, C. H., Zarzycka, E., She, C., Dobija, D., Krasodomska, J., & Bothello, J. (2024). Examining stakeholder reactions to corporate social irresponsibility: Evidence from social media. *European Management Journal*. ISSN 0263-2373. https://doi.org/10.1016/j.emj.2024.11.001

Ehrenfeld, T. (2018). *Lean lessons from Tesla*. https://www.lean.org/the-lean-post/articles/lean-lessons-from-tesla/

Fabric Editorial Team. (2024). *Tesla's digital strategy for becoming a trillion dollar company.* https://fabric.inc/blog/commerce/tesla-strategy

Fleureau, G. (2024). *Malgré des investissements croissants, les entreprises européennes peinent à atteindre la maturité numérique*, 12 septembre 2024. https://siecledigital.fr/2024/09/12/etude-malgre-des-investissements-croissants-les-entreprises-europeennes-peinent-a-atteindre-la-maturite-numerique/?

Kishwar, J., Mahfouz, A., Hayes, Y. (2023). *Strategy for growth and market leadership: The Netflix Case. AIMS International Journal of Management, 17*(2), 87. https://doi.org/10.26573/2023.17.2.2

Onggowidjojo, V., & Ellitan, L. (2023). Strategies for achieving competitive advantage: The case of Amazon. *International Journal of Research, 10*(09 September 2023). https://doi.org/10.5281/zenodo.8373836

Randall, T. (2018). Tesla's production problems spawn a legion pf Model 3 Stalkers. https://www.bloomberg.com

Thornhill, J. (2021). Trillion-dollar Tesla has redefined what carmaking means. Financial Times. https://www.ft.com/content/dd1b4aa6-5065-416a-8068-ca5e12cf32f3

Vinardi, C. (2023). *Business strategy with Hoshin Kanri*. SpringerBriefs in Applied Sciences and Technology. https://doi.org/10.1007/978-3-031-20963-5_1

Chapter 9
What Next?

Abstract In previous chapters, we discussed and illustrated the impact of Lean and Digital approaches, individually and in combination, for double sustainable performance. In an increasingly fast-paced world where information circulates rapidly, we observed that one solution, whether Lean or Digital, can sometimes replace another due to trends or a lack of knowledge put to use. However, with the current Digital solutions, we could transform organizations through well-defined, supported use cases. And what's next? This chapter looks towards the future and explores the concept of a post-Digital era, or the Digital practices of the Future.

When I wrote this book, my primary focus was finding the best ways to understand, integrate, and prioritize the key factors driving double sustainable performance. I intentionally chose not to delve deeply into the various use cases of Digital technology because an update would soon be necessary by the time I finished writing.

In the final chapter, I discuss the concepts that will either be confirmed or challenged in the coming years or decades. This uncertainty is inherent when we examine a subject from a forward-looking perspective. As a result, references to academic sources will be presented instead of interviews, except for my conversation with Luc Julia, whom I specifically interviewed on this topic.

? Has Digital technology as we know it today been fully developed beyond its current applications? What additional advancements are already emerging?

First, it should be noted that the proposed assessment framework is relevant for assessing future innovations. I have considered both current aspirations and future possibilities. Regardless of the future, it should be viewed this way. This approach has been effective in companies that successfully implement Lean, and will be necessary for Digital advancements and beyond.

When people and teams **lack trust and have doubts or questions, they often retreat into their familiar circles and stay with like-minded individuals** (Vinardi, 2016).

C. Vinardi, *Sustainable Performance in the Digital Age*,
https://doi.org/10.1007/978-3-032-01252-4_9

This cultural phenomenon is seen in politics and business, where crises often lead organizations to narrow diversity and reshape Executive Committees (ExCom) with members who are alike in terms of nationality, professional experience, education, etc.

As new generations seek deeper meaning, collaborative work, contribution, and respect for both individuals and the planet, future solutions must also take into account these emerging trends (see Sect. 4.3).

The landscape of future technological advancements is complex and challenging to understand. I have chosen to focus on the most promising developments based on current research. To do this, here is a broad classification with two categories: technologies inspired by nature and those part of natural science (De Castro, 2007):

- **Computing inspired by nature:** The main idea if this family of technologies is to develop computational tools (algorithms) by taking inspiration from nature for the solution of complex problems.
  - **AI:** based on Mathematics. It is particularly true for Deep Learning with ANN (Artificial Neuronal Network)
  - **General AI also called Artificial General AI**: is the ultimate AI that is capable of full human behavior.
- **Computing with nature or its connected laws:** The main idea is the use of novel natural materials and/or connected discipline (biology, chemistry, physics…) to perform computation. One notable category is referred to as **Molecular Computing,** which can be broadly defined as the use of (bio)molecules and biomolecular operations to solve problems and perform computation. It has at least three significant subcategories:
  - **Quantum Intelligence (QI) and Computing (QC)**: based on physics.
  - **Brain-Computer Interface (BRI) and Organoid Intelligence (OI)**: based on biology and neuroscience.
  - **DNA computing**: based on biology and chemistry.

Molecular computing can be broadly defined as the use of (bio)molecules and biomolecular operations to solve problems and perform computation (De Castro, 2007).

There are also various advancements involving **nanobots**—miniature robots used in nanomedicine. They are not used for **Computing with nature but for solving natural phenomena** (Abos et al., 2024). Nanobots are typically composed of a combination of nanomaterials, such as carbon nanotubes, metallic nanoparticles, and biomolecules, which are integrated into a microscale robotic system. The fabrication of nanobots often involves advanced techniques, such as self-assembly, 3D printing, and micro/nanofabrication *processes*,[1] to create the necessary components

[1] A more precise definition is found in the Glossary (See No. 2).

and integrate them into a functional device (Abos et al., 2024). While these developments have current applications and are expected to improve in the future, they are not the significant technological breakthroughs for what lies ahead. As a result, I will not consider them in this chapter.

We explored the first category (concerning AI), which is part of Digital Systems that we already looked at. However, it did not discuss **Edge AI** specifically. Edge AI is unique in that its management occurs not in data centers but on local devices—such as smartphones, connected devices, and industrial equipment, etc. These embedded applications are highly energy-efficient and can perform both inference and even learning phases locally.

Traditionally, computing and memory units are separate on a chip**, resulting in 80 to 90% of energy consumed by data transmission between the two** (Dauvé, 2025).

Edge AI is an interesting alternative to minimize environmental impact, but it will require innovations in electronic architecture.

Additionally, I have yet to address the concept of Artificial General Intelligence, which operates differently from standard AI. It will be addressed in Sect. 9.1 before discussing Quantum Intelligence & Computing, le Brain-Computer Interface and Organoïd Intelligence and finally DNA Computing.

9.1 Artificial General Intelligence

9.1.1 Definitions and Origin

Because human intelligence is general, human-like AI is often called Artificial General Intelligence (AGI) (Fjelland, 2020). The **fundamental goal of AGI is to mimic the core cognitive activities of human,** create machines with general intelligence comparable to that of humans (Fei et al., 2022).

AGI is a **type of artificial intelligence that surpasses current capabilities**. It can perform a wide range of tasks autonomously, much like humans. This concept is often referred to as "Strong-AI," in contrast to the current AI types, known as "Narrow-AI" (Goertzel, 2014).

"Narrow-AI" is designed to perform specific tasks with high efficiency, AGI aspires to replicate human-like intelligence, possessing the ability to understand, learn, and adapt across a wide range of domains (Rayhan et al., 2023).

To replicate human behavior, AGI should have the following characteristics (Roli et al., 2022; Rayhan et al., 2023):

- reasoning and problem-solving,
- learning, creating new representations of the knowledge acquired, transferring knowledge, skills acquired in one context to new, unfamiliar situations,
- inference-making,

- using common-sense knowledge,
- autonomously defining and adjusting goals,
- dealing with ambiguity and ill-defined situations.

Nobody knows if and when AGI could be achieved (Siegel, 2023). It is also worth noting that pioneers in this field had a relatively optimistic outlook (Mitchell, 2024):

- In 1965, H. Simon predicted that machines would be capable, within twenty years, of doing any work that a man can do.
- In 1970, M. Minsky declared that "in from three to eight years, we would have a machine with the general intelligence of an average human being".

However, it is clear that by 2025, we are still quite far from having a reasonable timeline for the realization of AGI, if it is indeed possible.

9.1.2 Interest and Potentials

Perhaps the most ambitious scientific quest in human history is the creation of general artificial intelligence, which roughly means AI that is as smart or smarter than humans (Clune, 2020).

Having "humanoids" around has always been the stuff of dreams and nightmares alike, for humans. The world's biggest tech companies and entire governments are making important decisions on the basis of what they think AGI will entail (Mitchell, 2024).

9.1.3 Limitations and Disadvantages

The **overestimation of technology is closely connected with the underestimation of humans** (Fjelland, 2020).

In addition, human intelligence must first be understood in order to replicate it, particularly how the brain functions. We are still far from uncovering all the mysteries and capabilities beyond. Luc Julia, a prominent AI expert at Renault, also emphasized this point:

> *"Understanding is needed to mimic nature, biology, physics, and the brain. However,* ***worldwide experts acknowledge that, despite our efforts, we only understand about 20 to 40% of how the brain functions****. Consequently, we are still far from being able to reproduce it.*
> *Regarding AGI, it's important to note that AI is based on algorithms and mathematics. Humans invented mathematics, and it can not explain*

> *everything so far. Given our present understanding and tools,* ***I do not believe we will achieve AGI for thousands of years, if it's even possible at all!***"
>
> From an interview of Luc Julia with Carine Vinardi

The history of AI has repeatedly disproved our intuitions about intelligence. Many early AI pioneers thought that machines programmed with logic would capture the full spectrum of human intelligence (Mitchell, 2024).

Many who study biological intelligence are also skeptical that so-called "cognitive" aspects of intelligence can be separated from its other modes and captured in a disembodied machine. Psychologists have shown that important aspects of human intelligence are grounded in one's **embodied physical and emotional experiences**. Evidence also shows that individual intelligence is **deeply reliant on one's participation in social and cultural environments**. The abilities to understand, coordinate with, and learn from other people are likely much more important to a person's success in accomplishing goals than is an individual's "optimization power" (Mitchell, 2024).

To **understand another person is not to look into the chemistry of that person's brain, not even into that person's "soul", but is rather to be in that person's "shoes".** It is to understand the person's lifeworld (Fjelland, 2020).

Moreover, the AGI's input knowledge must be explicitly transcribed in digital format. We have seen that this is **only a small fraction of human knowledge.** Physics, for example, is primarily an experimental science. To carry out experiments the physicist must be able to move around, to handle instruments, to read scales, and to communicate with other physicists (Fjelland, 2020).

Quelques challenges et obstacles à lever ont donc déjà été identifiées (Rayhan et al., 2023; Müller, 2014; Roli et al., 2022):

- **Algorithm limits:** It is impossible to list all possible goals, actions, or affordances of an organismic agent in advance
  - Algorithms cannot take or shift perspective and that is why they cannot leverage ambiguity for innovation in the way an organismic agent can. Algorithms cannot jury-rig Not all possible behaviors of an organismic agent can be formalized and performed by an algorithm.
  - AGI is impossible in the current algorithmic frame of research in AI and robotics, since algorithms cannot identify and exploit new affordances. As a direct corollary, truly open-ended evolution into the adjacent possible is impossible in algorithmic systems, since they cannot transcend their predefined space of possibilities.

- **Computational Limitations and Hardware Requirements**
- **Data Availability and Quality for Training**: procuring vast and high-quality datasets that encompass the breadth of human knowledge and experiences remains

a daunting task. Ethical considerations, privacy concerns, and legal restrictions further complicate the acquisition and utilization of such datasets.

- **Ethical Considerations**
  - Ensuring that AGI's actions align with human values and norms becomes paramount.
  - If high-level AI occurs, this would have a significant impact on humanity, especially on the ability of humans to control their fate on Earth. This loss of control is a significant risk, perhaps an existential risk for humanity.
- **Safety and Control Mechanisms**: Ensuring Safety and Control Mechanisms in AGI Systems Safety and control represent fundamental aspects of AGI development, as creating super-intelligent entities necessitates measures to avert unintended consequences or uncontrollable behaviors.
- **Job consideration:** it would raise concerns about job displacement and economic disparities, necessitating careful planning for a smooth societal transition.

The path to AGI traverses through a complex interplay of key components, each carrying its significance and intricacies: perception and understanding of the environment, learning and adaptation mechanisms, memory and knowledge representation and reasoning, planning, and decision-making, synergize to form the bedrock of AGI (Rayhan et al., 2023).

9.1.4 Applications and Perspectives

If AGI were to become a reality, its capabilities would help us explore **innovations beyond our current imagination**. It would allow us to find solutions to highly complex problems across various fields and help address workforce shortages, as we would have "machine" resources at our disposal.

9.2 Quantum Intelligence

9.2.1 Definitions and Origin

Quantum Intelligence (QI), like artificial intelligence (AI), is a form of intelligence based on **Quantum Computing** and the laws of **quantum physics**, dealing with atomic-scale phenomena. Quantum physics explains the behavior and interactions of particles and the force fields that influence them (Soulier, 2024). Here are the main fundamental concepts of quantum physics:

- **Wave-Particle Duality**: This concept describes the unique behavior of microscopic objects, such as electrons, having the properties of both particles and waves.

- **Entanglement**: This phenomenon occurs when two (or more) particles become so interconnected that they share the same properties, even when separated by large distances. For instance, if a measurement is made on one particle, the other particle's state changes instantaneously to stay compatible with the first measurement (Toffano, 2014).
- **Superposition**: This is a particle's ability to have multiple states simultaneously.
- **Quantization** refers to the discrete values of energy associated with electromagnetic waves. This energy is distributed into separate "packets" known as "quanta." Energy transitions occur as quantum "leaps".
- **Exclusion** is the concept that certain microscopic particles, such as electrons, cannot have the same quantum state simultaneously.
- **Indeterminism** highlights the probabilistic nature of measurements in quantum physics. Unlike classical physics, where results can be predicted precisely, quantum outcomes are inherently probabilistic. The result may not be the same when a quantum experiment is conducted under identical conditions.
- **Tunneling** describes the phenomenon where a particle can pass through an energy barrier, even if it doesn't have enough energy according to classical laws.
- **Uncertainty principle** states that it is impossible to know both a particle's position and velocity accurately. The more precisely one of these properties is measured, the less precisely the other can be determined.

The **laws of quantum physics date back over a century** and were not developed by a single individual or team. Starting in the early 1900s, several renowned contributors made significant advancements, including Max Planck and Albert Einstein, who worked on quantization and wave-particle duality; Wolfgang Pauli, known for the exclusion principle; Louis de Broglie and Erwin Schrödinger, who explored superposition; and Werner Heisenberg, who introduced the uncertainty principle.

Additionally, in 1982, Alain Aspect and his team conducted experiments that confirmed the accuracy of quantum theory, particularly regarding particle entanglement. During that period, in the early 1980s, **the second quantum revolution began**, leading to the development of Quantum Computing.

As early as the 1990s, researchers proposed algorithms for quantum computers. **If run on these machines, these algorithms have been mathematically shown to perform certain calculations at speeds that far exceed those of classical computers** (Bourdet, 2021).

Classical computing relies on bits representing binary states: either 0 or 1, meaning there are only two possible values.

In contrast, **quantum computing is based on qubits**, or quantum bits.

Qubits can exhibit superposition, simultaneously representing both 0 and 1. With two qubits, for example, there are four possible states. This capability, along with the principle of entanglement, means that qubits can have several interdependent states at the same time (Louvet, 2024). This fundamental difference gives quantum computing its extraordinary potential.

9.2.2 Interest and Potentials

First, let's return to quantum physics. Although the concepts require a solid scientific foundation and a high level of abstraction, they have led to applications that we already use widely, which I will discuss in Sect. 9.2.4.

Regarding Quantum Computing, its significance and potential are high:

- Qubit coding enables **much faster processing speeds and the ability to handle vast amounts of information, as it grows exponentially** compared to conventional memory (Soulier, 2024).
- Quantum computers could **overcome the mechanical limitations** of engraving that currently restrict the miniaturization and performance of today's chips (Toffano, 2014).
- **Quantum laws would allow quantum computers to carry out massively parallel calculations**, opening new possibilities for solving complex problems (Louvet, 2024).

The potential of quantum computing will be truly appreciated when we have a machine capable of working with a large number of qubits.

The ultimate goal is to develop quantum computers that can solve **real-world problems more quickly and efficiently than current supercomputers**. Here are some potential areas of application to consider (Louvet, 2024; Soulier, 2024):

- **Molecular Simulation**: This could be used to discover new drugs or optimize materials.
- **Network Optimization**: This has practical transportation, logistics, and energy systems applications.
- **Fundamental Research**: Quantum simulations could help towards a deeper understanding of the laws of physics.
- **Cybersecurity**: This field has a significant opportunity, particularly with **quantum cryptography**. Using photons can enhance security against hacking, although we will discuss potential drawbacks of quantum computers related to this subject in the next chapter.

Some believe that physics-based prospects are more realistic than those of AGI. Luc Julia from Renault also shares this opinion:

> *"Quantic (based on Physics) and Organoïd (based on Biology) Intelligence, may be promising* ***as it will not 'only' come from Mathematics****."*
>
> From an interview of Luc Julia with Carine Vinardi

9.2.3 *Limitations or Disadvantages*

The quantum computer, as a macroscopic object with a memory capacity comparable to the earliest binary computers and capable of tackling various problems, **does not yet exist**. Instead, we currently have specialized computers that have a small number of qubits (Bourdet, 2021). Moreover, using Windows on a quantum computer would not be feasible, and even if it were, it would be highly inefficient (Bourdet, 2021).

Currently, there are several obstacles to the implementation of the first quantum computer, with the main challenges being (Futura, 2024; Toffano, 2014; Bourdet, 2021; Soulier, 2024; Vicenti, 2024):

- **Entangling large numbers of Qubits**

  While quantum simulation has already produced results, increasing the number of qubits could lead to even more significant advancements.

- **Addressing Qubit Instability (Decoherence):** Qubits often lose their quantum information when they interact with their environment, which can include magnetic fields, light, and thermal agitation. This loss of coherence tends to accelerate as the number of qubits increases.

  The longer we maintain coherence with a large number of qubits, the greater the potential for computing power. To achieve this, we need to understand the environment surrounding the particles better, as no system is perfectly isolated. However, a key advantage of quantum simulation is that decoherence is not necessarily an obstacle; in fact, the systems we simulate are also affected by this phenomenon, meaning that perfect quantum computers may not be required for progress.

- **Countering probabilistic results** related to quantum laws.

  Classical computers yield predictive results.

- **Finding a more cost-effective way to design equipment. Current equipment** based on superconducting electrical circuits **must be cooled to -273 °C and shielded from vibrations and interference.** This is necessary due to the inherently unstable state of qubits.

Google, in collaboration with NASA, has acquired a quantum computer. This machine is not a traditional computer; rather, it is a quantum computer that can be programmed to perform only specific algorithms. This has sparked lively controversy, and opinions within the scientific community are divided.

- **Reducing the high error rate** is one of the main challenges in this field, particularly related to entanglement. The goal is to lower this error rate to below one in a million. The best error correction technologies can only achieve about one in a thousand.
- **Simplifying the programming of qubits.**

  Current programming requires a **deep understanding of quantum physics.**

As opportunities in cybersecurity are being explored, some researchers express concerns that Quantum Computing could pose a significant risk. Quantum computers have the potential to crack current encryption codes. With their immense computing power, these quantum machines could solve asymmetric encryption algorithms that secure electronic communications in just a few hours, a task that would take classical computers thousands of years to complete (Gaie et al., 2024).

So, is quantum computing an opportunity for advancements in cryptography or a threat? It is wise to consider all potential scenarios at this stage.

At the end of 2024, Google reached a significant milestone with the launch of Willow, their new quantum processor. This advancement marks a major step forward in overcoming the challenges of error correction. To achieve acceptable error levels, Google is also developing more reliable logical qubits, targeting error rates of less than one in a million. This ambitious objective will require approximately 1457 physical qubits for each logical qubit, relying on new developments in error correction and engineering (Louvet, 2024).

According to Google, Willow will be capable of solving a problem in five minutes, taking the fastest and most powerful computers today, 1024 years.

However, predicting when the first fully error-tolerant quantum computers will be available remains uncertain. Could they emerge in 10, 15, or even 20 years?

Addressing these challenges requires substantial investment, similar to what is needed for AI. However, the required technological components are being developed, and AI has enabled us to define the ecosystem (Virol, 2025). As Luc Julia from Renault points out, innovations must also have a globally coherent purpose in addition to overcoming technological hurdles:

> ***"The environmental impacts must be considered for all new technologies.*** *Quantum computers run at stable temperatures close to –273°C, which is power-intensive; otherwise, they become more unstable. As for General AI, we are still far from identifying the most sustainable and effective solutions—* ***if we ever do!***

*Additionally, **we need to answer better WHY** we need to develop these technologies."*

From an interview of Luc JUlia with Carine Vinardi

9.2.4 Applications and Perspectives

As we await advancements in Quantum Computing, quantum physics **is already a part of our daily lives in various technologies. For example** (Bourdet, 2021; Soulier, 2024):

- **Lasers**: This unique form of light is made up of photons, identical particles of light. Lasers are created by stimulating atoms to transmit energy in unison.
- **Optical Fibers**: These transmit data as light pulses over long distances
- **Semiconductors** are between a conductor and an insulator. It becomes conductive when a small electrical voltage is applied in a particular way.
- **LEDs (Light Emitting Diodes)** are created by combining controlled light emission with semiconductors.
- **Photovoltaic Panels** and CCD Sensors used in cameras are closely linked to the principle of wave-particle duality.

Many applications already have a high TRL.

Although the TRL of quantum computing is making progress for IT applications, it remains relatively low. However, several companies have successfully implemented quantum algorithms (Bughin, 2025):

- **Volkswagen** used quantum technology in 2019 to optimize bus routes in real time during the Web Summit, significantly reducing traffic congestion and demonstrating the potential of quantum technology for urban mobility.
- **JPMorgan Chase** has leveraged quantum computing to accelerate its financial models, achieving a fivefold increase in training speed compared to traditional portfolio optimization models.
- **Daimler AG** has focused on rapidly simulating over 150 molecular reactions related to lithium-sulfur batteries. This has shortened research and development times for electric car battery innovations from months to hours.

Some analysts estimate significant growth potential in the application market, which currently represents 15–20% of total spending and is projected to be worth between 5 and 20 billion USD in the next five years, with a sharp acceleration expected afterwards (Bughin, 2025).

At present, businesses have two options: either to rent Nvidia GPUs (and bear the cost) or invest in the quantum computing potential for a substantial leap forward (Bughin, 2025).

Major tech companies, known as GAFAM, are already making strides in this area. Google is developing its Willow chip, Amazon is enhancing its Amazon Braket service by collaborating with startups to integrate various quantum algorithm technologies, and Microsoft is expanding its Azure Quantum service.

Companies should strive to develop a strong internal quantum culture to prepare for upcoming applications. This includes training employees in AI and quantum technologies, understanding their interaction, and investing in hybrid quantum–classical platforms (Bughin, 2025).

9.3 Brain-Computer Interfaces (BCI) and Organoid Intelligence (OI)

9.3.1 Definitions and Origin

Organoid Intelligence (OI) combines two disruptive technologies based on biology, neuroscience, and digital technology (Bastien, 2024):

- **The 3D culture of human brain cells**, known as organoids, mimics the functioning of the human brain.
- **Brain-Computer Interfaces** (BCIs) explore using cell cultures, biological tissues, and even mini-brains to create intelligent systems.

A **Bioelectronic Interface** (BCI) is needed for brain organoids to interact with electronic components. This interface enables bidirectional communication between biological cells and modern electronic circuits (Bastien, 2024).

BCI is a **communication system that directly interfaces the brain and a device or computer**. BCIs are systems that translate a measure of a user 's brain activity into messages or commands for an interactive application (Lotte et al., 2018). It detects brain neural signals and analyzes and converts them to commands that are transmitted to output devices to perform specific tasks. This type of system creates an alternative pathway in contrast to neuromuscular transmission (Farzan, 2025).

For its part, OI requires scaling up current brain organoids into complex, durable 3D structures enriched with cells and genes associated with learning, and connecting these to next-generation input and output devices and AI/machine learning systems (Smirnova et al., 2023). Unlike the ongoing debate regarding "intelligence" in AI, this has a mini-brain developed in the lab that can learn.

Let's look at the origins of BCIs. The idea of an interface between the brain and machines isn't new; it was made possible by the German scientist Hans Berger who discovered that the human brain produces electrical currents in 1920.

The end of 80's and beginning of the 90's saw a handful of researchers from the USA and Europe pioneering the BCI field, by proposing the first real-time and working BCI implementations. The journal "Brain-Computer Interfaces" published its first issue in 2014. Today, the most advanced BCI equipment uses AI (at least machine learning ML techniques) (Lotte et al., 2018).

In contrast, OI was first described in 2013, but it gained significant public attention in 2022, thanks to the emergence of this new discipline spearheaded by Cortical Labs. This Australian start-up, affiliated with the University of Melbourne, made headlines when **human-made neurons were grown in the laboratory and connected to a two-dimensional microelectrode array**, allowing them to learn the video game Pong (which involves moving a paddle to hit a ball) in just five minutes. The formal term Organoid Intelligence (OI) was coined in 2023 by an international team led by American researchers (Théodule, 2023).

9.3.2 Interest and Potentials

In 2024, the interest in BCIs (Mann et al., 2024) emerged prominently in the healthcare sector. It is conceivable that, in the future, non-invasive BCIs—such as wearable brainwave monitors like Cogwear, Emotiv, or Muse—could communicate with AI design software.

As BCIs continue to evolve, the ability to express external, nonverbal facial cues will be needed. **The potential of BCIs to enhance synchronization between the human brain and AI presents exciting yet challenging** opportunities for human–computer collaboration.

The interest in this field was sparked by the physical constraints of classical computers, particularly the challenges associated with increasing the number of transistors on a small chip. In contrast, the human brain operates differently, with approximately 100 billion neurons having over 10^{15} synaptic connections.

There is a significant difference in power compared to our current technology (Bastien, 2024):

- **Data Processing**: Brain organoids can *process*[2] information similarly to the human brain, utilizing neural networks to perform cognitive tasks such as learning, pattern recognition, and sensory processing.
- **Learning and Adaptation**: Unlike traditional artificial intelligence systems, which require vast amounts of data and computing power for training, brain organoids can learn and adapt more efficiently. They do this by employing biological mechanisms of neuronal plasticity.

The advantages of OI are numerous (Bastien, 2024; Théodule, 2023):

- **Energy Efficiency**: Brain organoids consume significantly less energy compared to current supercomputers. The reactions occurring in organoids are chemical, while those in Digital systems rely on electricity. For instance, when operating at “full power,” the 1.4 kg human brain uses only 20 watts. In contrast, the Enterprise

[2] A more precise definition is found in the Glossary (See No. 2).

Frontier, currently the most powerful computer in the world, needs 21 megawatts to achieve speeds comparable to the human brain and occupies an area of 680 m^2. This figure does not account for the cost of its creation (approximately 600 million euros) or the running cost for its operation.

- **Speed and Efficiency**: Due to their biological neural networks, brain organoids can *process*[3] information more quickly and efficiently than systems based on microchips.

  OI could **enable to be faster, more efficient, and more powerful than silicon-based computing and AI, and only require a fraction of the energy**.
- **Advanced cognitive abilities:** Brain organoids **may surpass current artificial intelligence in understanding and interacting with their environment**, with their ability to mimic human cognitive functions.

OI research could also improve our understanding of brain development, learning, and memory, potentially helping to find treatments for neurological disorders such as dementia (Smirnova et al., 2023).

9.3.3 *Limitations and Disadvantages*

For BCIs, the main limits are ethical and societal, particularly regarding access to care (Mann et al., 2024):

- The **preservation of human autonomy**, which means that individuals should be able to make informed choices, **is a fundamental principle**.
- **Equitable access to these technologies is urgent**. If only certain privileged groups can benefit from brain-machine synchronization, it risks exacerbating existing societal divides.

In addition to the technical challenges of organoids and beyond, ethical limits are similar to those of BCIs, especially **considering the development of mini-brains in the laboratory**. Researchers advocate for a discussion involving legal and ethical representatives, researchers, and members of the public to advance and structure this work appropriately. However, one must also consider the potential military advancements that could arise, similar to those seen with AI.

9.3.4 *Applications and Perspectives*

The **first BCI applications are already being used in healthcare**. BCI technology captures neural signals from different brain regions to generate meaningful outputs

[3] A more precise definition is found in the Glossary (See No. 2).

for restoring function to the nervous system, particularly in conditions like stroke, neurodegeneration, and spinal cord injury (Farzan, 2025).

The most advanced versions incorporate AI. **In the near future, BCIs may enable the transformation of thoughts into digital commands and facilitate human communication**. This advancement will allow paralyzed individuals to control devices solely through their thoughts (Mann et al., 2024).

For OI, the current TRL is relatively low because it is still in the development phase.

It could serve as **biological hardware for future computers** by leveraging the human brain's information storage capacity, which is **estimated to be around 2500 terabytes** (Théodule, 2023).

In addition to its computing, storage, and learning applications, OI could contribute to medical research by enhancing our understanding of neurological disorders, particularly neurodegenerative diseases.

It could also be used for drug development, disease modeling, toxicology testing, and our basic understanding of the human brain (Smirnova & Hartung, 2024).

However, several additional steps are necessary before the first applications appear. Figure 9.1 presents a roadmap outlining key milestones.

9.4 DNA Computing

9.4.1 Definitions and Origin

DNA (Deoxyribonucleic Acid computing is a cutting-edge interdisciplinary field combining biology, chemistry, computer science and mathematics (Ma et al., 2021).

DNA computing is the performing of computations using biological molecules, rather than traditional silicon chips (Amos, 2019). It uses **DNA hybridization and replication *processes*[4] to solve computational problems** (Ratner & Ratner, 2002).

DNA computing leverages the self-assembly and programmable nature of DNA molecules to perform massively parallel computations. By encoding information in DNA sequences and using enzymatic reactions to manipulate and amplify the DNA (Abos et al., 2024).

The core idea of DNA computing is to use DNA's biochemical properties to perform computational tasks. Since it uses biomolecular rather than electronic and silicon components (Amos, 2019). DNA computing primarily **uses synthetic DNA that is custom-made in laboratories** (Amos, 2019).

A computation may be thought of as the execution of an algorithm, which itself may be defined as a step-by-step list of well-defined instructions that takes some input, *processes*[5] it, and produces a result. In DNA computing, information is represented

[4] A more precise definition is found in the Glossary (See No. 2).

[5] A more precise definition is found in the Glossary (See No. 2).

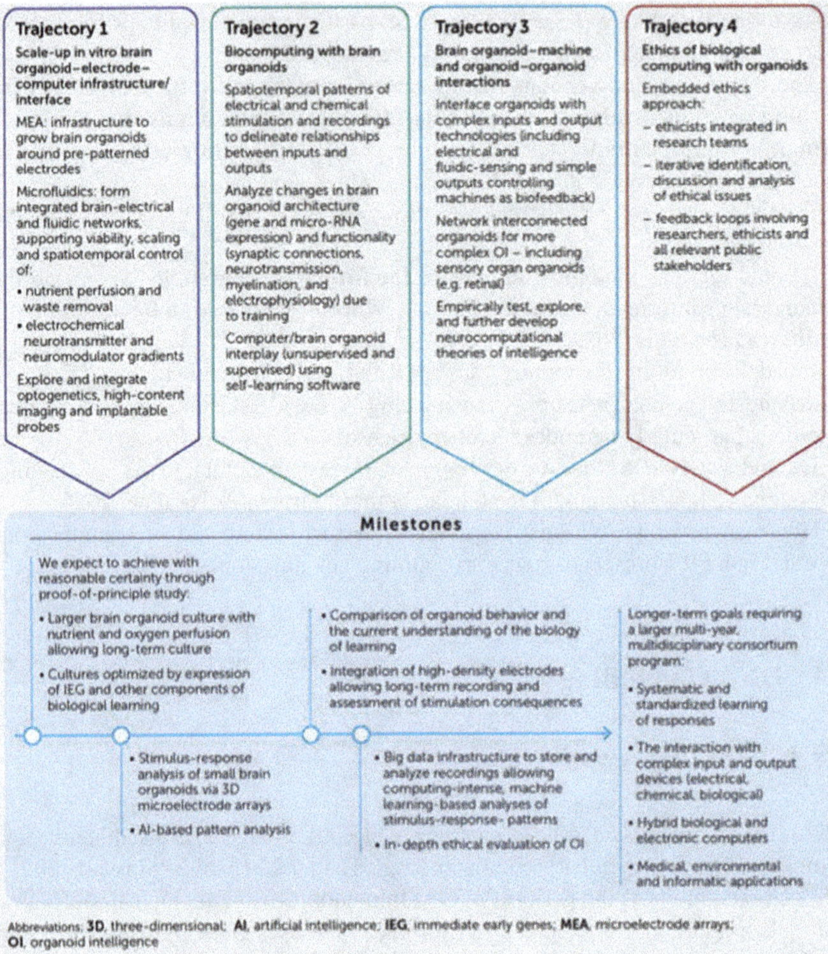

Fig. 9.1 OI roadmap with main milestones (Smirnova et al., 2023)

using the four- character genetic alphabet (A [adenine], G [guanine], C [cytosine], and T [thymine]), rather than the binary alphabet (1 and 0) used by traditional computers.

This is achievable because short DNA molecules of any arbitrary sequence may be synthesized to order (Amos, 2019). Think of **enzymes as hardware, DNA as software**, those computers would "computes" using enzymes that react with DNA strands, causing reactions (Trivedi, 2011).

The **idea that molecular systems can perform computations is not new** and was indeed more natural in the pre-transistor age (Ezziane, 2006). In 1994 L. Adleman demonstrated the possibility of using DNA to solve calculations. He proposed that the makeup of DNA and its multitude of possible combining nucleotides could have application in computational research techniques (Adleman, 1994).

9.4.2 Interest and Potential

The human body is, in many respects, an extremely efficient computer. One way in which it carries and *processes*[6] data is through DNA and its complex biochemistry (Ratner & Ratner, 2002).

Moreover, demands on faster information *processing* speed and denser data storage are catalyzing new computation modes. DNA Computing has shown its potential in information *processing* and storage due to its predictable base pairings and nanoscale size for programmable and high-throughput coding, as well as computing (Ma et al., 2021).

This technology has the potential to **store and *process*[7] vast amounts of information in a compact space while improving energy efficiency**. DNA computing could also lead to the development of **miniaturized biological computers** that are highly effective at solving specific **complex problems, such as optimization and cryptography**.

Main Potential/Advantages (Trivedi, 2011):

- Parallel Computing
- Incredibly light weight
- Low power Dissipation
- Solves Complex Problems quickly
- Gigantic Memory Capacity
- Clean, Cheap and Available

9.4.3 Limitations and Disadvantages

DNA computing is still in the research and development phase, and its Technology Readiness Level (TRL) remains low (Ma et al., 2021).

Main limits/disadvantages (Amos, 2019; Ma et al., 2021; Trivedi, 2011):

- Ethical & societal concerns (DNA use and related nanotechnology): privacy concerns, potential misuse, and the equitable access to these advanced therapies.
- Occasionally slower
- Human assistance is required
- Information can not be transferred/Lack of environmental compatibility
- Computing complexity
- Lack of universality: only specific problems
- No universal method of data representation and analysis
- Reliability Problems/Large amount of Errors
- High cost (time)

---

[6] A more precise definition is found in the Glossary (See No. 2).

[7] A more precise definition is found in the Glossary (See No. 2).

In addition, relatively small problems would require unmanageable volumes of DNA (on the order of large bathtubs) in order to represent all possible answers (Amos, 2019).

Effective strategies for problem-solving with this technology are yet to be defined.

9.4.4 Applications and Perspectives

Although DNA computing is still in the research and experimental phase, it promises a major breakthrough in the future. **Since its inception, it has shown potential for solving complex computational problems in multiple fields**, including medicine, cryptography, artificial intelligence, and others (Amos, 2019).

Scientists have created biosensors capable of diagnosing diseases by detecting specific genetic markers. The possible applications of this emerging technology could have an impact on many areas, including intelligent medical diagnostics and drug delivery, tissue engineering, energy, and the environment (Amos, 2019).

For example, DNA-based circuits could identify cancer cells by recognizing abnormal DNA sequences in a patient's body (Amos, 2019).

The engineering and programming of biochemical circuits would transform industries that use chemical and nanostructured materials. An important research aspect is the development of the communication interface between the biological materials and electronic components (Ezziane, 2006).

Expected applications (Ezziane, 2006; Trivedi, 2011):

- DNA chips: a key challenge remains of how to design and develop intelligent machine learning techniques so as to effectively explore such a vast amount of information. The problem of over fitting is a leading concern with machine learning approaches to DNA chip data.
- Genetic programming
- Pharmaceutical applications
- Cracking of coded messages

Like Quantum Computing, the applications of DNA computing synthesis techniques for making arbitrary sequences are becoming simpler, and custom DNA can already be ordered on a reasonable budget (Ratner & Ratner, 2002).

9.5 Hybrid Technologies

The future may see technologies combining rather than having just one. Scientists are already exploring several potential combinations:

- **Quantum AI (QAI)**

  Some experts believe artificial intelligence (AI) and quantum intelligence (QI) are the next frontiers in economic and technological innovation. There is also a promising yet uncertain market for hybrid AI and QI models (QAI) (Bughin, 2025).

  – AI could benefit QI in several ways:

    For instance, Machine Learning (ML) can further reduce quantum errors. By using AI pattern recognition, we can detect anomalies in qubit behavior, predict noise patterns in quantum systems, and optimize quantum error correction codes (Bughin, 2025).

  – Conversely, QI could help AI:

    Traditional AI models often exhibit hallucinations, an issue that could be addressed with quantum approaches. There is optimism surrounding the idea of complementarity (Virol, 2025). AI provides powerful, well-established algorithms that require high processing intensity, but they are limited by their significant energy consumption. This is where quantum computing could play a key role, driving advancements in AI. This concept gained traction around 2018-2019. Quantum computers could be particularly valuable for applications with limited inputs and outputs but they involve complex physics, linked to superconductivity or the simulation of chemical molecules.

The **current market value of QAI applications is essentially nonexistent.** Most of the spending currently allocated to QAI goes toward companies developing infrastructure and software platforms, such as Amazon and Nvidia, and native quantum players like Rigetti and D-Wave, leaders in developing quantum processors (Bughin, 2025).

Certain tipping points must be reached for broader adoption of QAI technology, including cost reduction, technological breakthroughs, and hybrid integration. An effective "no regrets" strategy would need fostering a quantum culture, exploring hybrid platforms, and conducting targeted experiments. **Ignoring the disruptive potential of quantum computing could be a significant strategic error** (Bughin, 2025).

- **Quantum OI**

  Organoid Intelligence (OI) could spark an unprecedented technological revolution through Biological Quantum Computers.

  This convergence of biology and quantum computing could redefine our understanding and use of technology (De Marval, 2023).

Whereas traditional AI depends on algorithms and computer calculations, Organoid Intelligence integrates biology into the *process*.[8] The combination of Organoid Intelligence with Quantum Computing means that the key components of these computers are based on biological molecules, which use quantum properties (De Marval, 2023).

- **AGI and OI**

  Incorporating organoid neural tissues could enhance or inspire AGI architectures, leading to more powerful and adaptive systems.

- **AI and DNA Computing**

  AI could be combined with this technology to leverage the vast amount of information available with DNA chips.

  The design of artificial intelligence techniques in DNA computing would serve to break the barrier of this brute-force method and get a final solution from a very small initial data pool, avoiding enumerating all candidate solutions.

  Arguably, forms of nano-computation, such as Quantum Computing or DNA computing may help unlock true AI (Ratner & Ratner, 2002).

This concludes our brief overview of the most promising technologies. **These were started over ten years ago,** and it is challenging to predict which ones will lead to practical applications for personal or professional use and over what timeframe.

Some pioneers are committing by investing in developing these technologies, while others are waiting for TRLs to increase. Determining which will succeed is difficult, and how we can scope these developments.

Issues of ethics and the protection of individuals are already being raised today, particularly concerning Digital technology and AI. **Achieving double sustainable performance will require a global consensus on a clear framework to address these risks and prevent potential solutions from becoming significant problems.** As Albert Einstein aptly pointed out:

> *"Technological progress is like an axe in the hands of a pathological criminal."*

At the company level, monitoring technical advances is more important than ever to seize upcoming opportunities. To achieve double sustainable performance, these innovations can be combined with traditional, proven methods such as Lean and Digital strategies. This approach allows for a holistic and pragmatic view, enabling the effective use of resources.

[8] A more precise definition is found in the Glossary (See No. 2).

References

Abos, P., Hamid, P., & Van Dam, S. (2024). *DNA computing and other biomolecular computation approaches*. https://www.researchgate.net/publication/383436500_DNA_computing_and_other_biomolecular_computation_approaches

Amos, M. (2019). *DNA computing*. Encyclopaedia Britannica. . https://www.britannica.comhttps://www.britannica.com/technology/DNA-computing

Adleman, L. M. (1994). Molecular computation of solutions to combinatorial problems. *Science, 266*, 1021–1024. https://doi.org/10.1126/science.7973651

Bastien, L. (2024). *Intelligence organoïde: ces ordinateurs à cerveau humain vont surpasser l'IA, tout savoir*. Lebigdata.fr, Intelligence artificielle https://www.lebigdata.fr/intelligence-organoide-oi

Bourdet, J. (2021). Ordinateur: les promesses de l'aube quantique. *CNRS Le Journal*. https://lejournal.cnrs.fr/articles/ordinateur-les-promesses-de-laube-quantique

Bughin, J. (2025). Le monde avant et après Willow: quelle stratégie d'entreprise liée au mariage quantique et IA? *Management et Datascience, 9*(1). https://doi.org/10.36863/mds.a.39967

Clune, J. (2020). *AI-GAs: AI-generating algorithms, an alternate paradigm for producing general artificial intelligence*. arXiv preprint arXiv:1905.10985 (2019)

Dauvé, S. (2025). *Semi-conducteurs: l'Europe peut-elle regagner du terrain?* Polytechnique Insights. https://www.polytechnique-insights.com

De Castro, L. N. (2007). Fundamentals of natural computing: An overview. *Physics of Life Reviews, 4*(1), 1–36. https://doi.org/10.1016/j.plrev.2006.10.002

De Marval, E. (2023). *Intelligence Organoïde: la révolution des ordinateurs biologiques*. Technologie & Société Numérique. https://sociologique.ch/intelligence-organoide-la-revolution-des-ordinateurs-quantiques-biologiques

Ezziane, Z. (2006). DNA computing: Applications and challenges. *Nanotechnology, 17*(2006), R27–R39. https://doi.org/10.1088/0957-4484/17/2/R01

Farzan, F. (2025). Brain-computer interface. In *Handbook of neural engineering*. Academic Press. https://doi.org/10.1016/B978-0-323-95730-4.00003-2

Fjelland, R. (2020). Why general artificial intelligence will not be realized. *Humanities and Social Sciences Communications, 7*(1), 1–9. https://doi.org/10.1057/s41599-020-0494-4

Fei, N., Lu, Z., Gao, Y., Yang, G., Huo, Y., Wen, J., et al. (2022). Towards artificial general intelligence via a multimodal foundation model. *Nature Communications, 13*(1), 3094. https://doi.org/10.1038/s41467-022-30761-2

Futura Web Site with ETX DAILY UP (2024). *L'informatique quantique prête à transformer l'humanité nous sommes à quelques pas de son moment « Spoutnik »*. https://www.futura-sciences.com/tech/actualites/ordinateur-quantique-informatique-quantique-prete-transformer-humanite-nous-sommes-quelques-pas-son-moment-spoutnik-115192/

Gaie, C., & Langlois-Berthelot, J. (2024). *Cryptographie: comment se protéger à l'ère quantique?* Polytechnique Insights. https://www.polytechnique-insights.com

Goertzel, T. (2014). The path to more general artificial intelligence. *Journal of Experimental & Theoretical Artificial Intelligence, 26*(3), 343–354 (pp. 1–12). https://doi.org/10.1080/0952813X.2014.895106

Lotte, F., Nam, C. S., & Nijholt, A. (2018). Introduction: Evolution of brain-computer interfaces. In *Brain-computer interfaces handbook: Technological and theoretical advance* (pp. 1–11). Taylor & Francis (CRC Press) (9781498773430, hal-01656743)

Louvet, B. (2024). Willow: le processeur quantique de Google qui redéfinit l'informatique du futur. https://sciencepost.fr/willow-processeur-quantique-de-google/

Ma, Q., Zhang, C., Zhang, M., Han, D., & Tan, W. (2021). DNA computing: Principle, construction, and applications in intelligent diagnostics. *Small Structures*, *2*(11). https://doi.org/10.1002/sstr.202100051

Mann, H., Walther, C. C., & Platt, M. (2024) *L'avenir de la synchronisation du cerveau et de la machine*. https://www.polytechnique-insights.com

Mitchell, M. (2024). Debates on the nature of artificial general intelligence. *Science 383*, eado7069. https://doi.org/10.1126/science.ado7069

Müller, V. C. (2014). Risks of general artificial intelligence. *Journal of Experimental & Theoretical Artificial Intelligence, 26*(3), 297–301. https://doi.org/10.1080/0952813X.2014.895110

Ratner, M., Ratner D. (2002) Nanotechnology: A Gentle Introduction to the Next Big Idea, https://doi.org/10.1016/S1369-7021(03)00236-0

Rayhan, A., Rayhan, R., & Rayhan, S. (2023). Artificial general intelligence: Roadmap to achieving human-level capabilities. *DOI, 10*, 13140.

Roli, A., Jaeger, J., & Kauffman, S. A. (2022). How organisms come to know the world: Fundamental limits on artificial general intelligence. *Frontiers in Ecology and Evolution,* 9, 806283. https://doi.org/10.3389/fevo.2021.806283

Siegel, E. (2023). *The AI Hype cycle is distracting companies.* Harvard Business Review. https://hbr.org/2023/06/the-ai-hype-cycle-is-distracting-companies

Singh, S. (2023). *The mega trends that will shape our future world.* Forbes. https://www.forbes.com/sites/sarwantsingh/2023/10/30/the-mega-trends-that-will-shape-our-future-world/

Smirnova, L., Caffo, B. S., Gracias, D. H., Huang, Q., Morales Pantoja, I. E., Tang, B., Zack, D. J., Berlinicke, C. A., Boyd, J. L., Harris, T. D., Johnson, E. C., Kagan, B. J., Kahn, J., Muotri, A. R., Paulhamus, B. L., Schwamborn, J. C., Plotkin, J., Szalay, A. S., Vogelstein, J. T., … Worley, P. F., & Hartung, T. (2023). *Organoid intelligence (OI): the new frontier in biocomputing and intelligence-in-a-dish.* https://doi.org/10.3389/fsci.2023.1017235

Smirnova, L., & Hartung, T. (2024). The promise and potential of brain organoids. *Advanced Healthcare Materials, 13*, 2302745. https://doi.org/10.1002/adhm.202302745

Soulier, L. (2024). *Démystifier l'IA générative: le vrai, le faux et l'incertain.* Polytechnique Insights. https://www.polytechnique-insights.com/tribunes/science/demystifier-lia-generative-le-vrai-le-faux-et-lincertain/

Théodule, M. L. (2023). *Après l'intelligence artificielle, l'intelligence organoïde.* Le Temps/Sciences. https://www.letemps.ch/sciences/apres-l-intelligence-artificielle-l-intelligence-organoide

Toffano, Z. (2014). *Intrication quantiqu: mythe ou réalité? Res-Systemica, Modélisation des Systèmes Complexes, 12* (article 14 (2014), hal-01264835v3)

Trivedi, D. (2011). *DNA Computing*. https://doi.org/10.13140/2.1.2789.2162

Vicentini, F. (2024). Informatique quantique et IA: moins compatibles que prévu?, https://www.polytechnique-insights.com

Vinardi, C. (2016). *Les équipes multiculturelles en entreprise: Les comprendre, les gérer, les développer*. Maxima Editions.

Virol, G. (2025). *Le calcul Quantique: à la poursuite de son moment ChatGPT* (pp. 12–13., Usine Nouvelle No. 3740, Mars 2025.

Chapter 10
Conclusion

Often referred to as our environmental capital, the planet must be cared for, and this is frequently discussed in the media. However, it is equally important to prioritize the well-being of employees and suppliers, known as human capital. This includes addressing their physical and psychological health for the best working conditions and aspirations. We have seen that the aspirations of younger generations, the driving force of the future, go beyond just environmental concerns.

- Companies must **invest with the same intensity as in environmental matters, in social and societal issues.**

I have focused on double sustainable performance because it combines traditional financial performance with a social, societal, and environmental responsibility commitment. This approach must be developed for long-term sustainability. We have seen a list of criteria and KPIs to evaluate this performance. Moreover, achieving this level of performance means selecting various actions that are meaningful to people:

- By **prioritizing strategies to meet customer needs effectively, we can optimize our consumption to just what is necessary.** These include **human resources, material resources** (like equipment, materials, products, energy and water) **and intangible resources** (like data and GHG emissions).
- By **minimizing waste across these human, material, and intangible resources.**
- At the same time, **ensuring safe and efficient working conditions.** Safety and ergonomics are consistent components required for sustainable performance.

This is why I became interested in the role of Digital solutions, combined with established methodologies like Lean, as well as best practices in management, leadership, and innovation. When Lean is implemented correctly, it naturally incorporates these aspects.

- **Proven approaches based on human dynamics are always relevant, even in a technological context.**

C. Vinardi, *Sustainable Performance in the Digital Age*,
https://doi.org/10.1007/978-3-032-01252-4_10

To highlight this, I would like to quote D. Acemoglu, Professor of Economics at MIT:

> *"We are in the age of AI but I remain convinced that* ***human productivity, human ingenuity and human resources broadly construed****, are still key to the* ***meaningful flourishing of countries.****"*

Environmental performance is increasingly becoming the focus of academic research. However, the concept of **double sustainable performance**—incorporating social and societal issues along with a timeline—**is rarely examined.** Moreover, while topics such as managerial practices, innovation, Lean and Digital transformation are frequently covered in specific publications, **these subjects are seldom explored holistically. The investigation of how these influence double sustainable performance, is even less common.** Given this gap, I found it worthwhile to adopt an empirical-deductive approach, which I have illustrated through interviews and current literature.

- Clearly, **this area of study warrants continued exploration.**

Before analyzing the combination, it was necessary to define the concepts of Lean and Digital separately. Lean is understood as a comprehensive system that adopts a holistic and pragmatic approach, incorporating the fundamentals of managerial and organizational performance. Digital encompasses various elements, including, but not limited to, Artificial Intelligence. This clarification allowed for alignment on terminology (establishing a common language) and substance (identifying key principles and components). **This was necessary to prevent errors or misinterpretations that could result in vastly different performance outcomes.** Additionally, this approach highlighted significant milestones in the development of Lean and Digital, helping to illustrate how these methodologies evolved over a long timespan. This also helped to distinguish regression phases, such as those experienced by AI and the Metaverse.

- With its proven methodologies, we have observed that **Lean significantly contributes to achieving double sustainable performance.** This is made possible because it **focuses on customer satisfaction alongside employee well-being, the effective use of necessary resources, a structured approach to waste elimination, and a practical and sustainable implementation of automation, particularly with Karakuri.**

Moreover, we have found that Digital technology can enhance the pursuit of double sustainable performance in a way that complements Lean. This is especially true regarding *process*[1] optimization and the development of new products or services. Therefore, **combining Lean and Digital can maximize double sustainable performance, though a few precautions should be considered:**

- Consider **deploying in the right order**, with the basics of Lean first. **Ensure that development is structured in coherence with maturity and organization.**
- Integrate **deployment with your business strategy.**
- Implement **decisive cybersecurity measures for digital deployment** to protect customers, suppliers, employees, and the company.
- **Focus on digital applications, particularly on AI, to reduce environmental impact and grow the social potential.**
- **Establish rigorous and comprehensive data management practices** to safeguard customers and employees while minimizing environmental effects.
- **Limit the environmental impact of data centers and reduce their carbon footprint, water consumption, and electricity waste.** Projections indicate that electricity consumption necessary for the development of Digital technology may exceed the current capacity of many states. This increased demand for electricity production requires significant infrastructure investments, while the widespread deployment of renewable energy sources is leading to instability in the electrical grid, which currently lacks a sustainable solution. Moreover, the issue of water scarcity is becoming increasingly pressing. This critical situation must not be overlooked any longer. If current consumption trends continue, by 2050, it is projected that two-thirds of the world's population will reside in water-stressed areas (Porcher, 2024).

Implementing a combined Lean and Digital deployment to achieve double sustainable performance is challenging but achievable. Beyond demonstrating individual contributions, engaging and developing organizations is decisive. To do this, we must:

- **Support change through all phases by identifying the most effective and appropriate paths to initiate, deploy, and sustain it.** This helps to foster a learning organization at every level.
- **Engage and develop leaders who can facilitate this change. These leaders must cultivate resilience** throughout the organization, inspire their teams, and build the trust necessary for high performance. **Additionally, they must be curious and capable of effectively monitoring and training their teams.**

I have ended this book with a brief overview of how to address double sustainable performance, particularly through the integration of Lean and Digital practices. I provided examples from both transactional and industrial companies selected based on their implementation of Lean and Digital strategies, as well as the ESG (Environmental, Social, and Governance) reports they have published. We observed that,

[1] A more precise definition is found in the Glossary (See No. 2).

for most companies (with the exception of Toyota), it remains challenging to clearly and simply communicate their *Strategic Objectives*[2] regarding the double sustainable performance of specifically Lean and Digital practices.

- I gave hypotheses regarding the possible causes of these difficulties, **which should be validated through a comprehensive diagnosis with the selected companies.**

Understanding and addressing challenges is recommended to fully leverage the combination of Lean and Digital strategies for achieving double sustainable performance. It is fundamental to know how to integrate performance actions effectively, especially since alternatives to Digital solutions can also shape the future. Four promising technologies have been addressed, and their future development could complement Lean and/or Digital approaches to enhance double sustainable performance.

- Now more than ever, **leaders at all levels must prioritize and support their teams in making informed decisions and taking action, rather than merely managing teams based on objectives alone. Therefore, the *Hoshin kanri*[3] approach has become more relevant than it was in the past.**

We must not overlook the importance of solutions, particularly of tools, as highlighted by French anthropologist A. Leroi-Gourhan, who said:

> *"The tools of men extend their bodies and touch/elevate their spirits."*

To achieve double sustainable performance, current and future technological solutions must **support the development of individuals and companies while giving a clear sense of purpose.**

Digital tools, especially AI, raise important questions about their contribution and sense of purpose, and we still have the opportunity to respond positively. **Pragmatism and desire for meaning should guide our choice of tools**, particularly digital ones, as emphasized by the renowned Spanish architect A. Gaudi:

> *"To do things right, first you need love, then technique."*

Reference

Porcher, S. (2024). *La fin de l'eau* (Fayard ed.).

[2] A more precise definition is found in the Glossary (See No. 5).

[3] A more precise definition is found in the Glossary (See No. 3).

Chapter 11
Glossary

(1) Benchmarking (Vinardi, 2013)

Benchmarking comes from marketing that needed comparisons with the competition regarding the market, services, and products.

This term is now used to understand the competitive environment in terms of results, processes, and organization (to adapt and reproduce them).

It is supplemented with competitive intelligence, which describes the state of the art, at time t, in terms of technology, process, or products.

(2) Process (Vinardi, 2013)

It is a series of coordinated actions performed by an individual or a group, focusing on inputs from suppliers and the outputs to customers of the said individual or group.

To describe this process, we often use the SIPOC table or diagram for Suppliers, Inputs, Processes, Outputs, and Customers. Here is a breakdown of each component:

- **Suppliers** are the providers in the process
- **Inputs** are the items required by the process, in the correct format
- ***Processes*** are the different steps and tasks, stating who is responsible for each
- **Outputs** are the generated items by the process, in the correct format
- **Customers** are the people/entity that benefits from the output

(3) Hoshin Kanri (Vinardi, 2023)

Hoshin Kanri is not a new method as the principal moments traced from Japan's official introduction in the 1960s.

For those who do not speak Japanese, the term "*Hoshin Kanri*" does not immediately reveal what it is all about. A breakdown of the Japanese words gives (Vinardi, 2013):

C. Vinardi, *Sustainable Performance in the Digital Age*,
https://doi.org/10.1007/978-3-032-01252-4_11

- "*ho*" is a form or method, and "*shin*" is a compass needle: so "*Hoshin*" would convey the idea of following the compass bearing, pointing a direction, and
- "*Kanri*" refers to "control, behavior, change, or management."

Keeping those notions in mind, here is a simple version of the definition:

"*Hoshin kanri* is a way of collecting and deploying strategic objectives via concrete and relevant actions on all levels and teams within a company." (Vinardi, 2013)

Hoshin Kanri is the art of defining the WHAT, and transforming the WHAT into HOW. Some companies, particularly Anglo-Saxon ones, have various terms of which here are examples: "Policy Deployment" or "Strategy Deployment," "Management By Policy," "Managing for results," "Hoshin Planning," "target-means deployment," or even "Managing into action".

(4) Vision (Vinardi, 2023)

The vision of the company: what it wants to be in the future, that is, in the distant but defined long-term. This relates to the market and the products and services the company for the future. **The vision** must be simple and inspiring.

(5) Strategic Objectives (Vinardi, 2023)

Objectives are concrete elements to target. **The strategic or breakthrough objectives** are connected to the strategy and have their own time scales:

- long-term (related to the vision),
- short-term (or annual): in line with the yearly breakdown of long-term performance.

Once the strategic objectives are set, the corresponding action(s) need to be defined**, which is why they are called strategic action**s. To attain performance that has never been reached, it is often necessary to engage in actions that have never been set. These are **called breakthrough actions**. However, large-scale continuous improvements can also generate results that have never been reached. Therefore, we will note that strategic actions are mainly (but not exclusively) breakthrough actions.

(6) Bottom-up (Vinardi, 2023)

"From bottom to top" describes the upward flow from the most operational levels to top executives of an organization or company. This flow concerns information, problems, ideas for improvements, and decisions.

(7) Top-Down (Vinardi, 2023)

"From top to bottom" describes the downward flow from top executives of an organization or company to the most operational levels. This flow concerns information, problems, ideas for improvements, and decisions.

(8) Blockchain/Cryptoasset (Bercy, 2024)

A Cryptocurrency is a type of cryptoasset created using cryptographic technology.

It is recorded on a blockchain, whose protocol allows transactions to be tracked in a decentralized, secure, and transparent manner. This blockchain enables users connected to a network to share data without intermediaries.

- In the case of Bitcoin, transactions are validated by miners. A miner is a person or entity who offers processing power to the blockchain network.
- Once a transaction is validated, it is encrypted (converted into computer language) and timestamped before being added to the blockchain.
- Miners, also known as nodes of the network, are compensated for each block they create. Once these cryptoassets are generated, they are stored in a digital wallet for safekeeping.

Despite their success, first-generation cryptoassets like Bitcoin and Ethereum do not qualify as legal tender and lack intrinsic value. These cryptoassets are often **mistakenly called cryptocurrencies. However, they do not fulfill the three primary functions of a currency:**

- They are not an instrument of exchange, which would allow for easy buying and selling of everyday transactions,
- They do not serve as a unit of account, which would enable simple comparison of the prices of various goods and services,
- They do not function as a store of value.

(9) Value Creation

Value creation means increasing a company's worth by improving its performance, particularly regarding productivity and future profitability. It identifies the intersection between the overlapping interests of customers, main stakeholders, and the organization itself. For companies anywhere in the world, creating long-term shareholder value requires satisfying the needs of your customers, suppliers, and employees.

(10) Ominichannel approach

Customers, including employees and suppliers, interact through various channels and interfaces, both physical and real. This is known as a multichannel approach.

The first step is to cross-reference data collection and interconnect interfaces to establish a cross-channel approach.

The ultimate goal is to achieve homogenization by unifying the customer experience, ensuring that the same services are available regardless of the interface. This unified experience is referred to as an omnichannel approach.

References

Bercy Infos. (2024). *Cryptoactifs, cryptomonnaies: comments'y retrouver?* Epargne et investissements mobiliers. https://www.economie.gouv.fr/particuliers/cryptomonnaies-cryptoactifs

Vinardi, C. (2013). *Le Lean: atouts, impacts et limites* (Vuibert ed.).

Vinardi, C. (2023). *Business atrategy with Hoshin Kanri.* SpringerBriefs in Applied Sciences and Technology. https://doi.org/10.1007/978-3-031-20963-5_1

GPSR Compliance
The European Union's (EU) General Product Safety Regulation (GPSR) is a set of rules that requires consumer products to be safe and our obligations to ensure this.

If you have any concerns about our products, you can contact us on

ProductSafety@springernature.com

In case Publisher is established outside the EU, the EU authorized representative is:

Springer Nature Customer Service Center GmbH
Europaplatz 3
69115 Heidelberg, Germany

www.ingramcontent.com/pod-product-compliance
Lightning Source LLC
Chambersburg PA
CBHW071128151225
R17539700001B/R175397PG36512CBX00015B/3
* 9 7 8 3 0 3 2 0 1 2 5 1 7 *